Plant Responses to Air Pollution

Umesh Kulshrestha • Pallavi Saxena
Editors

Plant Responses to Air Pollution

Springer

Editors
Umesh Kulshrestha
School of Environmental Sciences
Jawaharlal Nehru University
New Delhi, India

Pallavi Saxena
School of Environmental Sciences
Jawaharlal Nehru University
New Delhi, India

ISBN 978-981-10-1199-3 ISBN 978-981-10-1201-3 (eBook)
DOI 10.1007/978-981-10-1201-3

Library of Congress Control Number: 2016947418

Printed on acid-free paper

This Springer imprint is published by Springer Nature
The registered company is Springer Science+Business Media Singapore Pte Ltd.

Contents

About the Editors

Umesh Kulshrestha is a professor at the School of Environmental Sciences, Jawaharlal Nehru University, New Delhi. He is Deputy Director of South Asia Nitrogen Center. He has 22 years of experience in Environmental Sciences. His area of expertise includes aerosol and precipitation chemistry relevant to climate change, air pollution and plant health, reactive nitrogen, air pollution chemistry and transport, atmospheric depositions and biogeochemical cycle etc. Prof. Kulshrestha is a recipient of the START Young Scientist Award by IGBP-START, USA, Thomas Kuhn Honor Pin by IUAPPA and International Academy of Sciences, CSIR Young Scientist Award, and MS Krishnan Gold Medal by the Indian Geophysical Union (IGU). Also, he is a Fellow of IGU. He has published a number of papers which have been widely cited. He has been associated with various international and national programmes on atmospheric science including Indian Ocean Experiment (INDOEX), Asian Brown Cloud (ABC), IGAC-DEBITS-CAD (Composition of Asian Deposition), Composition of Asian Aerosol and Precipitation (CAAP), Surya, ISRO-GBP-Integrated Campaign on Aerosol Radiation Budget (ICARB), etc.

Pallavi Saxena is a Young Scientist at School of Environmental Sciences, Jawaharlal Nehru University, from 2014 onwards. She has done her postdoc from the Space and Atmospheric Sciences Division, Physical Research Laboratory (PRL), Ahmedabad, Gujarat. She has completed her doctor of philosophy in environmental science from the University of Delhi on the topic "Effect of Photochemical Pollutants on Plant Species" in 2013. Her area of interest is air pollution and plant physiology. She has been working in this area for the last 9 years. She has also been awarded University Grants Commission Junir Research Fellowship (UGC, JRF) (2007–2008), Jawaharlal Nehru Doctoral Scholarship (2009–2010) and Council of Scientific and Industrial Research Senior Research Fellowship (CSIR SRF) (2012–2013) during her doctoral studies. She has been a recipient of postdoctoral fellowships like Indian Space Research Organization (ISRO) fellowship in Physical Research Laboratory (PRL) (2013–2014) and DST Fast Track Young Scientist at School of Environmental Sciences to pursue her independent project (2014–2017). She has published 13 research papers in international and national journals with high impact factor. She has also been an invited speaker on "Variation in the Concentration of VOCs in Atmosphere at Selected Sites in Delhi", at the "bVOCs Monitoring and Modelling" workshop at Lancaster

Environment Centre, Lancaster University, UK (2011). She is also an expert of TOAR International Meeting, which focuses on tropospheric ozone and its impact on vegetation, from 2015 onwards. In addition to that, she has been a recipient of various awards like the Young Scientist Award from the Indian Society of Plant Physiology (2013) and DBT Bio-CARe Award (2014) and several travel grant awards from World Meteorological Organization (WMO), International Global Atmospheric Chemistry (IGAC) and National Oceanic and Atmospheric Administration (NOAA) to participate as an expert in the Ozone Pollution and Its Impact on Vegetation in Trpospheric Ozone Assessment Report (TOAR) Meeting (2015–2016). She has also reviewed research articles for reputed journals, like *Atmospheric Environment* (Elsevier) and *Mitigation and Adaptation of Strategies for Global Change* (Springer), *Environmental Technology* (Taylor and Francis), *Polish Journal of Environmental Studies*, *Environmental Monitoring and Assessment* (Springer) and *International Journal of Physical Sciences* (Academic Journal). She has participated in various national (10) and international conferences (8) and presented several papers. Dr. Saxena has also participated in various workshops and trainings (15) in the area of air pollution and plant physiology.

Introduction

1

Umesh Kulshrestha and Pallavi Saxena

Being a living community on the planet earth, utilization of natural resources is quite obvious by humans. However, the overexploitation of these resources during past two decades has resulted in major environmental problems. Increasing human population and energy demand has contributed different kinds of harmful chemical and biological species in the atmosphere. Recently, air pollution has become a burning issue of common concern. The process of air pollution can be defined as the atmospheric conditions having high levels of pollutants which may produce undesirable effects on materials, plants, and human health. The major air pollutants include sulfur dioxide, nitrogen oxides, carbon monoxides, hydrocarbons and particulate matters, etc. which play a very significant role in affecting the biochemical and physiological processes of the plants and ultimately lead to yield losses (Heck et al. 1988). In developing countries like India, a very drastic change in the quality of air has been observed during past two decades. Many cities in Asia, Africa, Latin America are facing major challenges of air pollution (Ashmore 2005). Delhi (the capital city of India) is also facing the problem of severe air pollution in spite of the implementation of CNG-driven public transport (Saxena et al. 2012). The air quality in Delhi has been the worst among 1,600 cities of the world (WHO 2014–2015). According to the estimates, around 1.5 million people are killed every year. India has the world's highest death rate from chronic respiratory diseases and asthma.

1.1 Air Pollution and Plant Health

Generally, atmospheric pollutants have a negative effect on the plants; they can have direct toxic effects or indirectly by changing soil pH followed by solubilization of toxic salts of metals like aluminum. The deposition of particulate matter covers the leaf blade reducing light penetration and blocking the opening of stomata. These impediments influence the process of photosynthesis due to which rate of photosynthesis and growth declines sharply. On the other hand, leaves of the trees have role in retaining particulate matter but they are much more affected when the dry depositions are increased (Jyothi and Jaya 2010). Polluting gases such as SO_2 and NO_x entered into the leaves through the stomata following the same diffusion pathway as CO_2. NO_x dissolved into the cells gives rise to nitrite ions (NO_2^-, which are toxic at high concentrations) and nitrate ions (NO_3^-) that enter into nitrogen metabolism as if they had been absorbed through the roots. In some cases, exposure to pollutant

U. Kulshrestha (✉) • P. Saxena
School of Environmental Sciences, Jawaharlal Nehru University, New Delhi 110067, India
e-mail: umeshkulshrestha@gmail.com

U. Kulshrestha, P. Saxena (eds.), *Plant Responses to Air Pollution*,
DOI 10.1007/978-981-10-1201-3_1

gases particularly SO_2 causes stomatal closure, which protects the leaf against further entry of the pollutant but also curtails photosynthesis. In the cells, SO_2 is dissolved to give rise to bisulfite and sulfite ions, the sulfite is toxic, but, at low concentrations, it is metabolized by chloroplasts to sulfate, which is not toxic. At sufficiently low concentrations, bisulfite and sulfite are effectively detoxified by plants. In such cases, the SO_2 air pollution is a source of sulfur nutrient for the plant. In urban areas these polluting gases may be present in such high concentrations that these cannot be detoxified rapidly enough to avoid injury. For example, ozone is presently considered to be the most damaging phytotoxic air pollutant in North America (Heagle 1989; Krupa et al. 1995). It has been estimated that whenever the mean daily O_3 concentration reaches 40, 50, or 60 ppb (parts per billion or per 109), the combined yields of soybean, maize, winter wheat, and cotton would be decreased by 5 %, 10 %, and 16 %, respectively. Ozone is a highly reactive gas. It combines with plasma membranes and alters the metabolism. As a result, stomatal apertures are poorly regulated, chloroplast thylakoid membranes are damaged, RuBisCO is degraded, and photosynthesis is inhibited. Ozone reacts with O_2 and produces reactive oxygen species, including hydrogen peroxide (H_2O_2), superoxide (O_2^-), singlet oxygen ($^1O_2$*), and hydroxyl radical (OH^-). These denature proteins and damage nucleic acids (thereby giving rise to mutations) cause lipid peroxidation, which breaks down lipids in membranes. Reactive oxygen species (ROS) formed in the absence of O_3 too, particularly in during electron transport process in the mitochondria and chloroplasts, when electrons can be donated to O_2. Cells are protected from reactive oxygen species by enzymatic and nonenzymatic defense mechanisms (Heterich and Heterich 1990; Cape 2003). Defense against ROS is provided by the scavenging properties of molecules, such as ascorbic acid, α-tocopherol, phenolic compounds, and glutathione. Superoxide dismutases (SODs) catalyze the reduction of superoxide to hydrogen peroxide. Hydrogen peroxide is then converted to H_2O by the action of catalases and peroxidases. Of particular importance is the ascorbate-specific peroxidase localized in the chloroplast. Acting in concert, ascorbate peroxidase, dehydroascorbate reductase, and glutathione reductase remove H_2O_2 in a series of reactions called the Halliwell–Asada pathway, named after its discoverers. Glutathione is a sulfur-containing tripeptide that, in its reduced form, reacts rapidly with dehydroascorbate and becomes oxidized in the process. Glutathione reductase catalyzes the regeneration of reduced glutathione (GSH) from its oxidized form (GSSG) in the following reaction:

$$GSSG + NADPH + H^+ \rightarrow 2GSH + NADP^+$$

Exposure of plants to reactive oxygen species stimulates the transcription and translation of genes that encode enzymes involved in protection mechanisms. In *Arabidopsis*, exposure for 6 h per day to low levels of O_3 induces the expression of several genes that encode enzymes associated with protection from reactive oxygen species, including SOD, glutathione S-transferase (which catalyzes detoxification reactions involving glutathione), and phenylalanine ammonia lyase (an important enzyme at the start of the phenylpropanoid pathway that leads to the synthesis of flavonoids and other phenolics) (Weiss et al. 1997).

Therefore, in relation to air pollution phenomena and physiological processes of plants, the present book covers all the important topics in order to justify the title "Air Pollution and Plant Health: Climate Change Perspectives." Chapter 1 deals with the general introduction. It basically provides general information about air pollution and its effect on plants. Chapter 2 summarizes the air quality trends along with global and regional emissions of particulate matter, SOx, and NO_x. Also, this chapter gives an overview of global sources of air pollutants and their future projections. Chapter 3 deals with the effect of these urban air pollutants on biota. It describes about the effects of air pollutants such as particulate matter, SOx, and NO_x on plants as well as on human health. This chapter highlights the need for imperative shift toward renewable energy sources for sustainable and environmental friendly solution for new technologies and industrial producers. Chapter 4 describes the characteristics and mechanisms of plant which help in the uptake of air pollutants giving insights about the distribution of pollutants inside the plant machinery. It also describes how the pine needles and mosses accumulate air pollutants demonstrating the role of plants as biomonitors. Chapter 5 opens up the discussion about biogenic VOC emissions and their role in atmospheric sciences. In this chapter, plant chemistry and the conditions under which it emits BVOCs are detailed. Besides, this chapter also provides information about BVOC linkages with global warming and plant signaling processes. Chapter 6 deals with the biochemical effects of air pollutants on plants. This chapter summarizes about the significant effects posed by air pollutants on plant health. Biochemical parameters such as chlorophyll, proline content, and other enzymatic activities act as bioindicators for determining the health of the plant. Chapter 7 deals with air pollutants and photosynthetic machinery of plants. It describes how high concentrations of air pollutants damage leaves and reduce leaf area, thereby affecting the photosynthetic efficiency of plants to a significant extent. The damage to the photosynthetic apparatus accounts for the decline in photosynthetic efficiency, but each pollutant differs in its mode of action for damaging the photosynthetic apparatus of leaves. Photosynthesis and respiration are interrelated processes involved in the basic processes of plants. Therefore, Chap. 8 deals with the effect of air pollutants on plant gaseous exchange process: effect on stomata and respiration. This chapter describes about how respiration acts as an indicator for environmental stress. The study also shows that leaf characters including cuticle, stomata, epidermal cells, and guard cells get affected due to stress induced by the air pollutants. This further affects the gaseous exchange as well as respiration in plants. Chapter 9 deals with the effect of air pollutants on biomass and yield. It also linkages how respiration acts as an indicator in the physiological mechanisms. It is to mention that any book of this theme is incomplete without the provided dedicated chapter on environmental stresses and plant health. Therefore, Chap. 10 deals with air pollution stress and plant response. This chapter describes about how the structural and physiological changes occur due to stress imposed by air pollutants. It is very important to control the effect so as to the quality of air. Hence, Chap. 11 deals with the biomonitoring and remediation by plants. This chapter describes about how respiration acts as an indicator for environmental stress. The study also shows that leaf characters including cuticle, stomata, epidermal cells and guard cells get affected due to stress induced by air pollutants. Chapter 12 deals with air pollution control by policies and laws. It deals with different laws or amendments made by government to curb air pollution and other mitigation measures. Chapter 13 deals with the pollution and plants in relation to changing policies. This chapter determines the current pollution scenario in India and various interventions made at policy levels in particular for plants. Chapter 14 deals with tropospheric ozone and effects on plants. Ozone is important to study because it is the only phytotoxic pollutant having high oxidative capacity. It can damage plant to a larger extent as compared to other air pollutants. This chapter summarizes the information available on plant responses to O_3 at physiological, cellular, and biochemical levels, crop yield, forest, and grassland communities at present concentrations and also under projected future concentrations.

1.2 Conclusion

The collated information based on several studies concludes that air pollutants not only affect the vegetation near the point sources and urban centers but also affect the crops depending on the environmental conditions in suburban and rural areas. The physiology and metabolism of plants are altered due to the oxidizing potential of the pollutants. In order to survive, the responses of plants vary between different species and their cultivars. Responses of plants to air pollutants also depend on type of pollutants, concentrations' duration, and its magnitude. There is a need to screen out sensitive and tolerant cultivars especially in developing countries and establish the exposure indices of all the important crops to reduce the crop loss. The detailed description of air pollutant effect on plants, sources of air pollution, and control policies is given in this book through different chapters.

References

Ashmore MR (2005) Assessing the future global impacts of ozone on vegetation. Plant Cell Environ 28:949–964

Cape JN (2003) Effects of airborne volatile organic compounds on plants. Environ Poll 122(1):145–157

Heagle AS (1989) Ozone and crop yield. Annu Rev Phytopathol 27:397–423

Heck WW, Taylor OC, Tingey DT (1988) Assessment of crop loss from air pollutants. Elsevier Applied Science, London

Heterich R, Heterich R (1990) Comparing the distribution of nitrated phenols in the atmosphere of two German hill sites. Environ Tech 11:961–972

Jyothi JJ, Jaya DS (2010) Evaluation of air pollution tolerance index of selected plant species along roadsides in Thiruvananthapuram, Kerala. J Environ Biol 31:379–386

Krupa SV, Gruenhage L, Jaeger HJ, Nosal M, Manning WJ, Legge AH, Hanewald K (1995) Ambient ozone (O_3) and adverse crop response: a unified view of cause and effect. Environ Poll 87:119–126

Saxena P, Bhardwaj R, Ghosh C (2012) Status of air pollutants after implementation of CNG in Delhi. Curr World Environ 7(1):109–115

Weiss P, Lorbeer G, Stephan C, Svabenicky F (1997) Short chain aliphatic halocarbon, trichloroacetic acid and nitrophenols in Spruce needles of Austrian background forest sites. In: Organic xenobiotics and plants; impact, metabolism, toxicology. Federal Environment agency, Vienna

WHO (2014–2015) Air quality guidelines for Europe, 2nd edn. European series. WHO Regional Publications, Copenhagen, p 91

Air Quality: Global and Regional Emissions of Particulate Matter, SOx, and NOx

2

Darpa Saurav Jyethi

Abstract

Poor air quality is known to have deleterious effects on the environment and human health. A variety of air pollutants are identified at unprecedented levels. Particulate matter and oxides of sulfur and nitrogen are common pollutants of the atmosphere. The emissions of these criteria pollutants have been studied widely. An overview of their global and regional emissions is helpful in assessing the status of contributions from various sectors and efficacy of control strategies over the years in both developed and developing nations of the world. Particulate matter levels in developed regions have been observed to have decreased substantially toward the end of the twentieth century, contrary to the trend of emissions in developing countries. Transportation and power generation sectors are key sources of PM emissions. Oxides of sulfur have been observed to have peaked in the 1970s and subsequently decreased thereafter on a global scale; however, the developing economies have registered a rise in emissions. Similar trend has been observed for oxides of nitrogen with decline of emissions from developed regions of the world and subsequent increase from developing countries in the earlier part of the twenty-first century. Timely intervention of suitable strategies to combat emissions in developing regions is crucial to nullify the increasing trend of emissions. The present chapter provides an overview of three criteria air pollutants – particulate matter, oxides of sulfur, and oxides of nitrogen with respect to their changing global and regional emission scenario in the past decades and sector-wise contributions.

Keywords

Air quality • Air pollutants • Particulate matter • Oxides of nitrogen • Sulphur • Nitrogen

D.S. Jyethi (✉)
Atmospheric Sciences Research Center (ASRC),
State University of New York, Albany, NY, USA
e-mail: dsjyethi@gmail.com

U. Kulshrestha, P. Saxena (eds.), *Plant Responses to Air Pollution*,
DOI 10.1007/978-981-10-1201-3_2

2.1 Introduction

Air quality has gained immense concern because of its wide-ranging implications. Poor air quality has been generally attributed to population explosion and subsequent overexploitation of natural resources, urbanization, and industrialization. Much of the burden in the present times is markedly high in the developing nations of the world. A thorough understanding of sources, global and regional emission scenarios, and spatiotemporal variations is crucial in order to develop and implement strategies suitable to curb regional and local sources of air pollution. Air pollution sources are basically classified as natural (events that are not the result of any human activities, e.g., forest fires, dust, volcanoes) and anthropogenic (resulting from human activities). Stationary sources (e.g., power plants, refineries, and industries) and mobile sources (e.g., cars, trucks, buses) are two broad categories of prevalent sources in urban areas. Outdoor or ambient air pollution is mostly attributed to combustion of fossil fuels; however, fuel evaporation has also been known to contribute to pollution load (Black et al. 1998). Indoor sources (e.g., building materials and activities such as cleaning, cooking, etc.) are also other sources of outdoor air pollution. The United States Clean Air Act has categorized "major" and "area" stationary sources, the former being sources that emit 10 tons per year of any of the listed toxic air pollutants or 25 tons per year of a mixture of air toxics. Equipment leaks when materials are transferred from one location to another or during discharge through emission stacks or vents are considered in the category. On the other hand, "area" sources are those that emit less than the aforementioned limit. Emissions from single area sources are apparently often relatively less; however, their emissions are alarming when they are considered in a collective scenario, for instance, in a densely populated urban area. Apart from sources of air pollution, meteorology and topography are two important factors that determine the prevalence of pollution (Seinfeld and Pandis 2006) and extent of exposure and subsequent effects (Goncalves et al. 2005). Meteorology governs the dispersion, chemical transformations, residence time, transport, and fate of pollutants in the atmosphere. The following section deals with the implications of air quality deterioration from various perspectives.

2.1.1 Air Quality and Its Implications

Air quality is recognized as an important indicator of environmental sustainability, health status, economic condition, and social well-being. The effects of air pollution are not restricted in the source region but are trans-boundary. Environmental implications include damages of life-sustaining systems such as atmosphere, hydrosphere, and pedosphere, primarily caused by introduction of toxic air pollutants and through formation of secondary pollutants such as acid rain and tropospheric ozone. These harm plants, forests, food crops, wildlife, aquatic bodies, and biota (DEFRA 2013). Air quality has been linked to ocean acidification, eutrophication, long-term changes in water quality, visibility degradation and haze, and alteration of Earth's radiation budget (Seinfeld and Pandis 2006); bioaccumulation, behavioral, neurological, and reproductive effects in fish, birds, and wildlife; species extinction (Welch 1998); altered soil chemistry; and consequently plant species composition, agricultural yield, and damage to historical built environment (Thomas 1961; Brimblecombe 2003).

Health ramifications due to poor air quality are generally complex. It includes a host of outcomes that have been linked to short- and/or long-term exposure to air pollutants. Short-term or acute exposure to pollutants results in mild outcomes such as eye and throat irritation, headaches, nausea, intense effects such as respiratory tract infections – bronchitis and pneumonia – allergic reactions, and worsening of asthmatics and other sensitive populations such as infants, children, and elderly. Incidences of lung cancer, cardiovascular disorders, and detrimental effects on the brain, nerves, liver, and kidneys are some of the chronic effects of long-term exposure to specific air pollutants (Kim et al. 2014 and references therein). The Organisation for Economic Co-operation and Development (OECD) has declared that exposure to outdoor air pollution is expected to become the top environmental cause of premature mortality globally by 2050 (OECD 2012). Cardiovascular, pulmonary, and respiratory systems are predominantly affected. It is estimated that the global burden due to outdoor air pollution manifested in seven million deaths in 2012 (WHO 2014). On a regional scale, the World Health Organization (WHO) Western Pacific and Southeast Asian regions bear most of the burden (WHO 2014). The International Agency for Research on Cancer (IARC), WHO's specialized cancer agency, has declared outdoor air pollution as carcinogenic (Group 1) to human beings. Among various air pollutants studied, particulate matter is most closely related with increased cancer cases, especially lung cancer. Additionally air pollution has been linked to increase in cancer of the urinary tract/bladder (IARC 2013). Air quality index (AQI) is commonly used as a tool to assess the local air quality on a daily basis. It is expressed for individual pollutant (particulate matter, sulfur dioxide, tropospheric ozone, and carbon monoxide) and uses numerical values and color codes to alert general population regarding the status of air quality and its health effects upon exposure (0 to 5 (good, green); 51 to 100 (moderate, yellow); 101 to 150 (unhealthy for sensitive groups, orange); 151 to 200 (unhealthy, red); 201 to 300 (very unhealthy, purple); 301 to 500 (hazardous, maroon) (USEPA 2014).

Air quality degradation has economic consequences. The estimated cost of the deaths and illness attributable to air pollution in OECD member countries was about USD 1.7 trillion in 2010 (OECD 2014). Significant amount of the burden has been found to be arising from vehicular tailpipe emissions. The costs of deteriorating air quality are generally higher for the weaker (in terms of socioeconomic status) sections of the society (Wong et al. 2008; ALA 2013). Studies have concluded that sensitive population (Bateson and Schwartz 2004; Sunyer et al. 2000) and socioeconomically weaker sections of society (Forastiere et al. 2006; Jerrett et al. 2004; Neidell 2004) are most strongly affected by air pollution. The reasons attributed to such an observation are cheaper costs of production and less stringent environmental regulations for the shift of polluting industries into poorer areas from wealthier areas (Pulido 2000).

2.1.2 Air Pollutants

There are a gamut of chemical substances in the atmosphere that are known to be harmful for the environment and human health. The Clean Air Act of the United States Environmental Protection Agency (USEPA) defines air pollutant as "any air pollution agent or combination of such agents, including any physical, chemical, biological, radioactive (including source material, special nuclear material, and byproduct material) substance or matter which is emitted into or otherwise enters the ambient air." Certain air pollutants are ubiquitous. The USEPA classifies air pollutants as criteria and non-criteria pollutants for monitoring and regulation purposes. Commonly found air pollutants that are known to harm environment and health are classified as criteria pollutants. The term "criteria" emphasizes the need for regulation and routine monitoring by developing human health-based and/or environmentally based permissible levels. Presently, six pollutants – particulate matter, tropospheric ozone, carbon monoxide, sulfur oxides, nitrogen oxides, and lead – are designated as criteria pollutants by the USEPA. These are used as key indicators of air quality all over the world. The USEPA has set guideline values for these criteria pollutants referred to as National Ambient Air Quality Standards (NAAQS). Two standards – primary and secondary – are specified. Primary standard values of each of these pollutants provide public health protection, including protecting the health of "sensitive" populations such as asthmatics, children, and the elderly. Secondary standards provide public welfare protection, including protection against decreased visibility and damage to animals, crops, vegetation, and buildings (USEPA 2005). Non-criteria pollutants, unlike criteria pollutants, do not have assigned guideline value. In addition to criteria and non-criteria pollutants, certain air pollutants are classified as "air toxics" or hazardous air pollutants (HAPs). These include those pollutants that are known to cause cancer and other serious health outcomes, such as reproductive, immune, neurological, and congenital defects. Adverse environmental and ecological effects are also considered in classifying pollutants as HAPs. Acetaldehyde, acetonitrile, asbestos, benzene, carbon tetrachloride, chloroform, dibenzofurans, phenol, polycyclic aromatic, nickel, arsenic, antimony, lead, mercury, and cadmium compounds, radionuclides, and pesticides are among 187 compounds that are listed as HAPs (USEPA 2015a). The USEPA regulates HAPs in two phases – firstly, based on emissions from best available technology and, secondly, on the basis of health risks.

Air pollutants are either primary or secondary in origin. Pollutants that are directly released into the air from the pollutant source are termed as primary pollutants. Once emitted in the atmosphere, some of the primary pollutants are chemically and/or photochemically transformed into secondary pollutants. Gaseous pollutants and particulate matter are two broad categories with marked differences in measurement techniques and control strategies. Tropospheric ozone, carbon monoxide, sulfur oxides, nitrogen oxides, ammonia, and benzene are some of the gaseous pollutants commonly found in ambient atmosphere. Particulate matter is a relatively complex entity suspended in the atmosphere with size-dependent atmospheric residence time. Size distribution of particulate matter in atmosphere along with number, mass concentrations, and chemical constituents is highly variable on a spatial as well as temporal scale.

2.2 Particulate Matter

Particulate matter (PM), otherwise known as atmospheric aerosol, is a criteria air pollutant. The term "aerosol" denotes a system of particles (solid or liquid) suspended in a gas (air). PM underscores the particle component of atmospheric aerosols. It is highly complex in composition. Size is an important factor of particulates and governs the transport and fate in the atmosphere. "Fine" and "coarse" are two broad categories of particulates. Fine particulates are those having an aerodynamic diameter of 2.5 micrometers (μm) or less, whereas particulates having aerodynamic diameter larger than 2.5 μm and smaller than 10 μm are termed as coarse. Coarse particulates are generally mechanically generated, whereas fine particulates are either directly emitted from sources or formed in the atmosphere from gas to particle conversion reactions. Fine particulates are of particular concern due to association with adverse health concerns as they are inhalable and are able to penetrate deep into the lungs. Fine particulates generally comprise the accumulation mode (0.1 μm < particle size <2.5 μm) and the nuclei mode (particle size <0.1 μm). Nuclei mode particles have relatively short lifetimes in the atmosphere and end up in the accumulation mode. Gas-phase condensations and secondary formations on the existing particles in atmosphere also play a crucial role on the chemistry of the particles. Concentrations of ultrafine particles (UFPs, particulate matter of less than 100 nm in aerodynamic diameter) are also known to be of particular relevance especially from inhalation exposure perspective. It is noteworthy that not only size of particulates but size distribution is also an important attribute in the study of PM. It has been reported that the methods used for measuring PM concentrations are affected by the size distribution. Capareda et al. (2004) found that size distribution of PM emitted in rural areas is significantly larger than that of PM present in urban areas. The authors reported that mass median diameter (MMD) of urban PM is generally less than 10 μm aerodynamic diameter, whereas agricultural PM will have an MMD larger than 10 μm. The size of PM also governs the process of removal from atmosphere. Fine particulates are generally removed through rainout and washout, whereas coarse particulates are mostly removed through sedimentation. Apart from size of particulates and size distribution, the composition of PM is highly variable and complex. PM is mostly composed of carbonaceous matter and includes inorganic and organic components. These components drive health outcomes upon inhalation and consequent deposition in the respiratory system. Certain chemical components trigger the production of reactive oxygen species and induce oxidative stress in lung cells (Tao et al. 2003). Other attributes that influence the environmental and health effects are mass concentrations and morphology. Size and chemical composition have been often debated to be the competing attributes of PM in relation to health effects. The USEPA, however, reports that the health outcomes due to PM exposure vary by size and composition with both bioavailable metals and organic chemicals playing a crucial role in the health outcomes. Studies have concluded that no known chemical substance is of sufficient toxicity given the current levels of exposure to particulate matter to explain the observed magnitude of health effects (USEPA 2002). Fine and coarse PM has been routinely monitored in various cities across the world. Anthropogenic exploitation of fossil fuels is known to generate significant amount of PM. Naturally occurring PM originates from volcanoes, dust storms, wild fires, and biogenic and sea spray. Long-term exposure to PM has been linked with morbidity and mortality (Dockery et al. 1993; Pope et al. 2004). PM exposure for a short term results inflammatory response in the upper respiratory tract and decreased lung functions in general population. Sensitive populations such as children and elderly, asthmatics, and people suffering from heart and lung ailments are extremely susceptible to initiation of cough, phlegm, wheezing, shortness of breath, bronchitis, increased asthma attacks, and aggravation of lung or heart disease (Cohen and Pope 1995). Ambient and occupational exposure to elevated levels of PM has been associated to chronic pulmonary obstructive diseases (COPD) such as bronchitis and emphysema.

Environmental concerns of PM include absorption or scattering of solar radiation, consequent alteration of the radiation budget of Earth, and haze formation in the lower troposphere. Composition and size of PM are important in the manifestation of these effects. For instance, black carbon or soot-dominated aerosol is absorbing and sulfate aerosol is scattering in nature. Exposure in plants is either through deposition on vegetative surfaces or soil–root pathway. Effects manifest in alteration of key physiological processes such as photosynthesis and respiration; foliar surface damage by acidic and alkaline component; nutrient uptake from soil; foliar nutrient leaching; reduced vigor, productivity, reproductive success, and resilience; and disturbed nutrient cycling, soil stabilization, water cycling, and even energy flux (Grantz et al. 2003). Other environmental effects of PM include acidification of aquatic and terrestrial ecosystems upon deposition and degradation of heritage monuments and built structures.

2.2.1 Global and Regional Emissions of Particulate Matter

PM is one of the most widely studied criteria air pollutants. Quantification of ambient PM load has been achieved by gravimetric and continuous or real-time measurements. Satellite-derived aerosol optical depth (AOD) has also been used to derive PM concentrations. Sampling and measurement techniques are subjected to significant uncertainty because of sampling artifacts, handling procedures, and by the fact that a significant portion of PM is semi-volatile and partition between gas and particle phase is temperature dependent (Van Dingenen et al. 2004). Satellite-derived PM also suffers from uncertainty due to aerosol vertical profile variability on a seasonal and regional scale and cloud fraction (Li et al. 2015).

van Donkelaar et al. (2015) used satellite observations as a surrogate to report an increased population-weighted fine particulate concentrations (2.1 %/year globally) from 1998 to 2012. The authors also found that higher emission trends in some developing countries influenced the trend on a global scale. Figure 2.1 depicts a world map exhibiting a measure of extinction of radiation or total aerosol optical depth at 550 nm due to aerosol scattering and absorption. It is

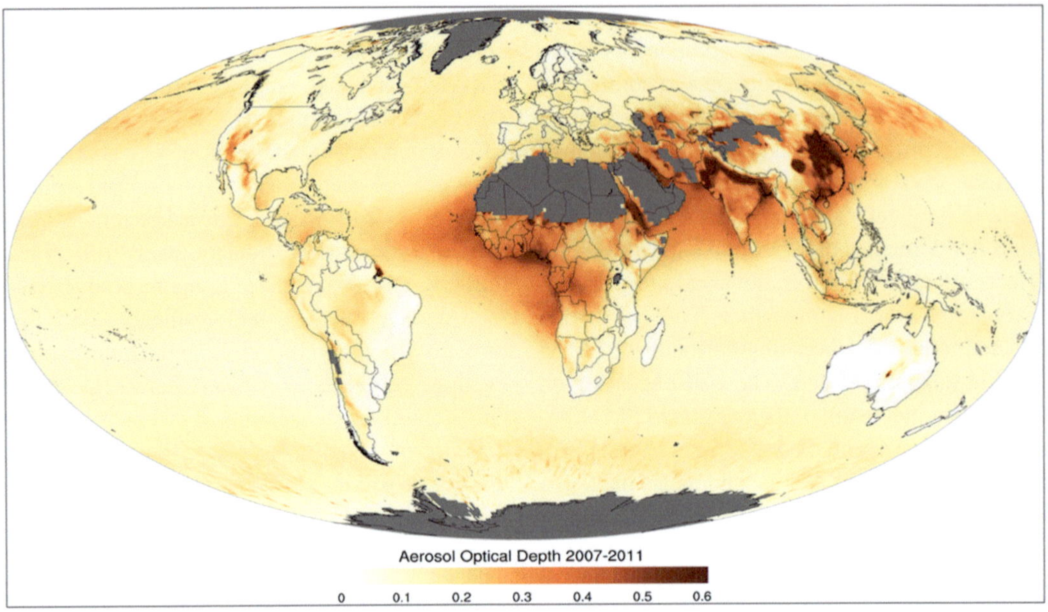

Fig. 2.1 World map showing aerosol optical thickness from Moderate Resolution Imaging Spectroradiometer (MODIS) on NASA's Terra satellite. *Dark brown* pixels show high aerosol concentrations, while tan pixels show lower concentrations, and *light yellow* areas show little or no aerosols. *Gray* denotes that the sensor could not make its measurement (Source: https://en.wikipedia.org/wiki/Particulates)

indicative of the PM pollution status of the regions in the world. The USEPA has reported 35 % and 30 % reduction in the national average of $PM_{2.5}$ and PM_{10} levels, respectively, from 2000 to 2014 in the USA (USEPA 2016). Upadhyay et al. (2011) studied PM levels in Southern Phoenix, Arizona, and found that soil and combustion processes explained the observed variance in PM_{10} levels, whereas combustion sources explained the observed $PM_{2.5}$ concentrations. The difference in urban and rural $PM_{2.5}$ levels during 2009–2012 was studied in the Midwestern region of the USA, and it was concluded that the observed levels in rural areas were generally lower (8.4–10.4 μg m^{-3}) compared to the urban sites (9.5–11.6 μg m^{-3}). Crustal materials dominated in rural areas, whereas secondary aerosols and combustion products dominated in urban areas (Kundu and Stone 2014). Karagulian et al. (2015) reported that 24 % and 30 % of the observed $PM_{2.5}$ and PM_{10}, respectively, in the USA can be attributed to traffic sources. It was also reported by the authors that secondary particle formation contributed as high as 46 % and 44 % to the $PM_{2.5}$ and PM_{10} levels, respectively. Annual mean levels based on daily averages of PM data monitored by the European Environmental Agency (EEA) in various countries of the European Union were found to be relatively higher in nations such as Turkey, Bulgaria, Slovakia, the Czech Republic, Poland, and parts of Italy as compared with the other member nations during the year 2012 (EEA 2014). Primary PM_{10}, $PM_{2.5}$, and $PM_{2.5-10}$ emissions during 1990–2010 have been reported to be decreased by 26 %, 28 %, and 21 %, respectively, across regions in Europe. Increase in emissions from road transport and agriculture since 1990 has been attributed as the reason for the resulting difference in the trends in $PM_{2.5}$ and $PM_{2.5-10}$. The transition of source (from coal to natural gas) for power generation and better pollution control strategies have been the key reasons for the decrease in PM_{10} levels (EEA 2012). Considerable variations with respect to sites and seasons in the contribution of non-combustion fraction to the observed PM were observed in the cities of Athens, Madrid, and London. PM_{10} levels in

Athens and Madrid were found to be higher in hot and dry season indicating toward secondary formations and crustal particles (Kassomenos et al. 2014). On a global scale, the highest contribution of natural dust to observed levels of $PM_{2.5}$ and PM_{10} has been found in the Middle East region (52 % and 44 %, respectively) (Karagulian et al. 2015). PM pollution is particularly worse in Asian cities. Mean PM_{10} concentrations across Northern China was found to be an order higher than those recorded in European Union and the USA. Coal and biomass fuel combustion for heating in winter contributed the highest to the observed levels in China (Luo et al. 2014). PM emissions in the Indian subcontinent have been considerably high in megacities. $PM_{2.5}$ levels in rural areas in the Indo-Gangetic plains are reported to be higher than some urban centers (Dey et al. 2012). Delhi, the capital of India, is one of the worst in the world with respect to particulate pollution. Marrapu et al. (2014) estimated the contribution of transportation, power generation, industrial activities, and domestic sources as 86.8 %, 7.9 %, 4.6 %, and 0.8 % and 52.6 %, 9.9 %, 15.3 %, and 22.2 % for PM_{10} and $PM_{2.5}$, respectively, for the city during 2010. On a global level, Karagulian et al (2015) found 25 % of urban ambient air pollution from $PM_{2.5}$ is contributed by traffic, 15 % by industrial activities, 20 % by domestic fuel burning, 22 % from unspecified sources of human origin, and 18 % from natural dust and salt. Takeshita (2011) examined the role of transport sector using REDGEM70 global energy system model and concluded that in a business as usual (BaU) situation (wherein an assumption is made that developing countries will adopt identical stringent regulations as in developed countries but with a lag of 20 years), global PM emissions from road vehicles will decrease by 93 % from 2000 to 2050. It is estimated that an early implementation of strict regulatory measures in developing countries can potentially alleviate the PM pollution scenario from road vehicles.

2.3 Oxides of Sulfur

Oxides of sulfur (SO_x) is a generic term for a group of compounds containing sulfur and oxygen molecules. SO_x emissions in the atmosphere predominantly result from anthropogenic sources, i.e., combustion of sulfur-containing fuels such as coal and oil used in electricity generation. Petroleum refineries and smelting of metal sulfide ores for metal manufacturing are other sources. Apart from anthropogenic sources, natural sources such as volcanoes and wildfires are also known to emit considerable amounts of SO_x into the atmosphere. Sulfur dioxide (SO_2) is the predominant form emitted from usual high-temperature combustion processes (Bowman 1991) and is considered a criteria air pollutant. Harmful effects of SO_2 are well evidenced on plant and human health (WHO 2000; ATSDR 1998; Chen et al. 2007). Once emitted in the atmosphere, SO_2 is oxidized to sulfate aerosol (Seinfeld and Pandis 2006). Saxena and Seigneur (1987) suggested that in addition to the gas-phase oxidation by hydroxyl (OH) radicals, SO_2 oxidation in aqueous aerosols may also contribute significantly to sulfate formation. Wet and dry deposition are the removal modes of SO_2 and its oxidation products. The most severe environmental effect associated with SO_2 is that of acid deposition with consequent damage to terrestrial and aquatic ecosystems. SO_2 dissolves readily in water present in the atmosphere to form sulfurous acid (H_2SO_3). Sulfur trioxide (SO_3), another oxide of sulfur, is either emitted directly into the atmosphere or produced from sulfur dioxide and is rapidly converted to sulfuric acid (H_2SO_4). Sulfate particles are responsible for scattering visible light, global cooling, and haze formation (IPCC 2007). Human health effects of SO_2 exposure have been associated with respiratory illness and exacerbation of existing cardiovascular and pulmonary diseases in children, the elderly, and asthmatics (Bremmer et al. 1999; Mar et al. 2000; Maynard and Ayres 2014). Plant exposure to elevated levels of SO_2 is manifested through serious effects such as foliar damage and reduced yield in crops (Varshney et al. 1979; Malhotra and Hocking 1976; Winner et al. 1985). Building materials are also damaged due to the corrosive action of the pollutant.

2.3.1 Global and Regional Emissions of Oxides of Sulfur

Global emissions of SO_2 during 1850–2005 were estimated by a bottom-up mass balance method, taking into account country-level inventory data with consideration of coal combustion, petroleum combustion, natural gas processing and combustion, petroleum processing, biomass combustion, shipping bunker fuels, metal smelting, pulp and paper processing, other industrial processes, and agricultural waste burning (AWB). It was found that global emissions peaked in the early 1970s and decreased until 2000, with an increase in recent years due to increased emissions in China, international shipping, and developing countries in general (Smith et al. 2011). The authors reported emissions from China and international shipping to be source of uncertainty. The authors have attributed the declining trend to changes in fossil fuel use, increases in the amount of sulfur removals from oil and nonferrous metals, and controls on coal-fired power plants. A comparative discussion on quantitative differences in various other estimates is outlined in Smith et al. (2004). It is also found that North America and Europe (including Russian Federation and Eastern Europe, Caucasus and Central Asia (EECCA) countries) were responsible for over 50 % of total global SO_2 emissions in 1990. The contribution was found to drop to less than 25 %, and Asian emissions represent more than 50 % of the total during 2010 (Klimont et al. 2013).

Hand et al. (2012) reported that the total SO_2 emissions have decreased from nearly 31 million tons in 1970 to 8 million tons in 2010 in the USA. It was also reported that regulation of power plant emissions led to the decline in the total emissions. Annual trend within the USA was found to be predominantly associated with the total emissions and concentrations in the Eastern United States. Western Europe was the highest contributor to the global emissions of SO_2 during 1850, a period marked by the end of the industrial revolution (Smith et al. 2011). The emissions were an order higher than those of the USA and Canada during the same period. Peak emissions in Western Europe were estimated to be during the 1970s. 2005 emission levels in Western Europe were comparable to 1890–1900 levels. Emissions from former USSR have been observed to increase steadily in 1940 and sharply decline after the collapse of USSR in 1990. Chinese emissions have also been observed to rise during the 1940s; however, the emissions were found to be on an all-time high in 2005. Indian emissions were found to be on a steep rise only in the 1980s. Overall, top countries/regions in terms of total SO_2 emissions during 2005 were found to be as China, the USA, Europe, South and East Asia, India, and Russia (Smith et al. 2011). Figure 2.2 represents a world map depicting SO_2 emissions across regions during 2010. It can be observed that Chinese emissions are the highest. Chinese emissions of SO_2 can be linked to the increase (~73 %) in the total coal consumption from 1995 to 2006, marked by rapid growth of energy-intensive industries (Gao et al. 2009). Lu et al. (2010) reported that Chinese SO_2 emission increased by 53 %, during 2000–2006 (annual growth rate of 7.3 %) with emissions from power plants as the dominant source. However, the emission growth rate slowed around 2005, and consequently decreased emissions were observed after 2006 basically because of policy interventions to curb power plant emissions. The authors also highlighted the transport of SO_2 from the Asian continent to southwestern areas of Japan despite the relatively short atmospheric lifetime of SO_2. The United States Energy Information Administration (USEIA) has reported that the reduction in SO_2 in the USA after 1990 was basically the outcome of coal-fired power plants employing strategies (flue gas desulfurization (FGD), scrubbers, the use of lower sulfur coal) to curb emissions, not due to reduction in the use of coal. Coal generation was

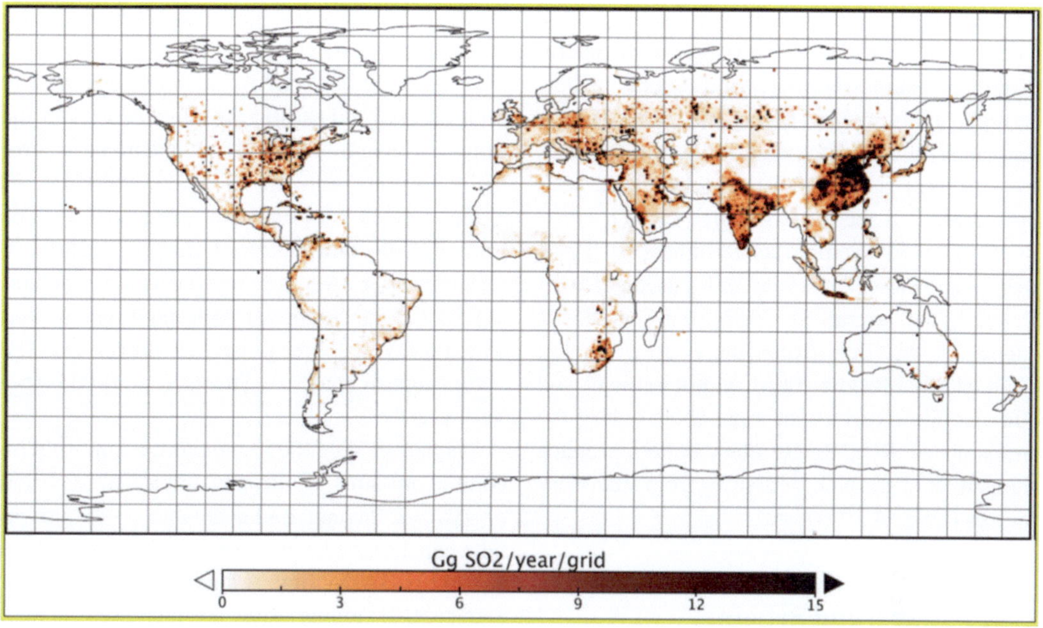

Fig. 2.2 Anthropogenic land-based emissions of SO_2 in 2010, gridded by 0.5×0.5°; Greenhouse Gas–Air Pollution Interactions and Synergies (GAINS) model (Adapted from Klimont et al. 2013)

high in 2007 when significant SO_2 reduction was already underway (USEIA 2013). Ray and Kim (2014) analyzed the long-term trend (1989–2010) of SO_2 levels in South Korea. The authors found that the period 2000–2010 recorded three to five times lower SO_2 levels than the period 1989–1999.

2.4 Oxides of Nitrogen

From an air quality perspective, oxides of nitrogen in the atmosphere include nitrous oxide (N_2O), nitrogen dioxide (NO_2), and nitric oxide (NO). N_2O (a greenhouse gas) is mostly emitted by natural sources and is usually not implied in the notation NO_x, which is expressed as the sum of NO_2 and NO. Interestingly the atmospheric residence of NO_x (NO + NO_2) is reported to be <0.01–0.03 years, whereas for N_2O it is around 120 years depending on photolysis rates and other chemical feedbacks that alter the atmospheric residence (IPCC 2001). NO is emitted from natural and anthropogenic sources, whereas NO_2 is emitted from combustion processes along with NO and atmospheric oxidation of NO. NO_3 and N_2O_5 are yet other oxides of nitrogen found in the atmosphere in comparatively lower concentrations and are primarily important from atmospheric chemistry perspectives. Health effects of NO_x, primarily from NO_2 exposure, include most notably respiratory effects, with a marked increase of outcomes in susceptible populations such as asthmatics, children, and elderly. ASTDR (2002) reports that low levels of nitrogen oxides in ambient air produce irritation of the eyes, nose, throat, and lungs, shortness of breath, tiredness, and nausea. High levels of nitrogen oxides have been associated with rapid burning, spasms, decreased oxygenation of body tissues, fluid accumulation in respiratory tract and lungs, and death. Secondarily, health effects due to inhalation exposure of particles (formed by reaction of NO_x with moisture, ammonia, and other compounds) and tropospheric ozone (formed by photochemical reaction of NO_x with volatile organic compounds) are also of serious concern (USEPA 2015b). Environmental impact of NO_2 is manifested in the form of acid rain (oxidation product nitric acid), photochemical smog initiation (nitrous acid is the major source of the highly reactive hydroxyl free radical), and tropospheric ozone formation (Seinfeld and Pandis 2006), eutrophication of water bodies, and deterioration of plants, animal life, and buildings (USEPA 1998). Fossil fuel combustion (emissions, 33.0 Tg N year^{-1}) is the largest source of NO_x emissions followed by biomass burning (7.1 Tg N year^{-1}), soils (5.6 Tg N year^{-1}), lightning (5.0 Tg N year^{-1}), and aircraft (0.7 Tg N year^{-1}) predominantly in the free troposphere at 8–12 km (Seinfield and Pandis 2006). Motor vehicles and other mobile sources contribute about half of the emissions of NO_x. Power plant boilers produce about 40 % of the NO_x emissions from stationary sources (USEPA 1999).

2.4.1 Global and Regional Emissions of Oxides of Nitrogen

One of the earliest estimates of NO_x emission sources by Logan (1983) indicates that combustion of fossil fuels contributes to ~40 % of the total emissions followed by biomass burning (~25 %) and the remaining from lightning and microbial activity in soils. The author also estimated that industrial and agricultural activities provide approximately two thirds of the global source for NO_x. The uncertainty in the estimations, especially for the contribution from fossil fuel combustion and industrial activities, was, however, around 30 %. Hameed and Dignon (1988) observed that the greatest emission rate during 1966–1980 was in the northern midlatitudes. However, the greatest increase in the period was recorded in the tropics. Jaeglé et al. (2005) used satellite observations from the Global Ozone Monitoring Experiment (GOME) instrument onboard the European Space Agency's ERS-2 satellite and found that fuel combustion dominates NO_x emissions at northern midlatitudes, while fires are a significant source in the tropics. Soil emissions over tropical savanna/woodland ecosystems (Africa), as well as over agricultural regions in the Western United States (Great Plains), Southern Europe (Spain, Greece, Turkey), and Asia (North China Plain and North India), were also evident. Summertime emissions over these regions were found to be significant at midlatitudes and during the rainy season in the tropics. The authors found soil and biomass burning emissions account for 22 % and 14 % of global surface NO_x emissions, respectively. The estimate for contribution of fuel combustion to NO_x emissions over Asia (8.8 TgN year^{-1}, including East Asia, Southeast Asia/India, and Japan) was, however, found to be 28 % larger than the bottom-up inventory of Streets et al. (2003) (6.9 TgN year^{-1}).

Richter et al. (2005) monitored tropospheric column amounts of nitrogen dioxide for a period of 1996–2004 and found notable reductions in nitrogen dioxide levels over some Europe regions and the USA. A highly significant increase of about 50 % along with an increased annual growth rate was found over the Chinese industrial areas. The authors used the data retrieved from two satellite instruments: GOME and Scanning Imaging Absorption Spectrometer for Atmospheric Chartography (SCIAMACHY) (launched on board ENVISAT). Figure 2.3 depicts a world map of nitrogen dioxide tropospheric column density (10^{15} molecules/cm^2) during 2011. High emission from Chinese region is clearly evident. Ghude et al. (2009) also used the data from the same source but for the period 1996–2006. The results exhibit that tropospheric NO(2) column amounts have increased in China (11 ± 2.6 %/year), South Asia (1.76 ± 1.1 %/year), the Middle East (2.3 ± 1 %/year), and South Africa (2.4 ± 2.2 %/year). Tropospheric NO(2) column amounts show some decrease over the Eastern United States (−2 ± 1.5 %/year) and Europe (0.9 ± 2.1 %/year). It was concluded that although tropospheric NO(2) column amounts decreased over the major developed regions in the past decade, the present tropospheric NO(2) column amounts over these regions are still significantly higher than those observed over newly and rapidly developing regions (except China). Tropospheric NO(2) column amounts show some decrease over South America and Central Africa, which are major biomass burning regions in the Southern Hemisphere. Ghude et al.

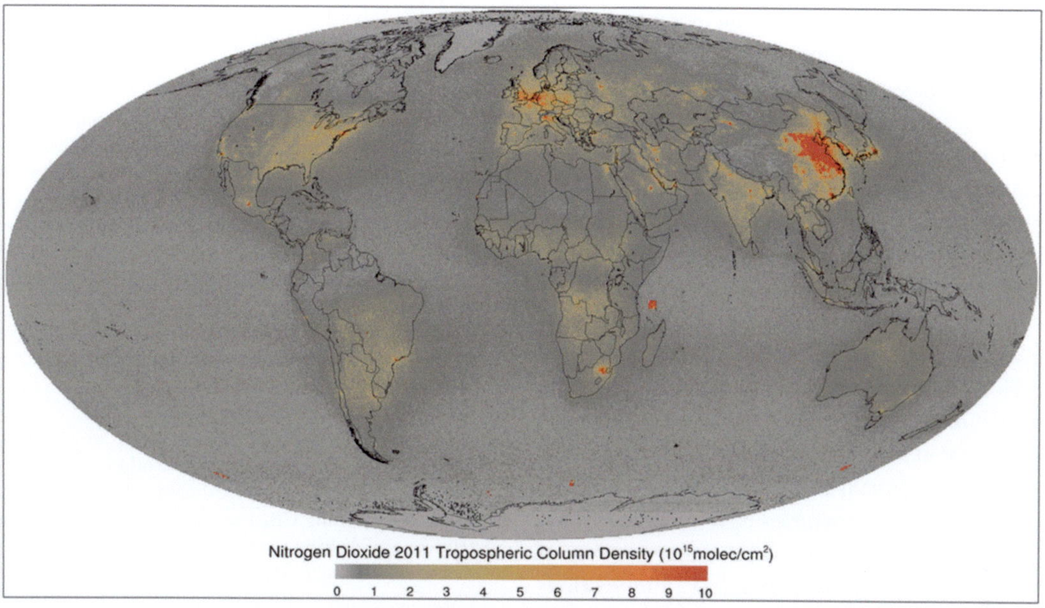

Fig. 2.3 World map of nitrogen dioxide tropospheric column density (10^{15} molecules/cm^2) based on Ozone Monitoring Instrument (OMI) onboard NASA's Aura satellite during 2011 (Source: https://en.wikipedia.org/wiki/Nitrogen_dioxide)

(2013) analyzed the scenario for India and found a growth rate of 3.8 % ± 2.2 % yr^{-1} between 2003 and 2011 for anthropogenic sources, which is clearly related to the growth in oil and coal consumption in India. Hilboll et al. (2013) measured tropospheric NO2 columns and found it to be strongly increasing over China, the Middle East, and India, with values over East Central China tripling from 1996 to 2011. Significant decrease in NO2 amounts during the period was observed in the USA, Western Europe, and Japan. In terms of city-wise missions, Dhaka, Bangladesh (+27.2 ± 3.9 % $year^{-1}$), and Baghdad, Iraq (+20.7 ± 1.9 % $year^{-1}$), were among the higher ones, whereas Los Angeles, USA (−6.00 ± 0.72 % $year^{-1}$), was among the lower ones. Megacities (population >10 million) in China, India, and the Middle East have registered increasing NO2 columns of +5–10 % $year^{-1}$. Lamsal et al. (2011) estimated an increase in anthropogenic NOx emissions over land by 9.2 % globally and by 18.8 % from East Asia during 2006–2009. North American emissions were found to decrease by 5.7 % during the same period. Delmas et al. (1997) attributed combustion of fossil fuel (~50 %) and biomass burning (~20 %) as the globally dominant sources of NO_x. Lightning and microbial activity in soils were found to be contributing less than 30 % of total emissions. Huntrieser et al. (2002) estimated higher NOx produced by lightning (mean 3 TgN $year^{-1}$) than produced by aircraft (0.6 TgN $year^{-1}$) for Europe as well as on a global scale.

2.5 Conclusions

The impact of ambient air quality on environmental and human health cannot be overemphasized. Numerous studies have established strong linkage between observed levels and deleterious effects. Of the many air pollutants, particulate matter and oxides of sulfur and nitrogen are ubiquitous and widely monitored for ambient levels, sector-wise contributions, regional emission inventories, and global trends over the past decades. Particulate matter is found to be on an increasing trend on a global scale despite decline in developed regions of the world around the start of the twenty-first century. PM emissions from the developing regions are on an increasing trend. Traffic sources including vehicular emissions and non-exhaust emissions, domestic fuel burning, and industrial activities are chief anthropogenic sources of PM. Prompt and early intervention in developing countries especially in the road transport sector to curb PM emissions through fuel switching, stricter emission norms, and newer fleet with latest emission abatement technology can potentially combat the increasing trend of PM emissions as is evident in the developed regions of the world. After industrial revolution, peak atmospheric levels of oxides of sulfur were witnessed in the 1970s which declined toward the end of the twentieth century. However, China and other developing countries have emerged in the present times as the dominant source regions of oxides of sulfur. Technological interventions rather than coal use in the developing countries which include controls on coal-fired power plants have been attributed to be a leading reason for significant decline in emissions. Similar trends have been observed for oxides of nitrogen with marked increase in Chinese emissions in the first decade of the twenty-first century. It has been found that although emissions of oxides of nitrogen from developed countries have reduced in the last few decades but it is still higher than emissions from developing countries, with the exception of China. Timely implementation of the emission control mechanisms is urgently needed to curb the future emissions from developing countries.

References

ALA (2013) American Lung Association, Disparities in the impact of air pollution. http://www.stateoftheair.org/2013/health-risks/health-risks-disparities.html. Accessed on 16 June 2014

ATSDR (1998) Agency for Toxic Substances and Disease Registry. Toxicological profile for sulfur dioxide. US Department of Health and Human Services, Public Health Service, Atlanta

ATSDR (2002) Agency for Toxic Substances and Disease Registry. ToxFAQs™ nitrogen oxides. US Department of Health and Human Services, Public Health Service, Atlanta

Bateson TF, Schwartz J (2004) Who is sensitive to the effects of particulate air pollution on mortality? A case-crossover analysis of effect modifiers. Epidemiology 15(2):143–149

Black F, Tejada S, Gurevich M (1998) Alternative fuel motor vehicle tailpipe and evaporative emissions composition and ozone potential. J Air Waste Manage Assoc 48(7):578–591

Bowman CT (1991) Chemistry of gaseous pollutant formation and destruction. In: Bartok W, Sarofim AF (eds) Fossil fuel combustion. Wiley, New York, pp 215–260

Bremmer SA, Anderson HR, Atkinson RW et al (1999) Short term association between outdoor air pollution and mortality in London 1992–4. Occup Environ Med 56(4):237–244

Brimblecombe P (ed) (2003) The effects of air pollution on the built environment, vol 2, Air Pollution Reviews. World Scientific Publishing Company, New Jersey

Capareda SC L, Wang CB, Parnell Jr, Shaw BW (2004) Particle size distribution of particulate matter emitted by agricultural operations: impacts on FRM PM10 and PM2.5 concentration measurements. In: Proceedings of the 2004 Beltwide Cotton Production conferences. National Cotton Council, Memphis, TN

Chen TM, Gokhale J, Shofer S, Kuschner WG (2007) Outdoor air pollution: nitrogen dioxide, sulfur dioxide, and carbon monoxide health effects. Am J Med Sci 333(4):249–256

Cohen AJ, Pope CA (1995) Lung cancer and air pollution. Environ Health Perspect 103(8):219–224

DEFRA (2013) Department for Environment Food & Rural Affairs, United Kingdom. http://uk-air.defra.gov.uk/air-pollution/effects. Accessed 13 June 2014

Delmas R, Serça D, Jambert C (1997) Global inventory of NOx sources. Nutr Cycl Agroecosyst 48(1):51–60

Dey S, Di Girolamo L, van Donkelaar A, Tripathi SN, Gupta T, Mohan M (2012) Variability of outdoor fine particulate (PM2.5) concentration in the Indian Subcontinent: a remote sensing approach. Remote Sens Environ 127:153–161

Dockery D, Pope CA, Xu X, Spengler J, Ware J, Fay M, Ferris B, Speizer F (1993) An association between air pollution and mortality in six U.S. cities. N Engl J Med 329:1753–1759

EEA (2012) Emissions of primary PM2.5 and PM10 particulate matter. European Environmental Agency. http://www.eea.europa.eu/data-and-maps/indicators/emissions-of-primary-particles-and-5. Accessed on 14 June 2014

EEA (2014) European Environmental Agency. http://www.eea.europa.eu/data-and-maps/explore-interactive-maps#c5=air&c0=5&b_start=0. Accessed on 14 June 2104

Forastiere F, Stafoggia M, Tasco C, Picciotto S, Agabiti N, Cesaroni G et al (2006) Socioeconomic status, particulate air pollution, and daily mortality: differential exposure or differential susceptibility. Am J Ind Med 50(3):208–216

Gao C, Yin H, Ai N, Huang Z (2009) Historical analysis of SO_2 pollution control policies in China. Environ Manage 43(3):447–457

Ghude SD, RJ V d A, Beig G, Fadnavis S, Polade SD (2009) Satellite derived trends in NO2 over the major global hotspot regions during the past decade and their inter-comparison. Environ Pollut 157(6):1873–1878

Ghude SD, Kulkarni SH, Jena C, Pfister GG, Beig G, Fadnavis S, van der A RJ (2013) Application of satellite observations for identifying regions of dominant sources of nitrogen oxides over the Indian Subcontinent. J Geophys Res 118(2):1–15

Goncalves FLT, Carvalho LMV, Conde FC, Latorre MRDO, Saldiva PHN, Braga ALF (2005) The effects of air pollution and meteorological parameters on respiratory morbidity during the summer in São Paulo City. Environ Int 31(3):343–349

Grantz DA, Garner JH, Johnson DW (2003) Ecological effects of particulate matter. Environ Int 29(2–3):213–239

Hameed S, Dignon J (1988) Changes in the geographical distributions of global emissions of NOx and SOx from fossil fuel combustion between 1966–1980. Atmos Environ 22(3):441–449

Hand JL, Schichtel BA, Malm WC, Pitchford ML (2012) Particulate sulfate ion concentration and SO2 emission trends in the United States from the early 1990s through 2010. Atmos Chem Phys 12(21):10353–10365

Hilboll A, Richter A, Burrows JP (2013) Long-term changes of tropospheric NO_2 over megacities derived from multiple satellite instruments. Atmos Chem Phys 13(8):4145–4169

Huntrieser H, Feigl C, Schroder F, Gerbig C, van Velthoven P, Flatoy F, Thery C, Petzold A, Holler H, Schumann A (2002) Contribution of lightning- produced NOx to the European and global NOx budget: results and estimates from airborne EULINOX measurements. J Geophys Res 2(107):4113

IARC (2013) International Agency for Research on Cancer, outdoor air pollution a leading environmental cause of cancer deaths. http://www.iarc.fr/en/media centre/iarcnews/pdf/pr221_E.pdf. Accessed 12 May 2014

IPCC (2001) Contribution of Working Group I to the Third Assessment Report of the Intergovernmental Panel on Climate Change. In: Houghton JT, Ding Y, Griggs DJ, Noguer M, van der Linden PJ, Dai X, Maskell K, Johnson CA (eds) Climate change 2001: the scientific basis. Cambridge University Press, Cambridge/New York, p 881

IPCC (2007) Contribution of Working Group I to the fourth assessment report of the Intergovernmental Panel on Climate Change. In: Solomon S, Qin D, Manning M, Chen Z, Marquis M, Averyt KB, Tignor M, Miller HL (eds) Climate change 2007: the physical science basis. Cambridge University Press, Cambridge/New York, p 996

Jaeglé L, Steinberger L, Martin RV, Chance K (2005) Global partitioning of NOx sources using satellite observations: relative roles of fossil fuel combustion, biomass burning and soil emissions. Faraday Discuss 130:407–423

Jerrett M, Burnett RT, Brook J, Kanaroglou P, Giovis C, Finkelstein N et al (2004) Do socioeconomic characteristics modify the short term association between air pollution and mortality? Evidence from a zonal time series in Hamilton, Canada. J Epidemiol Commun H 58(1):31–40

Karagulian F, Belis CA, Dora CFC, Prüss-Ustün AM, Bonjour S, Adair-Rohani H, Amann M (2015) Contributions to cities' ambient particulate matter (PM): a systematic review of local source contributions at global level. Atmos Environ 120:475–483

Kassomenos PA, Vardoulakis S, Chaloulakou A, Paschalidou AK, Grivas G, Borge R, Lumbreras J (2014) Study of PM10 and PM2.5 levels in three European cities: analysis of intra and inter urban variations. Atmos Environ 87:153–163

Kim JW, Park S, Lim CW, Lee K, Kim B (2014) The role of air pollutants in initiating liver disease. Toxicol Res 30(2):65–70

Klimont Z, Smith SJ, Cofala J (2013) The last decade of global anthropogenic sulfur dioxide: 2000–2011 emissions. Environ Res Lett 8:014003

Kundu S, Stone EA (2014) Composition and sources of fine particulate matter across urban and rural sites in the Midwestern United States. Environ Sci Proc Impacts 28(6):1360–1370

Lamsal LN, Martin RV, Padmanabhan A, van Donkelaar A, Zhang Q, Sioris CE, Chance K, Kurosu TP, Newchurch MJ (2011) Application of satellite observations for timely updates to global anthropogenic NOx emission inventories. Geophys Res Lett 38(5):L05810

Li J, Carlsona BE, Lacisa AA (2015) How well do satellite AOD observations represent the spatial and temporal variability of PM2.5 concentration for the United States? Atmos Environ 102:260–273

Logan JA (1983) Nitrogen oxides in the troposphere: global and regional budgets. J Geophys Res 88(C15):10785–10807

Lu Z, Streets DG, Zhang Q, Wang S, Carmichael GR, Cheng YF, Wei C, Chin M, Diehl T, Tan Q (2010) Sulfur dioxide emissions in China and sulfur trends in East Asia since 2000. Atmos Chem Phys 10:6311–6331

Luo XS, Ip CCM, Li W, Tao S, Li XD (2014) Spatial–temporal variations, sources, and transport of airborne inhalable metals (PM10) in urban and rural areas of northern China. Atmos Chem Phys Discus 14:13133–13165

Malhotra SS, Hocking D (1976) Biochemical and cytological effects of sulphur dioxide on plant metabolism. New Phytol 76(2):227–237

Mar TF, Norris GA, Loenig JQ et al (2000) Associations between air pollution and mortality in Phoenix, 1995–1997. Environ Health Perspect 108(4):347–353

Marrapu P, Cheng Y, Beig G, Sahu S, Srinivas R, Carmichael GR (2014) Air quality in Delhi during the Commonwealth Games. Atmos Chem Phys 14(19):10619–10630

Maynard RL, Ayres J (2014) Air pollution and health. In: Harrison RM (ed) Pollution: causes, effects and control, vol 5E. RSC Publishing, Cambridge

Neidell MJ (2004) Air pollution, health, and socio-economic status: the effect of outdoor air quality on childhood asthma. J Health Econ 23(6):1209–1236

OECD (2012) The Organization for Economic Co-operation and Development Environmental outlook to 2050: the consequences of inaction. http://www.oecd.org/env/indicators-modelling-outlooks/oecdenvironmentaloutlookto2050theconsequencesofinaction.htm. Accessed 10 July 2014

OECD (2014) The effects of air pollution on mortality in socially deprived urban areas in Hong Kong, China. Environ Health Perspect Organization for Economic Co-operation and Development. The Cost of Air Pollution: Health Impacts of Road Transport, OECD Publishing, Paris. doi:http://dx.doi.org/10.1787/9789264210448-en

Pope CA III, Burnett RT, Thurston GD, Thun MJ, Calle EE, Krewski D et al (2004) Cardiovascular mortality and long-term exposure to particulate air pollution: epidemiological evidence of general pathophysiological pathways of disease. Circulation 109(1):71–77

Pulido L (2000) Rethinking environmental racism: white privilege and urban development in southern California. Am Geogr 90(1):12–40

Ray S, Kim KH (2014) The pollution status of sulfur dioxide in major urban areas of Korea between 1989 and 2010. Atmos Res 147–148:101–110

Richter A, Burrows JP, Nüß H, Granier C, Niemeier U (2005) Increase in tropospheric nitrogen dioxide over China observed from space. Nature 437:129–132

Saxena P, Seigneur C (1987) On the oxidation of SO_2 to sulfate in atmospheric aerosols. Atmos Environ 21(4):807–812

Seinfeld JH, Pandis SN (2006) Atmospheric chemistry and physics: from air pollution to climate change. Wiley, Hoboken

Smith SJ, Andres R, Conception E, Lurz J (2004) Sulfur dioxide emissions: 1850–2000. Joint Global Change Research Institute report. Pacific Northwest National Laboratory-14537

Smith SJ, van Aardenne J, Klimont Z, Andres RJ, Volke A, Arias SD (2011) Anthropogenic sulfur dioxide emissions: 1850–2005. Atmos Chem Phys 11:1101–1116

Streets DG, Bond TC, Carmichael GR, Fernandes SD, Fu Q, He D, Klimont Z, Nelson SM, Tsai NY, Wang MQ, Woo J-H, Yarber KF (2003) An inventory of gaseous and primary aerosol emissions in Asia in the year 2000. J Geophys Res 108:8809

Sunyer J, Schwartz J, Tobias A, Macfarlane D, Garcia J, Anto JM (2000) Patients with chronic obstructive pulmonary disease are at increased risk of death associated with urban particle air pollution: a case-crossover analysis. Am J Epidemiol 151(1):50–56

Takeshita T (2011) Global scenarios of air pollutant emissions from road transport through to 2050. Int J Environ Res Public Health 8(7):3032–3062

Tao F, Gonzalez-Flecha B, Kobzik L (2003) Reactive oxygen species in pulmonary inflammation by ambient particulates. Free Radic Biol Med 35(4):327–340

Thomas MD (1961) Effects of air pollution on plants. In: Air pollution. Columbia University Press, New York, p 442

Upadhyay A, Clements FM, Herckes P (2011) Chemical speciation of $PM_{2.5}$ and PM_{10} in South Phoenix, Arizona. J Air Waste Manage Assoc 61(3):302–310

USEIA (2013) Power plant emissions of sulfur dioxide and nitrogen oxides continue to decline in 2012. United States Energy Information Administration https://www.eia.gov/todayinenergy/detail.cfm?id=10151. Accessed on 31 Dec 2015

USEPA (1998) United States Environmental Protection Agency. NO_x: how nitrogen oxides affect the way we live and breathe. Office of Air Quality Planning and Standards Research Triangle Park, NC. EPA-456/F-98-005

USEPA (1999) United States Environmental Protection Agency. Technical Bulletin "Nitrogen Oxides (NOx), Why and How they are controlled. Office of Air Quality Planning and Standards Research Triangle Park. EPA 456/F-99-006R

USEPA (2002) United States Environmental Protection Agency. Health Assessment Document for Diesel Emission. Office of Research and Development, Durham. EPA/600/8-90/057C

USEPA (2005) United States Environmental Protection Agency. Guideline for carcinogen risk assessment. Risk Assessment Forum, Washington, DC. EPA/630/P-03/001F

USEPA (2014) AQI- Air quality index- a guide to air quality and your health. U.S. Environmental Protection Agency Office of Air Quality Planning and Standards Outreach and Information Division Research Triangle Park, NC February 2014 EPA-456/F-14-002

USEPA (2015a) United States Environmental Protection Agency. http://www.epa.gov/haps/initial-list-hazardous-air-pollutants-modifications. Accessed on 23 Dec 2015

USEPA (2015b) United States Environmental Protection Agency. http://www3.epa.gov/airquality/nitrogenoxides/health.html. Accessed on 23 Dec 2015

USEPA (2016) United States Environmental Protection Agency. http://www3.epa.gov/airtrends/pm.html. Accessed 10 Jan 2016

Van Dingenen R, Raes F, Putaud JP, Baltensperger U, Charron A, Facchini MC, Decesari S, Fuzzi S, Gehrig R, Hansson HC, Harrison RM, Huglin C, Jones AM, Laj P, Lorbeer G, Maenhaut W, Palmgren F, Querol X, Rodriguez S, Schneider J, ten Brink H, Tunved P, Torseth K, Wehn B, Weingartner E, Wiedensohler A, Wahlin P (2004) A European aerosol phenomenology-1: physical characteristics of particulate matter at Kerbside, urban, rural and background sites in Europe. Atmos Environ 38(16):2561–2577

van Donkelaar A, Martin RV, Brauer M, Boys BL (2015) Use of satellite observations for long-term exposure

assessment of global concentrations of fine particulate matter. Environ Health Perspect 123(2):135–143
Varshney CK, Garg JK, Lauenroth WK, Heitschmidt RK (1979) Plant responses to sulfur dioxide pollution. Crit Rev Env Contr 9(1):27–49
Welch H (1998) Mercury accumulation in snow and sea ice. Synopsis of research conducted under the 1997/98 Northern Contaminants Program. Indian and Northern Affairs Canada, Ottawa
WHO (2000) World Health Organization. Effects of sulfur dioxide on vegetation: critical levels Air Quality Guidelines – Second Edition. WHO Regional Office for Europe, Copenhagen, Denmark
WHO (2014) World Health Organization. http://www.un.org/climatechange/blog/2014/03/7-million-premature-deaths-annually-linked-to-air-pollution/#content. Accessed 24 June 2014
Winner WE, Mooney HA, Goldstun RA (1985) Sulphur dioxide and vegetation: physiology, ecology, and policy issues. Standard Univ Press, Stanford
Wong CM, Ou CQ, Chan KP, Chau YK, Thach TQ, Yang L et al (2008) The effects of air pollution on mortality in socially deprived urban areas in Hong Kong, China. Environ Health Perspect 116(9):1189–1194

Urban Air Pollutants and Their Impact on Biota

3

Svetlana Stevović and Jelena Markovic

Abstract

This chapter presents the results of studies of changes in the environment caused by air pollution increasement, in urban areas in the function of time, Vranje town case study, southern Serbia region. Vranje is a typical town with increasing traffic and industry and consequently increasing air contamination. Emissions of pollutants CO, nitrogen oxides, and soot were measured, as well as consequent pollution of agricultural land and biota in the town and region. Air quality measurements were carried out at two locations in the town. Three hundred sixty-five samples of sulfur dioxide, 365 samples of soot, and 365 samples of nitrogen oxides were collected and analyzed by physical chemical methods. The emission of pollutants from stationary sources, which was maximal on two representative sites, is determined by measuring and calculating the emission parameters, based on the measurement results. Measuring emissions of pollutants is done using special devices that are calibrated in accordance with EU legislation. Results of measuring emissions were compared with emission limit values, which are given by the Regulation and comply with EU environmental legislation. Impact of air pollution is analyzed on the biota within the town and in the surrounding agriculture areas. The conclusion is that pollution of air and soil is more intense in the winter. Pollution arrives even to agricultural products that are grown in the region. Findings indicate the imperative shift toward renewable energy sources and sustainable environmentally friendly technical and industrial solutions, in order to decrease pollutions in the towns and to preserve the quality of the environment.

S. Stevović (✉)
Faculty for Ecology and Environmental Protection,
University Union Nikola Tesla,
Cara Dusana 62-64, Belgrade, Serbia
e-mail: svetlanas123@gmail.com

J. Markovic
College of Applied Professional Studies,
Filipa Filipovića 20, Vranje, Serbia

U. Kulshrestha, P. Saxena (eds.), *Plant Responses to Air Pollution*,
DOI 10.1007/978-981-10-1201-3_3

Keywords
Air pollutants • Urban environment • Environment • Region • Biota

3.1 Introduction

The air is a gas mixture of variable composition. It consists of about 78 % nitrogen, 20.95 % oxygen, 0.93 % of argon, and very small quantities of krypton, xenon, and neon. The existing gases and particles in free and indoor spaces are important for the analysis of air pollutants. The emission and immission concentrations are measured in free space. The emission concentrations of polluting substances are found in exhaust gases before their release and dispersion into the environment. The immission concentrations are those which express the quality of the environment at a given time and space after the dispersion of gases in the surrounding area. A large number of different substances, which are present in clean air in very low concentrations or not present at all, are found in polluted air. All the polluting substances have never been found at one site, and this is the reason why they are not measured at the same location (Grsic et al. 2014). The substances which are regularly measured do not have to be measured on a daily basis.

Air protection is achieved by measures of systematic monitoring of air quality; by reducing concentrations of air pollutants below the prescribed limit values; by taking appropriate technical, technological, and other measures required to reduce emissions; as well as by continuous monitoring of the impact of air pollution on human health and the environment.

Different models for selecting the appropriate representative location for air pollution measurement control are used in the world (Mazzeo and Venegas 2000; Noll et al. 1977). The supporting specific programs for data processing, i.e., for the measured values of pollution (Tseng and Chang 2001), are carried out at the same time. Designing monitoring networks (Lanstaff et al. 1987) are particularly important in urban areas because a large number of people are exposed to air pollution in cities. Therefore, the focus of researchers on the pollution in urban areas is reasonable and clear. Besides, the monitoring network in one city (Coxford and Penn 1998; Noll and Mitsutomi 1983) is not only locally important, it is also regionally important, because of transboundaring pollutions (Milutinovic and Popovic 2001).

This chapter presents the results of research in the field of monitoring pollutant emissions of CO, nitrogen oxides (NO_x) expressed as nitrogen dioxide (NO_2), and soot in the urban environment and surrounding region. Measurements were taken at two sites and these are the Department of Public Health Vranje and the primary school "Svetozar Markovic" in Vranje. Three hundred sixty-five samples of sulfur dioxide, 365 samples of soot, and 365 samples of nitrogen oxides were collected and analyzed and examined by physicochemical methods. The emission of pollutants from stationary sources, such as these sites were, was determined by measuring or calculating the emission parameters based on the measurement results. The measurement of emissions of pollutants was conducted by using special devices that have specific legislation regulated. The measurement results were compared with the emission limit values. Table 3.3 shows the maximum allowed value of soot.

The main objective of this paper is to examine air pollution in quantitative and qualitative terms and to indicate the direction and the imperative to seek solutions for the established system of air quality monitoring in the urban environment that is in the city of Vranje. Urban areas affect the natural landscape changes and changes in the environment (Borgstrom 2009). The changes in air pollution affect the quality of the environment in the region, and these changes are reflected in people's lives. The changes in temperature, which has been varying and changing lately, also affect the changes in air pollution (Homer et al. 2010).

3.2 Materials and Methods

The experiment of systematic monitoring of air pollution was conducted in a small urban environment, the city of Vranje. Vranje is a city located in southern Serbia and it belongs to the Pčinja District. It occupies an area of 860 km^2 (the agricultural area covers 44,721 ha and the forest area covers 32,478 ha). Vranje has about 60,000 inhabitants. It is located in the valley of Vranje, on the banks of the Vranjska River, near its confluence with the South Morava. The climate is moderate continental with a varied soil structure. There is no significant industry in the city center, so the biggest polluters are the boiler room and traffic. The heating plant has boiler rooms which have not been operating at full capacity. Boiler rooms use convectional fuel that significantly affects the quality of the environment.

They are not equipped with modern filters which are used for air purification, and also they are not constructed in accordance with the new standards. Therefore, it can be said that during the winter period, when the heating season is underway, air pollution increases. This affects boiler rooms to operate at reduced capacity, and therefore they cannot meet the heating needs of the entire city. Since Vranje is a town that still lags behind certain metropolitan areas, the individual air pollutants which increase the air pollution in the city by their frequent and constant fuelling are commonly found.

In this paper, the two locations are shown. One of the two is excluded from industrial pollution and traffic and this is the Institute of Public Health of Vranje, while the other takes into account the individual contaminants and is also exempt from the impact of industrial pollution and traffic.

The monitoring program, or the process of air quality monitoring, determines which pollutants and by what mode (permanently or temporarily) will be measured in the area where the air quality is tested. It can be said that an increased amount of air adversely affects the urban ecosystems and the present biodiversities (Kowarik 2011). These are usually those substances that may endanger the tested area and those which are known to occur in a particular area based on experience or based on previous measurements (Fig. 3.1).

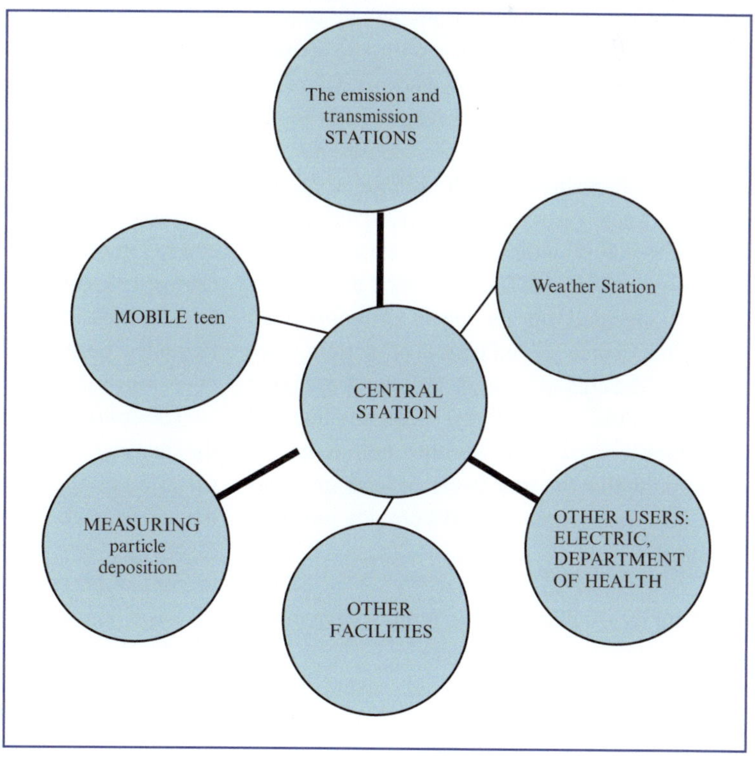

Fig. 3.1 Block diagram of the monitoring system for air quality monitoring

The concentration of gases in the air can be changed significantly in a short period of time. Therefore, the measurement of medium concentrations is common in a certain time period, usually 24 h.

One of the first measures of air quality control is an organization system for continuous monitoring and testing of the most harmful substances in the air. The system consists of a network of measuring stations and a monitoring system that, in addition to measurements, contains a subsystem for the collection, processing, storage, and distribution of data (Abellan-Nebot and Subiron 2010). The local Public Health Institute monitors air quality measurements and examines the amount of particles of SO_2, soot, and NO_2 present in the air. According to the Institute of Public Health of Vranje, regarding the examination of the amount of pollutants in the air, it can be said that air quality across the city varies. This means that air pollution changes over time. The findings reveal that air pollution has increased in recent years due to the development of industry and heavier traffic. It can also be said that the soil is exposed to pollution.

The purpose of monitoring is to determine the concentration of substances in the air and to compare the obtained values with the quality standards or with the ILV values (limit values of immission) and ELV (limit values of emission). The limit value of immission represents the maximum allowable concentration value of pollutants in waste gases from stationary and mobile sources of pollution that can be released into the air in a given period. The substance to be measured is decided on the basis of knowledge about the degree of influence of certain substances to the area of interest. Pollutants are most commonly measured in the area of interest. Concentrations of some specific pollutants are measured within the framework of local monitoring. Local monitoring networks provide information on:

- Existence of characteristic point source emissions (large energy facilities)
- The temperature regime of space (local overheating, ventilation)
- Meteorological parameters
- Characteristics of relief
- Traffic density
- Previous measurements

All the measurements of pollutants were carried out in the laboratory for testing emissions, noise, and wastewater into the environment at the Institute of Public Health in Vranje. The methods of chemical and physical tests were applied.

The limit value of emission (ELV) is the mass expressed in the form of specific parameters of concentration and/or level of an individual emission that is not allowed to be exceeded during one or more periods of time in accordance with special regulations (Law on Environmental Protection Official Gazette of RS 135/04. The limit value of emission is determined on the basis of characteristics of the installation, the geographical location, and the local environmental conditions.

Figure 3.2 presents the measuring points in the city of Vranje, the development of which has increased pollution in the environment of the entire region of South Serbia, Fig. 3.3.

The measurement and analysis of carbon monoxide (CO) and nitrogen oxides (NO_x) expressed as nitrogen dioxide (NO_2) were carried out within the framework of the conducted research; and the automatical determination of the smoke number and the contents of certain organic compounds were also conducted.

Based on the tests, it was shown that the concentration of nitrogen oxides (NO_x) expressed as nitrogen dioxide (NO_2) is a key component of phytotoxic emissions (Gupta et al. 2015) and it highlights the potential for adverse effects of vehicle emissions in urban ecosystems (Bell et al. 2011).

The following must be done in order to assist the measurement of air pollutant emissions: identification of all stationary sources of emissions in the air owned by the operator, identification of all discharges (emitters) at stationary sources, identification of pollutants and state parameters of waste gases to be measured at each discharge, identification method of measuring the emission and immission, and identification of the limit values of emission.

Fig. 3.2 The measuring points, the city of Vranje

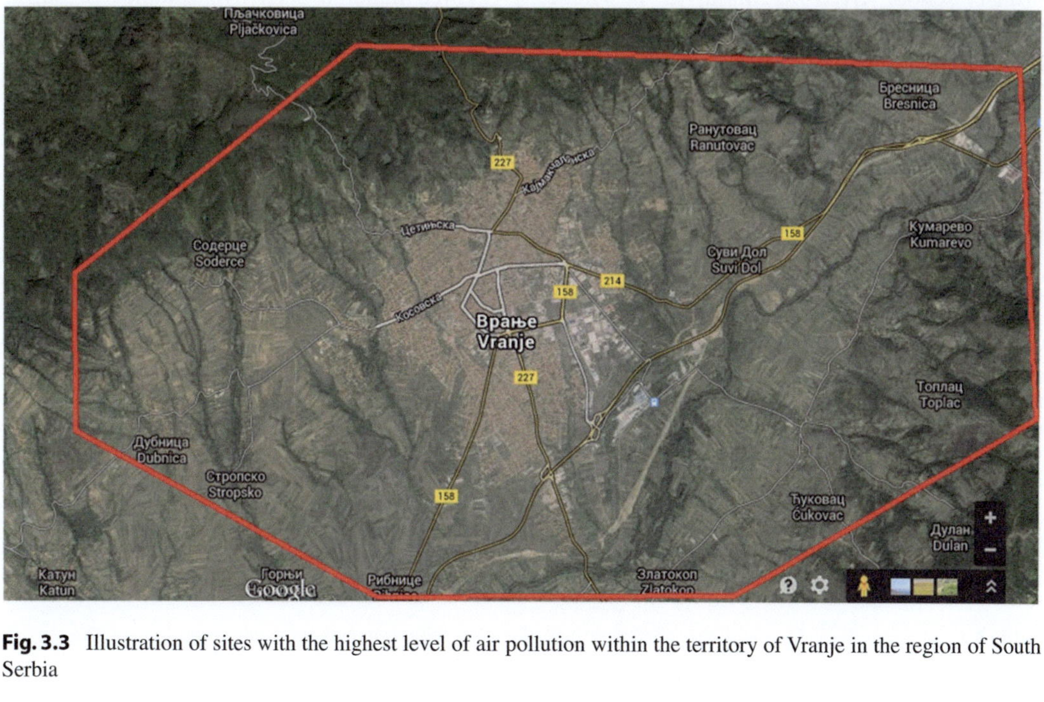

Fig. 3.3 Illustration of sites with the highest level of air pollution within the territory of Vranje in the region of South Serbia

The sampling of gaseous pollutants is conducted by Proekos brand devices (type AT-401) for air sampling, by absorption of contaminants from a known volume of air in the proper absorption solution.

The samples of soot are obtained by filtering a known volume of air through a filter paper. Refractometric measurements of soot index are performed on a device called a refractometer.

3.3 Results and Discussion

3.3.1 Results of Monitoring

The control of air quality in the city of Vranje is performed on two sites. One of these two is the Institute of Public Health of Vranje, which is located near the city center so the results would represent the air quality mainly from traffic impact. This measurement point is located within the health center, in close proximity to several busy roads. In order to monitor the degree of air pollution, the Institute of Public Health of Vranje measures the emission concentration of certain parameters of air pollution on the basis of the contract on the regulation of rights and obligations in performing air quality control and monitoring of the impact of air pollution in the city of Vranje. The second measuring point is within the primary school "Svetozar Markovic" in Vranje. This measuring point is in the direction of the dominant northeast wind in relation to the industrial zone, so that the results can maintain the impact of the industrial zone during the wind flow and the impact of local furnaces and local roads. The school is located in a residential area dominated mostly by individual residential buildings of low level, so air pollution occurs from individual furnaces and means of transport (Vosniakos et al. 2008).

The process of the measuring point selection took into account the following: the position and type of pollution, population density, topography, and meteorological conditions.

Air quality control is performed three times: after a month, after 3 months, and after a year (12 months). Daily systematic measurement of basic pollutants SO_2, NO_2, and soot and analysis of aero sediments in monthly precipitation samples are performed at both measuring points. The analysis of aero sediments determines the following: total sediment matter, pH, specific conductivity, sulfates, chlorides, ammonium ion, nitrate, nitrite, calcium, dissolved solids, combustible part, and the ash content. All the collected and analyzed samples of SO_2 and soot and the total sediment matter for 2013 were statistically analyzed and presented in accordance with the Law on Air Protection "Official Gazette of RS" No. 36/09, the Regulation on conditions and requirements for air quality monitoring "Official Gazette of RS" No. 11/10, and Regulation on elections and amendments and Regulation on conditions and requirements for air quality monitoring "Official Gazette of RS" No. 75/2010.

Three hundred sixty-five samples of sulfur dioxide, 365 samples of soot, and 365 samples of nitrogen oxides were collected and analyzed at the two measuring points, at the Institute of Public Health Vranje and the primary school "Svetozar Markovic" during 2013.

Table 3.1 shows the results of individual emission measurements of gaseous substances (CO, C_4H_{10}, C_8H_{18}, C_6H_{12}, C_7H_8, C_8H_{10}, C_2H_4O, C_2H_6O, C_3H_8O, $C_4H_{10}O$, $C_6H_{12}O$, CHOH, C_3H_6O, HF, C_5H_{12}, C_4H_8, and organic substances, expressed as total carbon), as well as the limit values of emission (ELV).

Based on measurements of emissions, which are shown in Table 3.1, it can be concluded that the measured values of pollutant emissions (CO and NO_2) and the smoke number are within the

Table 3.1 Measurement of gas emission concentrations of CO and NO_x is expressed as concentrations of NO_2

Measurement parameters	The unit of measurement	The measured value ± measurement uncertainty: The 1st measurement	The 2nd measurement	The 3rd measurement	ELV
Concentration of CO	mg/m^3	832,50±2,2 %	968,75±2,2 %	925,00±2,2 %	1000
Concentration of nitrogen oxides (NO_x)[a]	mg/m^3	8,20±3,2 %	8,20±3,2 %	10,25±3,2 %	250
The smoke number		1±0,2	1±0,2	1±0,2	≤1

[a]Expressed as nitrogen dioxide (NO_2)

permitted emission limits. Also the concentration of CO, individual organic matter, and organic substances expressed as total carbon are within the permissible limits of normal. Tables 3.2 and 3.3 show the results of the limit values and tolerance values of pollutants SO_2, NO_2, and the maximum permissible value of soot and total sediment matter.

Sulfur dioxide (SO_2) is a mandatory component of polluted air in urban areas. It is the product of the combustion of fossil fuels and other fuels, especially those that are rich in sulfur. It can be found in the air as a gas or dissolved in water droplets. In conditions of increased humidity, sulfur dioxide oxidizes and it partially transfers to the sulfuric or sulfurous acid. The concentration of sulfur dioxide in the air depends on temperature, air movement, humidity, and atmospheric pressure. The process of combustion of fuel is accompanied by the appearance of smoke (black smoke), which, depending on the efficiency of combustion, may contain more or less solid particles. Soot contains a large number of organic polycyclic aromatic compounds, whose particle size is around 5 μ and they remain in the air in the form of aerosols.

Based on the obtained results, it can be concluded that the total concentration of the above pollutants does not exceed the limit value allowed by the current Regulation.

These tables show that the averaging time was 24 h, and the results are shown as mean monthly, mean annual, and minimum and maximum values. Results in Tables 3.2 and 3.3 show that the measured emission of pollutants is within the acceptable limits, but there were days when the measured values exceeded the limit value, tolerance value, and the maximum allowable value. This occurs mainly in winter due to increased concentrations of soot obtained by fuelling.

Data source from the Table 3.4 marked by number 1 indicates that the data were obtained from the Institute of Public Health in Vranje. The methods of measurements marked by M in the Table 3.4 mean that they were carried out manually, and those marked by A mean that they were carried out automatically.

Table 3.2 Limit values and tolerance values of pollutants SO_2 and NO_2

The pollutant	The averaging period	Limit value (μg/m^3)	Tolerance value (μg/m^3)
SO_2	1 day	125	125
	Calendar year	50	50
NO_2	1 day	85	125
	Calendar year	40	60

Table 3.3 Maximum permissible value of pollutants (soot and total sediment matter)

The pollutant	The averaging period	The maximum allowed value (μg/m^3)
Soot	1 day	50
	Calendar year	50
Total sediment matter	1 day	450
	Calendar year	200

Table 3.4 Mean annual concentration of soot, SO_2, and NO_2, the number of days when the daily values exceeded the ELV and categories of air quality in 2014 determined on the basis of mean annual values

	Soot(μg/m^3)	SO_2(μg/m^3)	NO_2(μg/m^3)
The measuring point	Vranje IPH	Vranje IPH	Vranje IPH
The mean value	13	14	25
The number of days	7	1	2
Max daily value	89	79	86
Data source	1	1	1
The method of measurement	M	A	A
Availability (%)	96	87	97

The measurements obtained in accordance with the Art. 21 of the Law on Air Protection show that the concentration of air quality in Vranje is classified into the first category – slightly polluted air or clean air.

However, changes in the concentration of soot, nitrogen oxide, and sulfur dioxide greatly disturb the soil quality, as well as plants. Although the concentration of sulfur dioxide is within the normal range, these studies show that even small changes in the concentrations have a negative effect on the soil. Analyses carried out in the Institute of Agriculture, concerning the concentration of sulfur dioxide, have shown that gas negatively affects the quality of the soil because of the discoloration of plants within this period followed by a collapse of leaves and their eventual dying off.

The concentration of nitrogen oxides in the air, which is most commonly a result of exhaust gases, also adversely affects the quality of the soil. Since nitrogen is an important component of plants, and they absorb it mostly from the soil, polluted soil will cause the plants to be of poor quality.

The concentration of soot, which is here within the range of allowed values, reflects negatively on the flora and fauna especially in the fall due to stubble burning after which the soil is covered in soot in a greater degree. Table 3.5 shows the values of the emission concentration of soot, nitrogen oxide, and sulfur dioxide, which are harmful to the soil or disturb the quality parameters, which are in fact indicators of the unpolluted soil.

Increased urbanization in South Serbia has had a negative effect on the change of air temperature as well. Urbanization, which has increasingly become dominant in recent years, constitutes a major source of anthropogenic carbon dioxide emissions from the burning of fuels for heating purposes, from industrial processes, and from transportation. The magnitude of urban warming is highly variable in terms of time and space (Mitchell et al. 2002). Urban air temperatures are on average by 1–3 °C warmer although air temperatures in winter warm up to 10 °C due to constant fuelling when compared to rural areas. Air temperatures differ across the urban city. Temperatures from one to the other side of the street, from the park to the industrial zone differ significantly (Svirejeva-Hopkins et al. 2004). The largest intra-urban differences are associated with the open sky and reduced velocity of the wind, and they are spotted 2–3 h after sunset. Table 3.6 shows the changes in air temperature per years. It can be seen that the temperature has risen along with the growth of urbanization, industry, and transport, and it is shown graphically in Fig. 3.4.

Table 3.5 Parameter values of the emission concentration of soot, nitrogen oxide, and sulfur dioxide, which are harmful to the soil

Pollutant	The concentration of gas emissions from traffic expressed in %	The concentration of gas emissions from furnaces expressed in %	Changes in soil quality in values from 0 % to 100 %
SO_2	53	46	78
NO	27	21	56
NO_2	39	32	63
Soot	20	40	58

Table 3.6 Changes in the value of air temperature (average) per year from 2003 to 2014

Year	Temperature (°C)
2003	18,9
2004	19,05
2005	20,3
2006	20,9
2007	23,5
2008	24,04
2009	25,18
2010	27,9
2011	28,37
2012	28,78
2013	29,45
2014	30,23

The magnitude of the temperature range within the city can be very large, according to the season and depending on whether these are surface or air temperatures. The temperature range is larger in surface air temperatures, while air temperatures are of smaller scale because of their tendency to mix and collide with other surrounding air temperatures.

Figure 3.5 illustrates that increased urbanization leads to the increased concentrations of SO_2, NO_2, and soot, and thus the air quality becomes worse, that is, the air is more polluted. So the first three columns, marked with no 1, show that the air is heavily polluted, while the last three

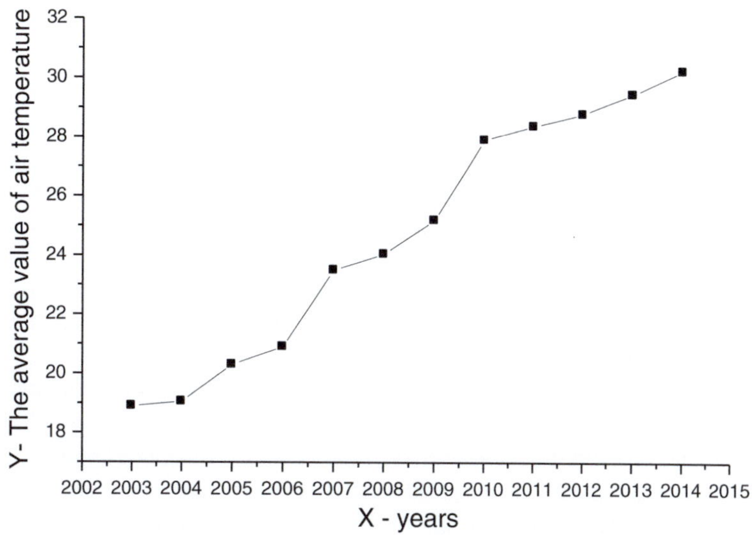

Fig. 3.4 Graphical representation of temperature changes per years from 2003 to 2014

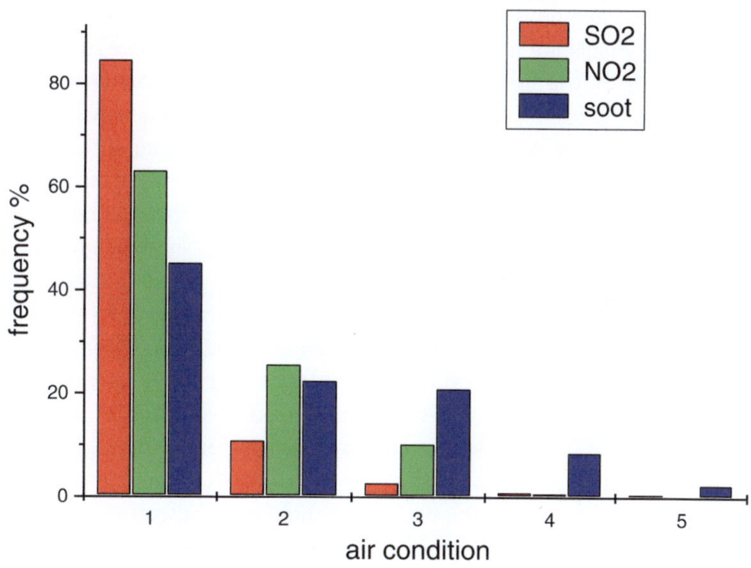

Fig. 3.5 The air quality in Vranje measured in 2010 under the influence of SO_2, NO_2, and soot

columns,marked with no 5, that is hardly noticeable, which means that the air is good (excellent), almost no pollution. The last three columns show the period before the start of global urbanization in the southern region. The numbers 1, 2, 3, 4, and 5 on chart 5 show how the air is polluted. Listing 1 shows that the air is highly polluted, number 2 the air is polluted, number 3 the air is acceptable, number 4 the air is good (no major pollution), and number 5 the air is not very polluted (it is excellent).

3.3.2 The Influence of Air Pollution on Plants

In this study it was shown that increased urbanization affects the increased concentration of harmful gases in the environment. These harmful gases, primarily SO_2, NO_2, and CO_2, reflect harmfully on plants that grow in this environment. Considering an urban environment, where no vegetables are grown, but it has woody and decorative plants, the consequences of harmful effects of these gases are not so noticeable. The tops of tall trees that grow in the city almost cannot feel the adverse effects caused by polluted air.

Exhaust gases and particles pollute the atmosphere of a narrower or broader area, to a lesser or greater extent, sometimes with critical consequences. The spread of pollutants around the sources of pollution depends on many elements: the type and strength of the wind, the position of the pollutants, the quality of particles, the character of buffer zones, etc. If the distance from the source of pollution increases, the amount of polluting particles in the air will reduce.

Plants are very important for the biotope and play an important role in the performance of photosynthesis. The process of photosynthesis that plants carry is multiplex, and the most important is that with the process of photosynthesis, plants produce food and prolong the life of other living organisms. Plants are a source of oxygen on the planet earth. If you disturb the natural balance, and that can happen with a variety of contaminants that are more and more frequent, it comes to many changes that reflect poorly on wildlife and mostly on plants. Air pollution today is the biggest problem. Nowadays there is less pure and more polluted air. Those are the consequences of increased industrialization, the development of various technologies, the increased number of vehicles, and fewer biological filters and air purifiers (Biočanin and Obhođaš 2011). The impact of air pollution on plant species has a negative impact on their quality and quantity. Air pollution is changing some of the abiotic factors such as light, temperature, humidity, the optimal amount of oxygen and carbon dioxide, as well as edaphic (soil) factors, which can indirectly change. Air pollution affects the anatomical and morphological structure, growth, development, and production of organic plants.

Air pollution reflects on plants in two ways

1. Sedimentation of solid particles on the surface of the leaves and minimize the doping of light and perform the process of photosynthesis, thus also minimizing the gas exchange.
2. Harmful gases together with CO and oxygen enter the internal structure of the leaves through the pores of plants, which causes the damage to the plant cells. Plants are less exposed to the danger at night in the summer and much more in a period of vegetation and during the day.

As the air is more polluted in winter due to increased smog, it can be said that in this period the plants are "protected" because they are in the phase of hibernation, and the great amount of harmful gases cannot affect them "badly." Deciduous plants, as well as annuals, are more at risk of the effects of air pollution than conifers and perennial plants. Deciduous and annuals cannot fight against the impact of harmful substances and gases contained in the air, because they have more delicate leaves, softer tree trunks, more open pores than evergreen and perennial trees ((Biochanin and Obhodjash 2011). Tests show that the needles of conifers retain up to 30 times the amount of particles than certain broadleaf species. In addition, there is the certain regularity in the content of pollutants under the crowns of trees, depending on the season of the year.

The greatest reduction of particle feels in September (38 %) and the lowest in May (about 20 %). During the period of vegetation, the low content of polluting elements in the air is 42 % lower than in the surrounding area. The sanitary importance of trees in the collection of particles is very important in the winter. The canopy of trees without leaves retain contaminants in great numbers (on average 37 % of their total amount in the air).

Based on the research, it has been calculated that during one vegetation period, adult specimens of trees and bushes can retain the following amounts of mechanical particles: elm tree, 28 kg mechanical particles per vegetation period; weeping willow, 38 kg; horse chestnut, 16 kg; jelly, 28 kg; maple, 33 kg; Canadian poplar, 34 kg; and black currant, 0.5 kg.

The accumulation of contaminant particles on the leaves of trees and bushes is even more important if the green space is larger.

Plants have certain mechanisms that protect themselves from the effects of polluting elements. Tests show that most of the plants can handle with no visible damage the amount of sediment of 0.75–1.50 g/m^2/day, especially if rain washes most of the deposited particles over a short period of time. The researchers of the pollution with lead come to the similar conclusions. Despite the extremely high concentrations of lead on the leaves of plants along the roads, there are no reports of their visible defects so far.

Plants filter the air using so-called vertical air purification. Wetter and cooler air above the green areas continuously replaces the air above the open space, thus taking up gas pollutions.

The choice of plants depends on local conditions, and some of the most resistant species in our country are *Ailanthus glandulosa*, *Celtis australis*, *Acer rubrum*, *Celtis occidentalis*, *Cornus mas*, *Corylus colurna*, *Platanus* sp., *Robinia pseudoacacia*, *Juniperus* sp., *Quercus*, *Rosa canina*, *Hedera helix*, *Thuja occidentalis*, *Juglans nigra*, *Acer platanoides*, etc.

The following measures are proposed for protection in cases of high air pollution: planting shelterbelts in the direction of dominant wind, combining resistant plants and plants with tenuous crown with plants that have dense and compact crown, concentrating plants as close to the source of pollution as possible, and forming wider belts, that is to say, larger green areas.

Table 3.7 shows the results of the values of parameters of concentrations of soot emissions, nitrogen oxides, and sulfur dioxide that have adverse effects on woody plants, ornamental plants, and fruit cultures.

Table 3.7 showed that the percentage of harmful gases has different effects on different plants.

Table 3.7 Values of the concentration of soot emissions, nitrogen oxides, and sulfur dioxide which adversely affect woody plants, ornamental plants, and fruit crops

Pollutant	On woody plants in %	On vegetables in %	On fruit in %
SO_2	25	61	52
NO	45	48	46
NO_2	36	46	48
Soot	10	20	18

In winter when the concentration of smog is increased in the air, the impact of smog is higher, provided that in this period the plants are at the phase of hibernation so that it cannot do much harm to them. But in the summer when the concentration of exhaust gases is increased (specifically the concentration of nitrogen is increased), the plants differently absorb the gas. Increased amounts of nitrogen oxides are differently reflected plants. Vegetable plants due to increased amounts of this gas can change the color of the leaves and can have a reduced yield, and due to increased amounts of SO2, fruits can be wizened and dry.

If you look at the surrounding arable land in the region of South Serbia, where different vegetable species can be grown as well as cereals, you can notice that there are no major changes to their trunks, leaves, and fruit when it comes to the impact of these harmful emissions. Due to the moderate continental climate that dominates in this region, there are no strong winds and no increased greenhouse effect, so that there is no large air pollution on plants. Plants grow on 80 % of healthy soil, so that the small amount of harmful gases in the environment cannot do much to change their physiognomy, unlike some cities and regions where air pollution drastically affects the quality of the plants. For example, the City of Pancevo where the refinery is located is subjected to the air pollution, and thus the quality of the plants is weaker.

3.4 Conclusion

This paper presents the research results of the conducted monitoring system of air quality and its impact on biota in urban areas such as the city of Vranje. The measurement of pollutant emissions, obtained as products of combustion and in the form of exhaust gases, was carried out. After the measurement, the averaging time was 24 h and certain limit and total values were established. It is shown in Tables 3.2 and 3.3 that these values do not differ much for CO_2 and NO_2. The most important thing is that all these values are in the range of normal. Although there are variations, these are special conditions or special periods during the year, when the measured emissions of air pollutants are increased. The increased measured emissions coincide with the winter period. The increased emissions are in direct correlation with operating of boiler rooms that use coal. The emissions from these boiler rooms which use coal reflect negatively on the quality of air, soil, and the entire city environment and on the surrounding agricultural sites within the region by reducing its quality parameters. This opens up new possibilities for testing and demonstration, as well as the conclusions about the necessity of seeking modern solutions for heating which produce less pollution to the environment, for example, from renewable energy sources. Consequently, the environment will be cleaner, and the soil will maintain the level of quality required for the cultivation of agricultural crops which is the goal of the region.

If the urbanization of cities were somehow regulated (e.g., introduction of ecological fuel), any type of pollution would be reduced, air temperatures would not also rise rapidly, the soil would not be exposed to harmful particles, and air quality would improve.

References

Abellan-Nebot JV, Subiron FR (2010) A review of machining monitoring systems based on artificial intelligence process models. Int J Adv Manuf Technol 47(1):237–257

Bell JNB, Honour SL, Power SA (2011) Effects of vehicle exhaust emissions on urban wild plant species. Environ Pollut 159(8–9):1984–1990

Biočanin R, Obhođaš S (2011) Environmental pollutants. International University of Travnik. Faculty for Civil Engineering, Novi Pazar, pp 322–323

BorgstrÖm ST (2009) Patterns and challenges of urban nature conservation-a study of southern Sweden. Environ Plan A 41(11):2671–2685

Coxford B, Penn A (1998) Siting considerations for urban pollution monitors. Atmos Environ 32:1049

Grsic Z, Dramlic D, Arbutina D, Miljevic N, Dramlic S, Milutinovic P, Kaljevic J, Pavlovic S, Joksimovic D (2014) Representativity of air quality control in limited number of grid points. J Environ Prot Ecol 15(1):1–6

Gupta GP, Kumar B, Singh S, Kulshrestha U (2015) Urban climate and its effect on biochemical and morphological characteristics of Arjun (Terminalia arjuna) plant in National Capital Region Delhi. Chem Ecol 31:524–538

Homer V, Ramis C, Romeo R, Alonso S (2010) Recent trends in temperature and precipitation over the Balearic Islands (Spain). Clim Change 98(1):199–211

Kowarik I (2011) Novel urban ecosystems, biodiversity, and conservation. Environ Pollut 159(8–9):1974–1983

Langstaff J, Seigneur C, Liu MK, Behar J, McElroy JL (1987) Design of an optimum air monitoring network for exposure assessments. Atmos Environ 21:1393

Mazzeo NA, Venegas LE (2000) Practical use of the ISCST3 model to select monitoring site locations for air pollution control. Int J Environ Pollut 14:246

Milutinovic P, Popovic M (2001) Importance of the local monitoring network for more accurate prediction of transboundary pollution. J Environ Prot Ecol 2(2):356

Mitchell VG, Mein RG, Mcmahon TA (2002) Utilising stormwater and wastewater resources in urban areas. Aust J Water Res 6:31

Noll KE, Mitsutomi S (1983) Design methodology for optimum dosage monitoring air site selection. Atmos Environ 17:2583

Noll KE, Miller TL, Norco JE, Raufer RK (1977) An objective air monitoring site selection methodology for large point sources. Atmos Environ 11:1051

Svirejeva-Hopkins A, Schellnuber HJ, Pomoz VL (2004) Urbanised territories as a specific component of the global carbon cycle. Ecol Model 173:295

Tseng CC, Chang NB (2001) Assessing relocation strategies of urban air quality monitoring stations by GA-based compromise programming. Environ Int 26:523

Vosniakos F, Triandafyllis J, Prapas D, Karyda A, Mentzelou P, Vasilikiotis G, Karagiannis D, Lazou P, Papastamou A, Vosniakos K, Argiriadis V (2008) Urban air pollution due to vehicles in katerini city on comparison basis (1999–2006). J Environ Prot Ecol 9(3):485

Mechanisms of Plant Pollutant Uptake as Related to Effective Biomonitoring

4

Yoshitaka Oishi

Abstract

Biomonitoring is a method that uses the responses of plants or animals to their surroundings to evaluate the status of an environment. Among taxonomic groups, pine needles and mosses are widely used for biomonitoring, especially for atmospheric environments. However, several studies have indicated that each of these plants reacts differently to changes in their habitat. Here, we characterized these contrasting responses and investigated the causes of these differences by comparing atmospheric pollutants (polycyclic hydrocarbons: PAHs) that accumulated in pine needles and mosses. Our results revealed that pine needles absorbed lower molecular weight PAHs, whereas mosses preferentially accumulated higher molecular weight PAHs. Furthermore, the comparison of their PAH isomer ratios showed that the pollution sources were not identical, even though the plant samples were collected from nearly the same sites. These differences can be explained by their distinct leaf structures and uptake mechanisms, as well as the influence of soil particles. Our novel results suggest that both pine needles and mosses can be used as bioindicators to assess PAH pollution multi-directionally.

Keywords

Air pollution • Biomonitoring • Plant uptake • Accumulation • Mechanisms

4.1 Introduction

In this chapter, we discuss the mechanisms of plant pollutant uptake and how this process can be applied to environmental evaluation. More specifically, this chapter first introduces an environmental evaluation method that utilizes the responses of living organisms to their surroundings, which is called as "biomonitoring". We then

Y. Oishi (✉)
Fukui Prefectural University, 4-1-1 Kenjojima, Matsuoka, Eiheiji-cho, Yoshida-gun, Fukui 910-1195, Japan
e-mail: oishiy@fpu.ac.jp

U. Kulshrestha, P. Saxena (eds.), *Plant Responses to Air Pollution*,
DOI 10.1007/978-981-10-1201-3_4

focus on two plant groups widely used for biomonitoring (pine trees and mosses) and explore the differences in how these species accumulate polycyclic aromatic hydrocarbons (PAHs), a hazardous atmospheric pollutant (Sect. 4.2). Furthermore, we evaluate the causes of plant-to-plant differences observed in PAH accumulation (Sect. 4.3). Finally, based on these findings, we demonstrate how to efficiently use biomonitoring to evaluate atmospheric environments (Sect. 4.4). These results highlight the importance of the understanding of plant uptake mechanisms when attempting to establish effective biomonitoring programs.

Air Pollution and Biomonitoring

Environmental problems have intensified because of an increase in human activities, and, among these, air pollution is one of the most concerning threats (Gurjar et al. 2010; Kopáček and Posch 2011; OECD 2012; WMO/IGAC 2012; Shibata et al. 2014). Atmospheric pollutants include a variety of substances (e.g., metals, sulfur oxide, nitrogen, and organic compounds) that can easily move over wide areas, resulting in transboundary pollution. Therefore, monitoring programs for atmospheric environments have been instituted in many parts of the world (e.g., Schröder et al. 2010; Harmens et al. 2015).

Air pollution has both direct and indirect effects on plants and animals, and therefore changes in these organisms can be correlated with the level of air contamination. For example, severe air pollution can cause a disappearance of epiphytes such as bryophytes and lichens, and an index based on the sensitivity of these plants to air pollution can indicate the level of contamination in the atmosphere (LeBlanc and De Sloover 1970). This type of environmental evaluation that uses the responses of living organisms to their surroundings has been termed "biomonitoring," and the subject organisms are known as "bioindicators."

Environmental conditions are generally evaluated with measuring devices (e.g., a stack gas analyzer), which would presumably produce more exact results than those of biomonitoring. Are there any advantages then to using biomonitoring for environmental evaluation? Of course, the answer is YES. One of the most important benefits of biomonitoring is that they can assess environments at a low cost and on a large scale. Imagine that we are measuring the concentration of atmospheric pollutants, which are affected by many factors such as human activities and weather and can even change drastically in a day. In order to obtain reliable data, we must therefore measure the concentration repeatedly with measuring devices. In contrast, if we were to perform this same environmental assessment using bioindicators, we can eliminate the steps of repeated measuring because by their nature, bioindicators have been exposed continuously to atmospheric pollutants, reflecting the time-integral effects of air pollution. These valuable characteristics enable the evaluation of air pollution on a large scale.

Plants as Bioindicators

Vegetation has been utilized as a bioindicator to identify point sources of pollutants and evaluate regional and global contamination patterns (Simonich and Hites 1995), particularly in atmospheric environments. Each plant group has its own morphology and life-strategy, qualities that are directly correlated with their usefulness as bioindicators. Among plant groups, pine trees (specifically their needles) (Tremolada et al. 1996; Piccardo et al. 2005; Klánová et al. 2009; Lehndorff and Schwark 2009; Wang et al. 2009; Ratola et al. 2010, 2011) and mosses (Holoubek et al. 2000; Gerdol et al. 2002; Migaszewski et al. 2002; Ötvös et al. 2004; Liu et al. 2005; Gałuszka 2007; Krommer et al. 2007; Skert et al. 2010; Oishi 2012, 2013) are valuable bioindicators for airborne contaminants because of their unique ecological characteristics (Figs. 4.1 and 4.2). Here, we describe the advantages of these plant groups for the evaluation of air pollution.

Pine needles are one of the most well-known bioindicators for atmospheric environments.

Fig. 4.1 Pine needles and mosses. (**a**) *Pinus densiflora* Sieb. et Zucc., (**b**) *Pinus pumila* (Pall.) Regel, (**c**) *Hypnum plumaeforme* Wilson, (**d**) *Racomitrium lanuginosum* (Hedw.) Brid. *P. densiflora* and *H. plumaeforme* (**a**, **c**) are distributed mainly in lowland areas, whereas *P. pumila* and *R. lanuginosum* (**b**, **d**) are found in mountainous areas. All species are used as bioindicators for air pollutants

Their leaves can persist for several years, and pine trees are widely distributed from urban to rural areas. A notable characteristic of pine needles is that their age can be determined easily, which enables us to calculate how long the leaves have been exposed to air pollution. Therefore, we can evaluate temporal trends of air pollution by analyzing different populations of same-age needles. Furthermore, the surface of the leaf is covered with a waxy cuticle that accumulates lipophilic organic contaminants from the air (Piccardo et al. 2005).

Bryophytes are characterized by a lack of vascular bundles and waxy cuticle layers. They absorb water and nutrients through their leaf cells, which allows them to grow on surfaces without soil, such as rocks and tree trunks (Fig. 4.3). As they take in pollutants efficiently from atmospheric environments, their pollutant contents are indicative of contamination by atmospheric fallout.

4.2 Comparison of PAH Accumulation in Plants

PAH Accumulation in Pine Needles and Mosses

Pine trees and mosses belong to different taxa, and their morphology and ecology is distinct. How then, and to what extent, can these contrasting characteristics affect their use as bioindicators? Here, we refer to the study by Holoubek et al. (2000) that described the accumulation of PAHs in pine needles and mosses in several parts of the Czech Republic. The results indicated that

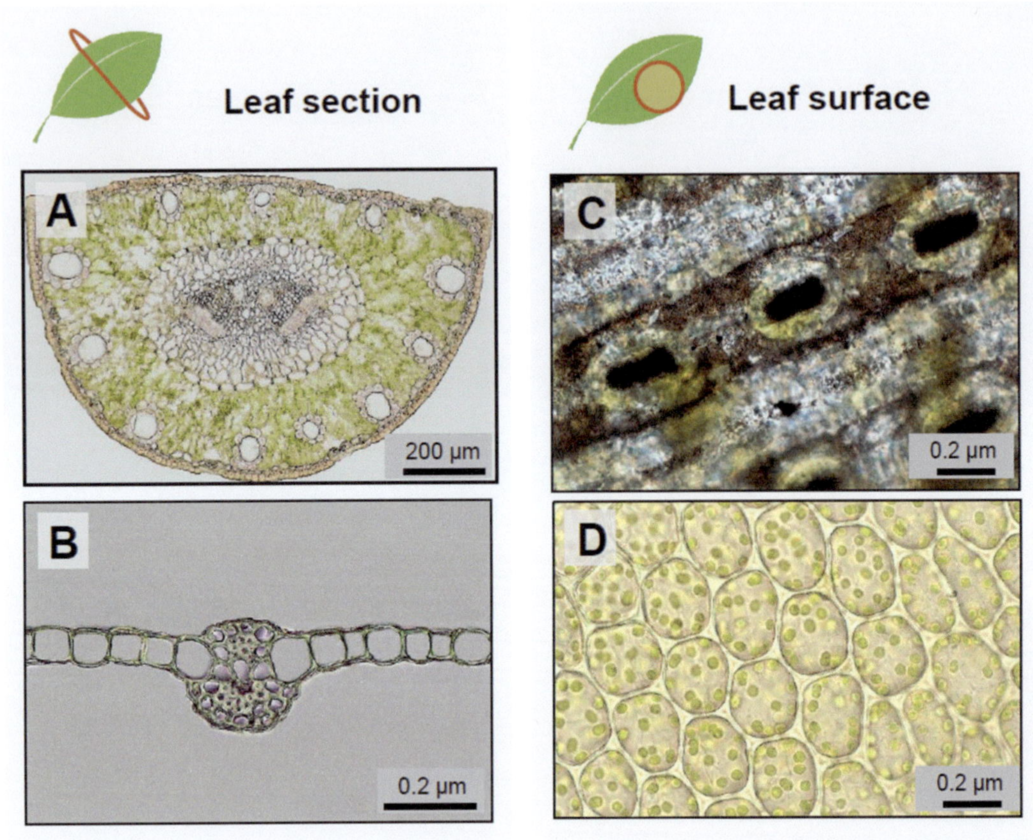

Fig. 4.2 Leaf section and leaf surface of pine needles and mosses. (**a**) Leaf section of pine needle (*Pinus densiflora*: photo by Azuma, W.), (**b**) leaf section of moss [*Plagiomnium acutum* (Lindb.) T.J. Kop.], (**c**) leaf surface of pine needle, (**d**) leaf surface of moss. The leaf section of moss (**b**) shows a simpler structure compared to that of a pine needle (**a**). Stomata are clearly identified in the leaf surface of pine needles (**c**), whereas mosses lack stomata (**d**)

the pattern of PAH accumulation in these plants was different.

Why did this contrast occur, and how did the differences between the structures of these plants affect the results? To answer these questions, one must understand how plants absorb air pollutants. Such knowledge is also essential in order to propose effective plant biomonitoring strategies. For these reasons, we investigated the characteristics and pollution uptake mechanisms of pine needles and mosses so that we could evaluate the most effective means of instituting biomonitoring. The questions we sought to answer were as follows:

1. Are the differences in PAH accumulation in Holoubek et al. (2000) also observed in our study site?
2. If so, why do these differences occur?
3. How can these results be applied to effective biomonitoring?

Characteristics of PAHs

PAHs are organic compounds with two or more fused aromatic rings (Fig. 4.4). They are emitted into the atmosphere through incomplete combustion from both anthropogenic and natural sources and are ubiquitous environmental contaminants (Maliszewska-Kordybach 1999). The predominant human-related sources of PAHs are activities that generate energy, such as vehicular movement, domestic heating, industrial processes, and electric power generation (Mastral and Callén 2000). Among them, motor vehicle exhaust is one of the major sources of PAHs in urban areas

Fig. 4.3 Mosses on substrates without soil layers. (**a**) *Leucobryum juniperoideum* (Brid.) Müll. Hal., (**b**) *Venturiella sinensis* (Vent.) Müll. Hal., (**c**) A stone figure (Ojizo-sama) covered with mosses, (**d**) *Grimmia pilifera* P. Beauv. Mosses absorb water and nutriment from the surrounding environment (e.g., rain, dews, and fogs) through the surface of their leaves. Therefore, they can grow on tree trunks or rocks that have scant soil layers

(Piccardo et al. 2005). PAHs are hazardous to human health and several have mutagenic and carcinogenic properties (Maliszewska-Kordybach 1999; Aas et al. 2001). For these reasons, there is increasing concern regarding the monitoring and regulation of PAHs in ambient air.

Aerosolized PAHs can exist in either a gaseous or a particle-bound phase. These phases are determined by several factors such as air temperature, the physicochemical characteristics of the compound, and the types of the absorbing surface (Pankow 1987). In general, PAHs with two to three aromatic rings exist primarily in the gas phase of the atmosphere because of their relatively low molecular weight (LMW). In contrast, PAHs with five to six rings have a relatively high molecular weight (HMW) and are more likely to be present in the particle-bound phase (Bidleman 1988; Maliszewska-Kordybach 1999). A temperature-dependent gas/particle phase partitioning occurs at intermediate vapor pressures with four-ring PAHs (Bidleman 1988; Liu et al. 2005; Wang et al. 2009).

An interesting property of PAHs is that their isomer ratio differs according to their source and the processes that they experienced. Using these properties, we can determine the source of a PAH based on the concept that isomeric PAHs behave similarly and may also experience comparable environmental transformations during their atmospheric movement (Yunker et al. 2002; Bucheli et al. 2004).

Differences in PAH Accumulation in Pine Needles and Mosses

In order to examine the mechanism of pollution uptake in plants, we compared accumulated

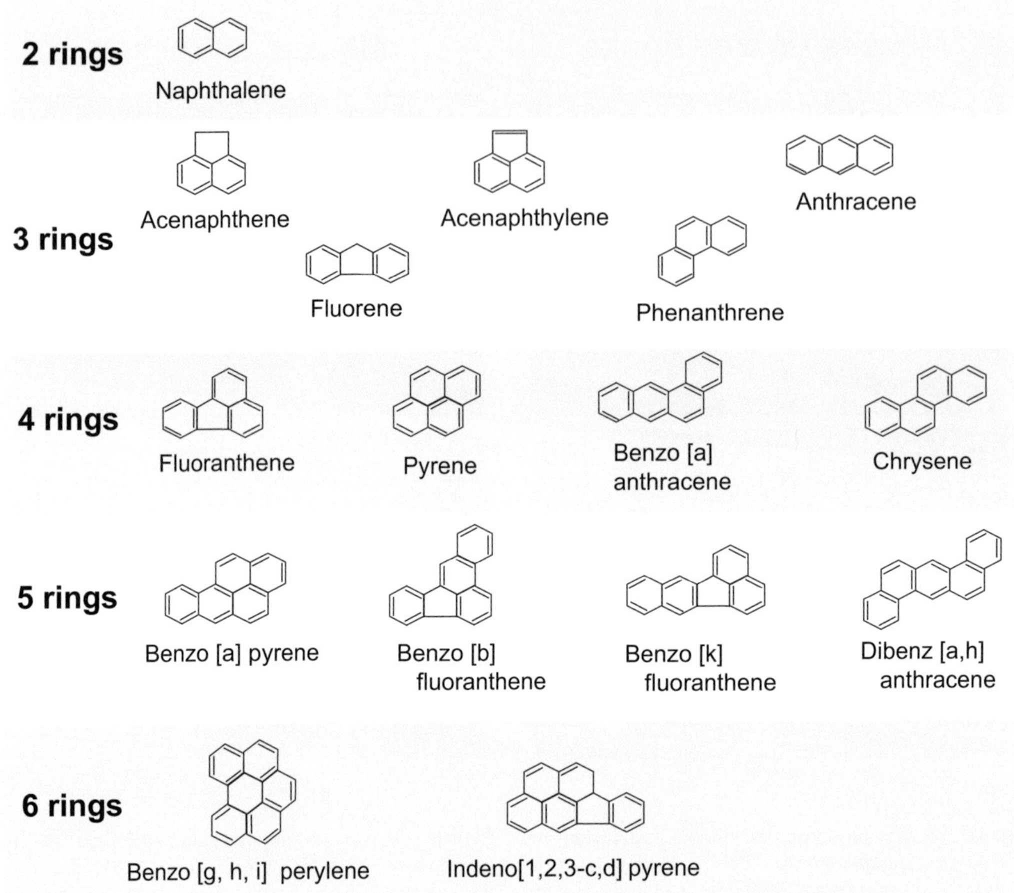

Fig. 4.4 Sixteen polycyclic aromatic hydrocarbons (PAHs) analyzed in this study. PAHs are characterized by the presence of two or more aromatic rings. PAHs with two to three rings exist mainly in the gas phase, whereas PAHs with five to six rings are in the particle-bound phase

PAHs in pine needles and mosses by collecting five sets of both pine needles (*Pinus thunbergii* Parl.) and moss (*Hypnum plumaeforme* Wilson) samples in a green area of Kyoto city. Each set of samples grew within 2 m diameter circular plots. The PAHs analyzed were as the follows: naphthalene (NAP), acenaphthene (ACE), acenaphthylene (ACL), anthracene (ANT), fluorene (FLU), phenanthrene (PHE), benz[a]anthracene (BaA), chrysene (CHR), fluoranthene (FLR), pyrene (PYR), benzo[a]pyrene (BaP), benzo[b]fluoranthene (BbF), benzo[k]fluoranthene (BkF), dibenz[a,h]anthracene (DBA), benzo[g,h,i]perylene (BPE), and indeno[1,2,3-cd]pyrene (INP).

In this section, we show two main results of this comparison: "Differences in PAH proportions in pine needles and mosses" and "Differences in PAH isomer ratios." These results are adapted from Oishi (2013).

PAH Proportions

The PAH analysis indicated that the total PAH content was 122.6±50.5 ng g^{-1} dry weight (mean ± SD) in pine needles and 44.5 ± 10.7 ng g^{-1} in the moss samples, respectively. The PAH content was significantly higher in pine needles than in the mosses (d.f. = 8, t-value = 3.0, $p = 0.016$).

The percentage contribution to the total PAH content by each individual compound is shown in Fig. 4.5. NAP was the most predominant PAH (29.5 %) in the pine needles, followed by PHE (26.8 %), FLU (16.3 %), and FLR (10.7 %). The

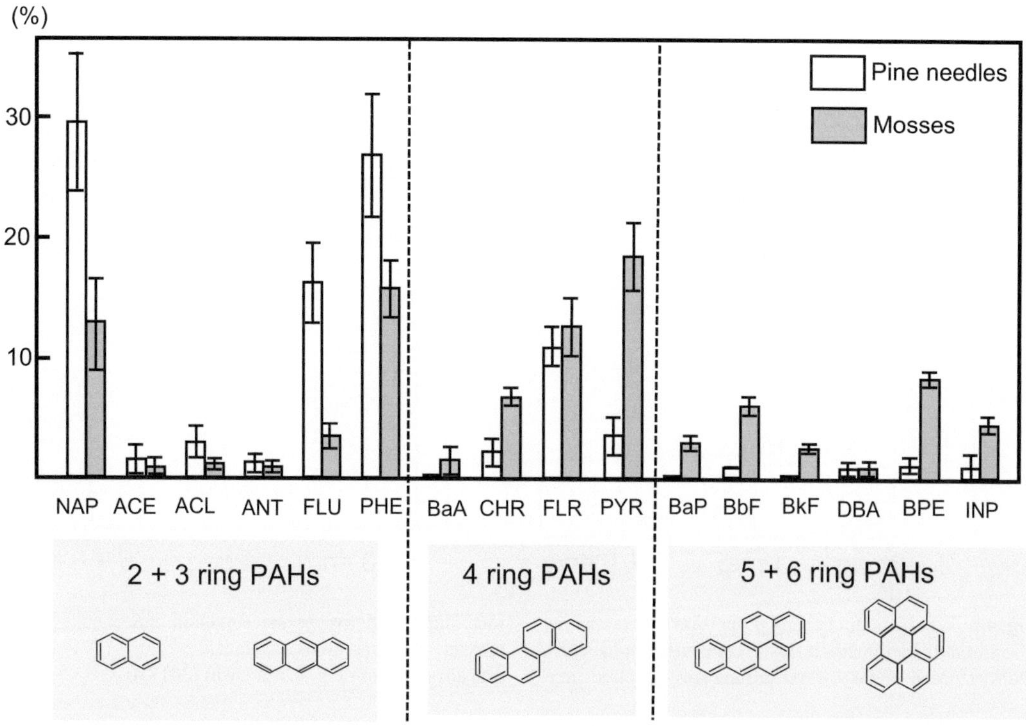

Fig. 4.5 Proportion of total polycyclic aromatic hydrocarbon (PAH) concentration attributable to each of three PAH groups (two to three rings, four rings, and five to six rings) in pine needles and mosses. *Bars* represent standard deviations (this figure was adapted from Fig. 4.2 in Oishi (2013)). Abbreviations of the 16 PAHs are as follows: *NAP* naphthalene, *ACE* acenaphthene, *ACL* acenaphthylene, *ANT* anthracene, *FLU* fluorene, *PHE* phenanthrene, *BaA* benzo[a]anthracene, *CHR* chrysene, *FLR* fluoranthene, *PYR* pyrene, *BaP* benzo[a]pyrene, *BbF* benzo[b]fluoranthene, *BkF* benzo[k]fluoranthene, *DBA* dibenz[a,h]anthracene, *BPE* benzo[g,h,i]perylene, *INP* indeno[1,2,3-cd]pyrene

concentrations of PYR, PHE, FLR, and NAP were relatively higher (18.4 %, 15.7 %, 13.0 %, and 12.6 %, respectively) in the mosses compared to other PAHs. Notably, NAP, ACL, ACE, FLU, and PHE were primarily found in the pine needle samples, whereas BaA, PYR, BaP, BbF, BkF, BPE, and INP were predominantly detected in the moss samples. We also found that in general, pine needles preferentially accumulated LMW PAHs and few HMW PAHs, as compared to mosses. These comparisons indicate that the accumulation patterns of pine needles and mosses are dissimilar.

To distinguish differences in PAH accumulations, we grouped the PAHs into three types (two to three rings, four rings, and five to six rings), according to their phase in the atmosphere (gas, intermediate, and particle bound), and compared the total amounts and proportions of each type (Fig. 4.6). The LMW PAHs were preferentially accumulated in pine needles, whereas the HMW PAHs were more often found in the moss samples (Fig. 4.6). The proportions of two + three rings, four rings, and five + six rings for pine needles were 78.5 ± 4.8 % (mean ± SD), 17.2 ± 2.6 %, and 4.3 ± 2.9 %, respectively (Fig. 4.6a), whereas those for mosses were 35.4 ± 6.8 %, 39.5 ± 4.5 %, and 25.1 ± 3.3 %, respectively (Fig. 4.6b). In this way, the proportion of each PAH group to the total decreased as the number of aromatic rings in the pine needles increased. Mosses showed a similar but less distinct decreasing trend. These differences were statically significant; the proportions of two + three rings in pine needles were significantly higher (d.f. = 8, *t*-value = 10.4, $p < 0.01$), whereas those of four rings and five + six rings were significantly higher in the moss

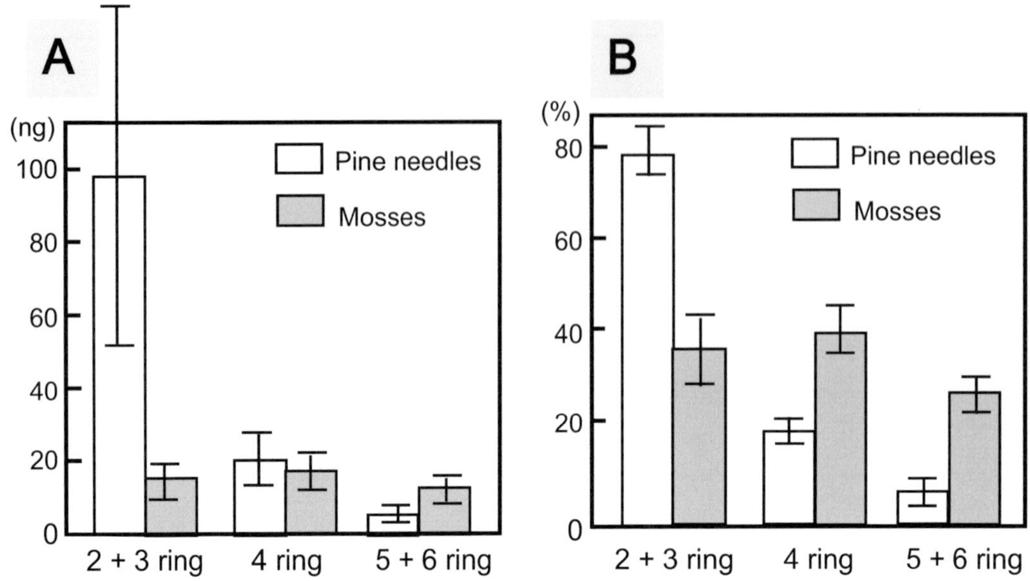

Fig. 4.6 Total amount (**a**) and proportion of total polycyclic aromatic hydrocarbon (PAH) concentration (**b**) attributable to each of three PAH groups (two to three rings, four rings, and five to six rings) in pine needles and mosses. *Bars* represent standard deviations (This figure was adapted from Fig. 4.3 in Oishi (2013))

samples (d.f. = 8, t-value = −8.6, $p < 0.01$; d.f. = 8, t-value = −9.5, $p < 0.01$).

PAH Isomer Ratios

Next, we used PAH ratios to examine the differences in PAH sources between pine needles and mosses. We compared the ratio of ANT and PHE and the ratio of FLR and PYR. The plots of the ANT/(ANT + PHE) versus the FLR/(FLR + PYR) ratios for pine needle and moss samples are shown in Fig. 4.7. The ANT/(ANT + PHE) ratio for all samples except one was <0.1 [0.05 ± 0.02 (mean ± SD) for pine needles; 0.07 ± 0.02 for mosses]. The FLR/(FLR + PYR) ratio for pine needles was approximately 0.7 [0.74 ± 0.06 (mean ± SD)] and approximately 0.40 [0.40 ± 0.07 (mean ± SD)] in the moss samples.

According to Yunker et al. (2002), the ANT/(ANT + PHE) ratios of most samples fell predominantly in the range of petrogenic area (<0.1), although the values of mosses tended to be higher than those of pine needles. The high FLR/(FLR + PYR) ratios in pine needles (>0.5) indicate that pine needles accumulated PAHs released by the combustion of coal and biomass. In contrast, the FLR/(FLR + PYR) ratios in mosses showed that they accumulated PAHs produced by petroleum or petroleum combustion.

In summary, these results indicate that pine needles and mosses do not accumulate PAHs from the same sources, even though they grow in similar regions.

4.3 Influence of Plant Uptake Mechanisms on Their PAH Accumulation

What Causes the Differences in the Accumulation of PAHs?

Our results are in agreement with previous research that reported high concentrations of LMW PAHs in pine needles (Simonich and Hites 1995; Wang et al. 2009) and high concentrations of HMW PAHs in mosses (Holoubek et al. 2000; Migaszewski et al. 2002; Liu et al. 2005). We will now discuss why these differences were observed from the following three viewpoints: (a) Leaf structure, (b) Uptake mechanism, and (c) Influence of soil particles. A graphic summary of these differences is shown in Fig. 4.8.

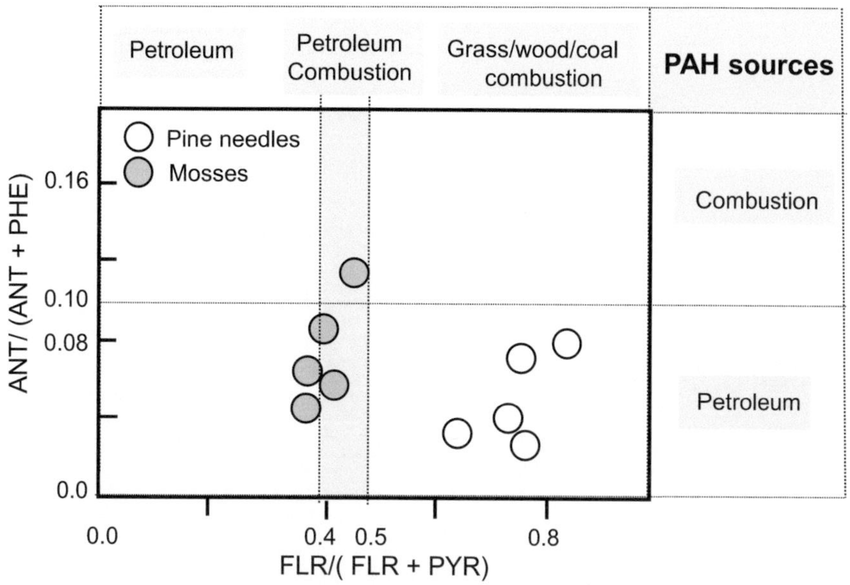

Fig. 4.7 Cross-plot for ANT/(ANT + PHE) and FLR/(FLR + PYR) ratios for pine needle and moss samples (This figure was adapted from Fig. 4.4 in Oishi (2013)). Abbreviations: *ANT* anthracene, *FLR* fluoranthene, *PHE* phenanthrene, *PYR* pyrene

Leaf Structure

External leaf properties of plant bioindicators greatly influence the characteristics of their PAH profiles (Howsam et al. 2000; Jouraeva et al. 2002; Niu et al. 2003; Piccardo et al. 2005; Wang et al. 2005). For example, the surface area of leaves directly affects the efficiency of PAH uptake; the larger the surface area is, the more PAHs it can absorb (Simonich and Hites 1995). In addition, according to Howsam et al. (2000), hairs, or trichomes, on the leaf surface can effectively trap PAHs. The presence of a waxy cuticle can also affect the uptake of organic pollutants (Simonich and Hites 1995; Piccardo et al. 2005).

Based on these previous studies, we conclude that the lack of a waxy cuticle layer on moss leaves may be a major factor in the differences in PAH accumulation between pine needles and mosses. As Piccardo et al. (2005) showed, LMW PAHs diffuse and accumulate in the tissues of pine needles either through the stomata or by diffusion through the cuticle. However, HMW PAHs tend to remain on the surface of the cuticle because of their strong interactions with the constituents of this waxy layer, making them more susceptible to external environmental factors (e.g., rain, temperature, ozone, and solar radiation). These dynamics may cause the loss of HMW PAHs from the leaves of pine needles (Jouraeva et al. 2002; Piccardo et al. 2005). In contrast, mosses lack the cuticle layers that facilitate the selective uptake of LMW PAHs, a distinction that can increase HMW PAH ratios in mosses compared to pine needles.

The presence of a waxy cuticle layer can also affect the total amount of PAHs accumulated in pine needles. We again focus on the comparison of the total amount of PAHs absorbed by pine needles and mosses in Fig. 4.6a. The pine needles examined in this study accumulated a significantly greater total PAH concentration than mosses. As Fig. 4.6b shows, this high PAH content can be attributed to the high content of LMW PAHs preferentially absorbed by pine needles.

Uptake Mechanisms

The cross-plots of the PAH isomer ratios (Fig. 4.7) show a clear distinction between the PAH sources in pine needles and mosses. Why did these differences occur? One potential explanation is that there is a stronger influence of wet deposition in mosses. Whereas mosses take up

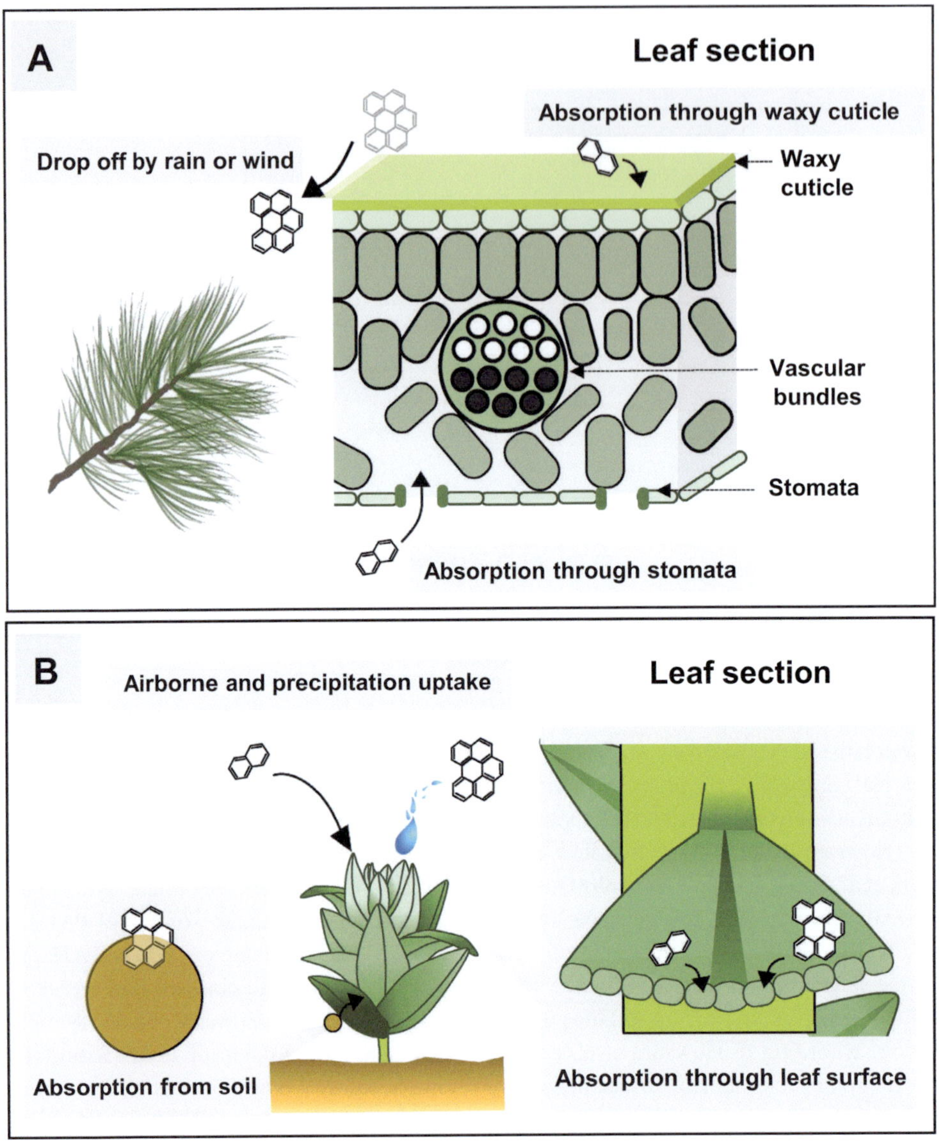

Fig. 4.8 Graphic summary of polycyclic aromatic hydrocarbon (PAH) uptake by pine needles and mosses. Pine needles (**a**) uptake low molecular weight (LMW) PAHs via stomata or diffusion across waxy cuticles. However, high molecular weight (HMW) PAHs are not as effectively absorbed because of the strong interaction between HMW PAHs and the waxy cuticle. In contrast, mosses (**b**) efficiently absorb HMW PAHs because they lack cuticle layers. Mosses on the ground can also uptake PAHs partially through soil particles

dissolved pollutants from precipitation (Thomas 1986), pine needles predominantly absorb gaseous PAHs via their stomata or diffusion (Lehndorff and Schwark 2004). This distinction in the uptake of pollutants can explain the differences in PAH sources and isomer ratios between these two plant types.

Influence of Soil Particles

In addition to the differences in leaf structures and uptake mechanisms, the relatively high concentration of HMW PAHs in mosses may partly be influenced by their ability to take up PAHs through soil particles (Migaszewski et al. 2009). Although this absorption route has not been proven experimentally, the possibility is supported by a previous study by Kłos et al. (2012), who used radioactive markers to determine that mosses absorb soil particles along with the heavy metals adhering to them. HMW PAHs exist mainly as particles and therefore can easily be absorbed into the soil (Bozlaker et al. 2008; Wang et al. 2009). Therefore, mosses can uptake HMW PAHs through soil particles in the same manner as they absorb heavy metals from soil particles.

4.4 Conclusion

Our results showed that pine needles and mosses absorb different types of PAHs, a contrast that can be explained by their unique pollutant uptake mechanisms. From the perspective of biomonitoring, these findings indicate that we can multidirectionally assess PAH pollution by using both pine needles and mosses as bioindicators. More specifically, pine needles are reliable indicators of airborne LMW PAH pollution, whereas mosses can be used to evaluate complex HMW PAH pollution in atmospheric and soil environments. Utilizing a combination of bioindicators for a more comprehensive environmental evaluation is a novel concept that can contribute to the effective biomonitoring.

Acknowledgments This research was supported by the Global COE Program, "Global Center for Education and Research on Human Security Engineering for Asian Megacities," MEXT, Japan, and by Grant-in-Aid for Young Scientists (B) No. 26870241 from the Japan Society for the Promotion of Science.

References

Aas E, Beyer J, Jonsson G, Reichert WL, Andersen OK (2001) Evidence of uptake, biotransformation and DNA binding of polyaromatic hydrocarbons in Atlantic cod and corkwing wrasse caught in the vicinity of an aluminium works. Mar Environ Res 52:213–229

Bidleman TF (1988) Atmospheric processes. Wet and dry deposition of organic compounds are controlled by their vapor-particle partitioning. Environ Sci Technol 22:361–367

Bozlaker A, Muezzinoglu A, Odabasi M (2008) Atmospheric concentrations, dry deposition and air-soil exchange of polycyclic aromatic hydrocarbons (PAHs) in an industrial region in Turkey. J Hazard Mater 153:1093–1102

Bucheli TD, Blum F, Desaules A, Gustafsson O (2004) Polycyclic aromatic hydrocarbons, black carbon, and molecular markers in soils of Switzerland. Chemosphere 56:1061–1076

Gałuszka A (2007) Distribution patterns of PAHs and trace elements in mosses *Hylocomium splendens* (Hedw.) B.S.G. and *Pleurozium schreberi* (Brid.) Mitt. from different forest communities: a case study, south-central Poland. Chemosphere 67:1415–1422

Gerdol R, Bragazza L, Marchesini R, Medici A, Pedrini P, Benedetti S, Bovolenta A, Coppi S (2002) Use of moss (*Tortula muralis* Hedw.) for monitoring organic and inorganic pollution in urban and rural sites in Northern Italy. Atmos Environ 36:4069–4075

Gurjar BR, Jain A, Sharma A, Agarwal A, Gupta P, Nagpure AS, Lelieveld J (2010) Human health risks in megacities due to air pollution. Atmos Environ 44:4606–4613

Harmens H, Mills G, Hayes F, Norris D, The participants of the ICP Vegetation (2015) Air pollution and vegetation. ICP vegetation annual report 2014/2015

Holoubek I, Korínek P, Seda Z, Schneiderová E, Holoubková I, Pacl A, Tríska J, Cudlín P, Cáslavský J (2000) The use of mosses and pine needles to detect persistent organic pollutants at local and regional scales. Environ Pollut 109:283–292

Howsam M, Jones KC, Ineson P (2000) PAHs associated with the leaves of tree species. I – concentrations and profiles. Environ Pollut 108:413–424

Jouraeva VA, Johnson DL, Hasset JP, Nowak DJ (2002) Differences in accumulation of PAHs and metals on the leaves of *Tilia* x *euchlora* and *Pyrus calleryana*. Environ Pollut 120:331–338

Klánová J, Čupr P, Baráková D, Šeda Z, Anděl P, Holoubek I (2009) Can pine needles indicate trends in the air pollution levels at remote sites? Environ Pollut 157:3248–3254

Kłos A, Czora M, Rajfur M, Wacławek M (2012) Mechanisms for translocation of heavy metals from soil to epigeal mosses. Water Air Soil Pollut 223:1829–1836

Kopáček J, Posch M (2011) Anthropogenic nitrogen emissions during the Holocene and their possible effects on remote ecosystems. Global Biogeochem Cycles 25:GB2017. doi:10.1029/2010GB003779

Krommer V, Zechmeister HG, Roder I, Scharf S, Hanus-Illnar A (2007) Monitoring atmospheric pollutants in the biosphere Wienerwald by a combined approach of biomonitoring methods and technical measurements. Chemosphere 67:1956–1966

LeBlanc F, De Sloover J (1970) Relation between industrialization and the distribution and growth of epiphytic lichens and mosses in Montreal. Can J Bot 48:1485–1496

Lehndorff E, Schwark L (2004) Biomonitoring of air quality in the Cologne Conurbation using pine needles as a passive sampler – Part II: polycyclic aromatic hydrocarbons (PAH). Atmos Environ 38:3793–3808

Lehndorff E, Schwark L (2009) Biomonitoring airborne parent and alkylated three-ring PAHs in the Greater Cologne Conurbation I: temporal accumulation patterns. Environ Pollut 157:1323–1331

Liu X, Zhang G, Jones KC, Li X, Peng X, Qi S (2005) Compositional fractionation of polycyclic aromatic hydrocarbons (PAHs) in mosses (*Hypnum plumaeformae* WILS.) from the northern slope of Nanling Mountains, South China. Atmos Environ 39:5490–5499

Maliszewska-Kordybach B (1999) Sources, concentrations, fate and effects of polycyclic aromatic hydrocarbons (PAHs) in the environment. Part A: PAHs in air. Pol J Environ Stud 8:131–136

Mastral AM, Callén MS (2000) A review on polycyclic aromatic hydrocarbon (PAH) emission from energy generation. Environ Sci Technol 34:3051–3057

Migaszewski ZM, Gałuszka A, Pasławski P (2002) Polynuclear aromatic hydrocarbons, phenols, and trace metals in selected soil profiles and plant bioindicators in the Holy Cross Mountains, South-Central Poland. Environ Int 28:303–313

Migaszewski ZM, Gałuszka A, Crock JG, Lamothe PJ, Dołegowska S (2009) Interspecies and interregional comparisons of the chemistry of PAHs and trace elements in mosses *Hylocomium splendens* (Hedw.) B.S.G. and *Pleurozium schreberi* (Brid.) Mitt. from Poland and Alaska. Atmos Environ 43:1464–1473

Niu J, Chen J, Martens D, Quan X, Yang F, Kettrup A, Schramm K (2003) Photolysis of polycyclic aromatic hydrocarbons adsorbed on spruce [*Picea abies* (L.) Karst.] needles under sunlight irradiation. Environ Pollut 123:39–45

OECD (2012) OECD environmental outlook to 2050: the consequences of inaction. doi:10.1787/9789264122246-en

Oishi Y (2012) Does uptake of Polycyclic Aromatic Hydrocarbons (PAHs) differ between pine needles and mosses? J Environ Inf Sci 40:31–36

Oishi Y (2013) Comparison of pine needles and mosses as bio-indicators for polycyclic aromatic hydrocarbons. J Environ Prot 4:106–113

Ötvös E, Kozák IO, Fekete J, Sharma VK, Tuba Z (2004) Atmospheric deposition of polycyclic aromatic hydrocarbons (PAHs) in mosses (*Hypnum cupressiforme*) in Hungary. Sci Total Environ 330:89–99

Pankow JF (1987) Review and comparative analysis of the theories on partitioning between the gas and aerosol particulate phases in the atmosphere. Atmos Environ 21:2275–2283

Piccardo MT, Pala M, Bonaccurso B, Stella A, Redaelli A, Paola G, Valério F (2005) *Pinus nigra* and *Pinus pinaster* needles as passive samplers of polycyclic aromatic hydrocarbons. Environ Pollut 133:293–301

Ratola N, Amigo JM, Alves A (2010) Levels and sources of PAHs in selected sites from Portugal: biomonitoring with *Pinus pinea* and *Pinus pinaster* needles. Arch Environ Contam Toxicol 58:631–647

Ratola N, Alves A, Psillakis E (2011) Biomonitoring of polycyclic aromatic hydrocarbons contamination in the island of Crete using pine needles. Water Air Soil Pollut 215:189–203

Schröder W, Holy M, Pesch R, Harmens H, Fagerli H, Alber R et al (2010) First Europe-wide correlation analysis identifying factors best explaining the total nitrogen concentration in mosses. Atmos Environ 44:3485–3491

Shibata H, Branquinho C, McDowell WH, Mitchell MJ, Monteith DT, Tang J et al (2014) Consequences of altered nitrogen cycles in the coupled human and ecological system under changing climate: the need for long-term and site-based research. Ambio 44:178–193

Simonich SL, Hites RA (1995) Organic pollutant accumulation in vegetation. Environ Sci Technol 29:2905–2914

Skert N, Falomo J, Giorgini L, Acquavita A, Capriglia L, Grahonja R, Miani N (2010) Biological and artificial matrixes as PAH accumulators: an experimental comparative study. Water Air Soil Pollut 206:95–103

Thomas W (1986) Representativity of mosses as biomonitor organisms for the accumulation of environmental chemicals in plants and soils. Ecotoxicol Environ Saf 11:339–346

Tremolada P, Burnett V, Calamari D, Jones KC (1996) Spatial distribution of PAHs in the UK atmosphere using pine needles. Environ Sci Technol 30:3570–3577

Wang D, Chen J, Xu Z, Qiao X, Huang L (2005) Disappearance of polycyclic aromatic hydrocarbons sorbed on surfaces of pine [*Pinus thunbergii*] needles under irradiation of sunlight: volatilization and photolysis. Atmos Environ 39:4583–4591

Wang Z, Chen J, Yang P, Tian F, Qiao X, Bian H, Ge L (2009) Distribution of PAHs in pine (*Pinus thunbergii*) needles and soils correlates with their gas-particle partitioning. Environ Sci Technol 43:1336–1341

WMO/IGAC (2012) Impacts of megacities on air pollution and climate. http://www.wmo.int/pages/prog/arep/gaw/documents/Final_GAW_205.pdf. Accessed 2 Feb 2016

Yunker MB, Macdonald RW, Vingarzan R, Mitchell RH, Goyette D, Sylvestre S (2002) PAHs in the Fraser River basin: a critical appraisal of PAH ratios as indicators of PAH source and composition. Org Geochem 33:489–515

Role of Global Warming and Plant Signaling in BVOC Emissions

5

Saurabh Sonwani, Pallavi Saxena,
and Umesh Kulshrestha

Abstract

Plants emit a substantial amount of biogenic volatile organic compounds (BVOCs) into the atmosphere having significant effects on atmospheric chemistry, physics, and the organisms. In this chapter, important facts about BVOCs' production mechanism, storage, and emissions due to various abiotic and biotic factors have been discussed. The role of BVOCs in defense system through plant signaling has also been discussed. The chapter continues with the discussions about the importance of BVOCs in the formation of tropospheric ozone and secondary organic aerosols and ultimately strengthens our understanding about global change research.

Keywords

BVOCs • Abiotic • Biotic • Plant signalling • Defense systems • Tropospheric ozone • Secondary organic aerosol

5.1 Introduction

Generally, any organic compound having vapor pressure high enough under normal conditions to be vaporized into the atmosphere is termed as a volatile organic compound (VOC). VOCs emitted by plants are widely known as biogenic volatile organic compounds (BVOCs). BVOCs are classified according to their structure and biosynthetic origin (Pichersky et al. 2006). BVOCs represent a group of organic trace gases (except carbon dioxide and monoxide) released into the atmosphere from the plants and soils. Biogenic VOCs include isoprenoids (isoprene and monoterpenes) as well as alkanes, alkenes, carbonyls, esters, ethers, alcohols, and acids. The lifetime of BVOCs in the atmosphere varies from minutes to several days (Kesselmeier and Staudt 1999).

Isoprene (C_5H_8) is the most widely studied single BVOC which accounts for >90 % of the total BVOC emissions of a particular plant species (Blande et al. 2007). The total annual isoprene emissions account for 440–660 TgC (Guenther et al. 2006), which is a large fraction of annual global BVOC emissions (700–1000

S. Sonwani (✉) • P. Saxena • U. Kulshrestha
School of Environmental Sciences, Jawaharlal Nehru University, New Delhi 110067, India
e-mail: sonwani.s19@gmail.com

U. Kulshrestha, P. Saxena (eds.), *Plant Responses to Air Pollution*,
DOI 10.1007/978-981-10-1201-3_5

TgC) (Laothawornkitkul et al. 2009). Isoprene belongs to the biochemical class of terpenoids; other BVOCs under the same class are monoterpenes ($C_{10}H_{16}$) and sesquiterpenes ($C_{15}H_{24}$). BVOCs produced by plants are involved in plant reproduction (by attracting pollinator), growth, wound healing, development, and defense against herbivores. They are also involved in a communication process between plants and insects, within plants, and within plant communities (Pichersky and Gershenzon 2002; Peñuelas et al. 1995; Shulaev et al. 1997) (Fig. 5.1). BVOCs can also protect plants against high temperatures. But, on the other hand, BVOC emissions increase with warming and might produce both negative and positive feedback on global warming through aerosol formation as well as direct and indirect greenhouse effects (Fig. 5.2).

Several authors have reported that plants re-emit a substantial fraction of their assimilated carbon into the atmosphere as BVOCs that can affect chemical and physical properties of the atmosphere (Peñuelas and Llusià 2001; Kesselmeier and Staudt 1999).

A large fraction of carbon can be lost during BVOC emission and it was estimated to be around ~10 % of the total carbon stored by plants through photosynthesis (Peñuelas Llusià 2003). Apart from abovementioned functions, there is another function of BVOCs that has dragged scientist attention given global warming. Recently, it is found that production and emission of certain BVOCs such as isoprene and monoterpenes, which contribute a large fraction of total BVOCs, confer protection against high temperature. There are various factors responsible for BVOC emissions and biological and physicochemical processes (Figs. 5.1 and 5.2). Several internal (biochemical and genetic) and external factors control emission rates of different BVOCs by altering their synthesis, diffusion, or vapor pressure to the atmosphere. The external factor includes abiotic (light, wind, water availability, temperature, and ozone) and biotic (animal, plant, and microorganism interactions) factors

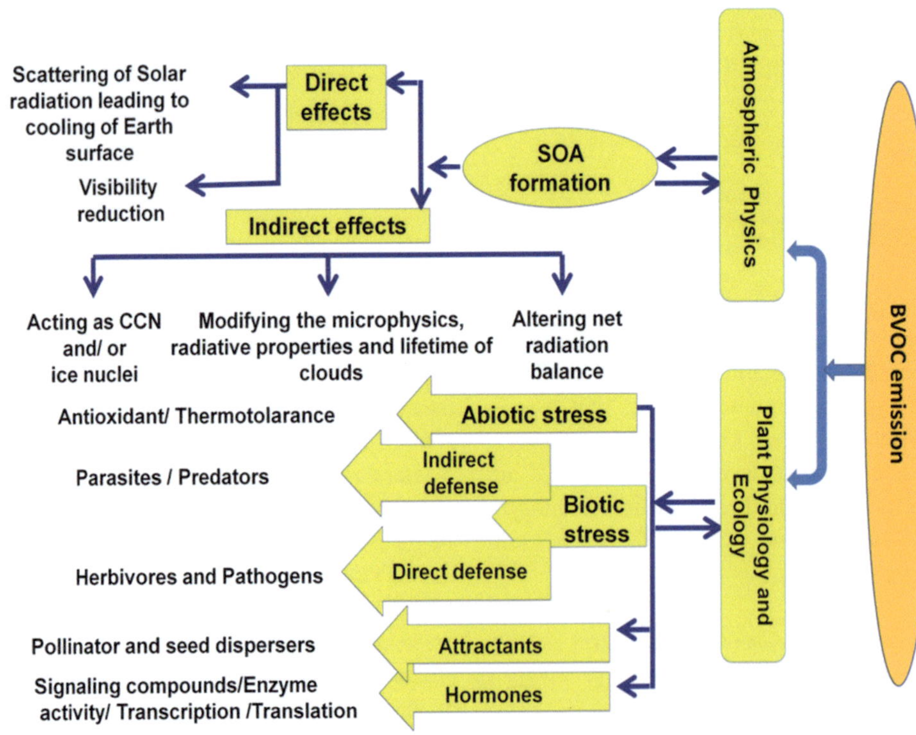

Fig. 5.1 Role of BVOCs in atmospheric physics and plant physiology

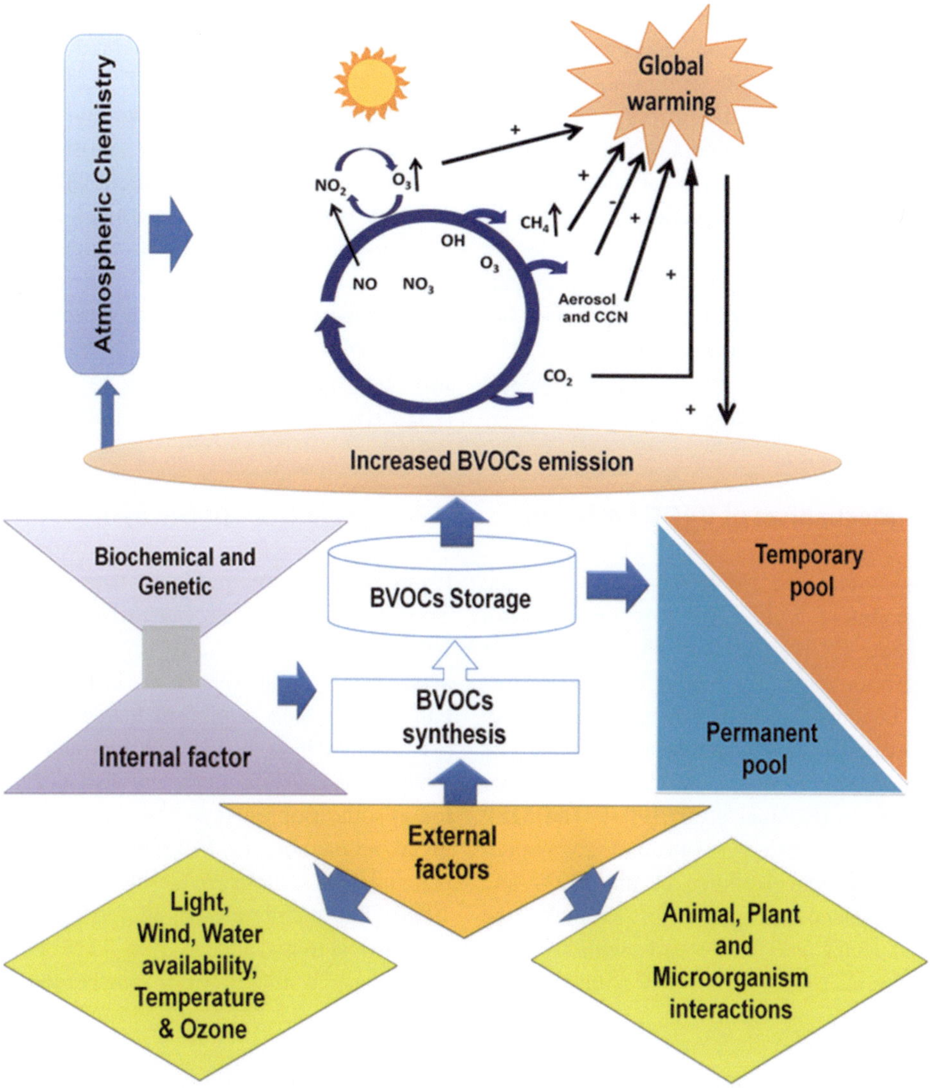

Fig. 5.2 BVOCs' synthesis, storage, and role in atmospheric chemistry

(Fig. 5.2). The complex net of these factors and their interactions and the different responses of the different BVOCs produce large qualitative and quantitative and spatial and temporal variability of emissions and frequent deviations from current standard emission models, mostly based on temperature and light response (Peñuelas and Llusià 2001). Plant emissions of BVOCs have strong relevance for plant physiology and plant ecology, atmospheric chemistry, and climate.

5.2 BVOC Emissions: Why and How

It is now well reported by several authors that a wide variety of volatile non-methane organic compounds (referred to hereafter as biogenic volatile organic compounds (BVOCs)) are emitted into the atmosphere from vegetation (Fall et al. 1999; Geron et al. 2000; Guenther et al. 1995, 2000; Fuentes et al. 2000). BVOCs are stored in specialized structures like glandular trichomes, glands, and resin ducts/resin canals (Turner et al. 2000; Franceschi et al. 2005) and are tightly separated by the surrounding cells by an impermeable layer, generally cuticular cells (Gershenzon et al. 2000). On the other hand, if plants have no storage structures for BVOCs, they have some temporary pools/nonspecific storage (observed in both conifers and broad-leaved trees) in mesophyll that freely diffuse out of the leaf along their concentration gradient. The only determining factor for this process of gas diffusion is stomatal conductance.

5.2.1 Mechanism of BVOC Emission

The mevalonic acid (MVA) pathway and the 2-deoxyxylulose 5-phosphate/2-methylerythritol 4-phosphate (MEP) pathway (Fig. 5.3) are the two metabolic routs responsible for DMAPP formation. Leucine metabolism is another pathway which works as a source of isoprenoids. The source of DMAPP within the chloroplast is the MEP pathway (Rohmer et al. 1993). The MEP pathway begins with pyruvate and glyceraldehyde 3-phosphate (Rohmer et al. 1993) and involves skeletal rearrangement to make the branched chain and finally ends with isoprene (Arigoni et al. 1997). In the MEP pathway, each carbon in the starting intermediate costs three ATPs plus two NADPH if at the redox level of a triose phosphate with CTP equivalent to ATP. According to Lichtenthaler and coworkers, this pathway is responsible for most of the isoprenoids made in plastids (Lichtenthaler et al. 1997). It is proved that this pathway is responsible for isoprene synthesis (Schwender et al. 1997; Zeidler et al. 1997). The MEP pathway is more efficient (cost is only six carbon atoms, 20 ATPs, and 14 NADPH) as compared to the MVA pathway (cost is nine carbon atoms, 24 ATPs, and 14 NADPH).

5.2.2 Role of Isoprene Synthase

Plants emit isoprene produced from DMAPP by isoprene synthase (enzyme) (Silver and Fall 1991) which works at a relatively high pH and needs Mg^{2+} (Schnitzler et al. 1996; Silver and Fall 1995) which is present inside chloroplasts (Mgaloblishvili et al. 1979; Wildermuth and Fall 1996, 1998). Wildermuth and Fall (1998) also found some isoprene synthase activity bound to thylakoid membranes. Isoprene synthase activity is found to be soluble (Wildermuth and Fall 1996, 1998). The membrane-bound form of isoprene synthase activity appeared to have the same kinetics as the soluble form. Whether the soluble

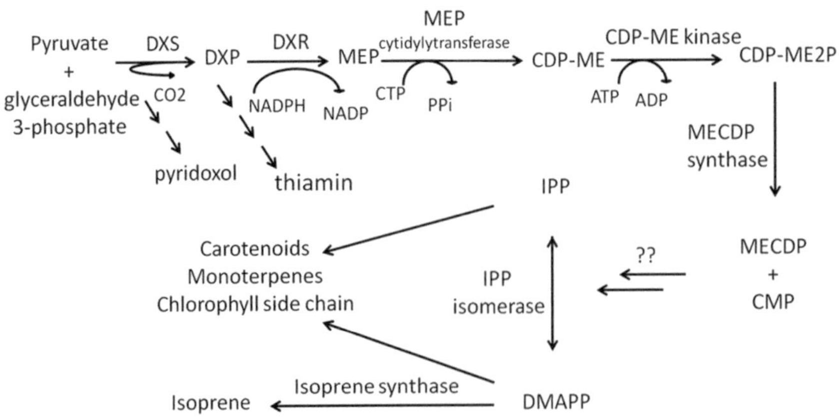

Fig. 5.3 BVOC synthesis; MEP pathway. Note: *CDP-ME2P*2-phospho-4-(cytidine50-diphospho)-2-C-methyl-D-erythritol, *CDP-ME* 4-(cytidine 50-diphospho)-2-C-methyl-D-erythritol, *DXR* deoxyxylulose-5-phosphate reductoisomerase, *DXS* deoxyxylulose-5-phosphate synthase, *MECDP* 2-C-methyl-D-erythritol 2,4-cyclodiphosphate, *IPP* isopentenyl pyrophosphate, *DMAPP* dimethylallyl pyrophosphate

form is converted to the bound form or vice versa and what effect this might have on the activity of isoprene synthase is not known. Thylakoid-bound isoprene synthase activity could be stimulated threefold by the addition of GTP and palmitoyl CoA (Wildermuth 1997). The conversion of DMAPP to isoprene is catalyzed by isoprene synthase. Perhaps isoprene is a product of the breakdown of higher isoprenoids in *E. coli*. In bacteria (*Bacillus subtilis*), isoprene emission occurs during different phases of growth (Wagner et al. 1999).

5.3 Atmospheric Chemistry and Physics of BVOCs

5.3.1 Production of Tropospheric Ozone

According to Fehsenfeld et al. (1992) and Thompson (1992), isoprene emission from plants plays a significant role in atmospheric chemistry (Fig. 5.2). BVOCs such as isoprene can alter atmospheric chemistry affecting the residence times of gases that contribute to the greenhouse effect. Isoprene oxidation in the atmosphere can give rise to ozone and smog if nitrogen oxides are present in the atmosphere (Daum et al. 2000; Haagen-Smit 1952). In the high-isoprene emission zone, like a zone with extensive tropical forest, isoprene oxidation can account for up to 71 % of the reduction in OH (compared to only 11 % for CO oxidation and 5 % for CH_4 oxidation) (Grosjean 1995). Hence, it is very important to understand the role of BVOCs in atmospheric chemistry (Thompson et al. 1992; Fehsenfeld et al. 1992). BVOC emissions can indirectly increase the concentrations of other important greenhouse gases such as methane resulting in reduced atmospheric oxidation capacity due to depletion of OH radicals (Lelieveld et al. 2008). Hydroxyl radicals are very reactive oxidants and act as the atmospheric detergent (Lelieveld et al. 2008). After the study of chemistry and transport model by Wang et al. (1998), it is clear that global emissions of BVOCs, particularly isoprene, can cause a 15 % increase in the background levels of O_3 and a 20 % decrease in the mean OH concentrations (Wang et al. 1998). Another study suggested that global VOC emissions (isoprene as a large fraction) cause an 18 % increase in the global tropospheric O_3 concentration, a 16 % decrease in the global OH concentrations, and a 20 % increase in the tropospheric lifetime of CH_4 (Poisson et al. 2000). Even the NOx spatial distribution and deposition are affected by BVOC interactions.

The oxidation of BVOCs by OH in the presence of sufficient NOx leads to the formation of O_3 in the troposphere by disruption of the photochemical steady state of O_3 (i.e., allows the oxidation of NO to NO_2 without removal of an O_3 molecule) and so causes elevated O_3 concentrations. NOx emissions may result from fossil fuel combustion, fertilizer application, and biomass burning, as well as natural production by lightning. As tropospheric photochemistry is highly nonlinear with respect to the emissions of O_3 precursors, modeling is required to determine the effects of BVOC emissions on O_3 concentrations in the troposphere (Fowler et al. 2008).

At night, NOx is present in the atmosphere as NO_2. According to Monson and Holland (2001), sunlight photolyzes NO_2 which leads to the formula of one molecule of ozone per one molecule of NO_2. In the absence of hydrocarbons, the total NOx level determines the amount of ozone in the atmosphere.

$$RH_2 + OH + O_2 \rightarrow RO_2 + H_2O \quad (5.1)$$

$$RO_2 + NO \rightarrow RO + NO_2 \quad (5.2)$$

$$RO + O_2 \rightarrow R`CHO + HO_2 \quad (5.3)$$

$$HO_2 + NO \rightarrow OH + NO_2 \quad (5.4)$$

$$2\left(NO_2 + O_2 \rightarrow NO + O_3\right) \quad (5.5)$$

$$NET: RH + 4O_2 \rightarrow R`HCO + 2O_3 + H_2O \quad (5.6)$$

This cycle is continued as long as there are C–H bonds available which enable the formation of ozone. At a very high NOx level, one isoprene molecule can result in multiple molecules of ozone. But at a low NOx level, isoprene emission can reduce ozone level via different set of reactions (Trainer et al. 1987). Isoprene also plays a role in the growth of aerosols in the atmosphere. Though aerosol's yield per molecule in the atmosphere is much lower than for monoterpenes, due to its large emission, isoprene became a significant source of secondary organic aerosol (Claeys et al. 2004; Edney et al. 2005; Kroll et al. 2005, 2006; Ng et al. 2006; Olcese et al. 2007). The resulting changes in climate may affect BVOC emission rates, providing a positive feedback in the climate system. The oxidation of BVOCs also contributes atmospheric CO (Hatakeyama et al. 1991; Fehsenfeld et al. 1992; Bergamaschi et al. 2000; Griffin et al. 2007) which impacts the oxidative capacity of the atmosphere similar to isoprene through acting as a sink for OH (Logan et al. 1981). Thereby, the oxidation of CO is a source or sink of O_3, depending on the NOx availability. Atmospheric oxidation of BVOCs can also result in the formation of peroxyacetyl nitrates (PANs) (Fehsenfeld et al. 1992). PANs have longer residence time and can be transported to a larger distance. PANs are a source of NOx when they are thermally decomposed in warmer air (Fehsenfeld et al. 1992; Poisson et al. 2000), contributing a high level of NOx in the regime without local NOx emissions which ultimately lead to higher O_3 levels in remote areas.

5.3.2 Secondary Aerosol Formation

According to various laboratory and field studies, BVOCs are a significant source of secondary organic aerosol (SOA) (Fig. 5.1) (Leaitch et al. 1999; Joutsensaari et al. 2005). Through the mechanism SOA formation via BVOC oxidation is still not fully explained (Kulmala 2003), but it is understood that BVOC oxidation results in lower vapor pressure compounds which triggers their condensation on preexisting atmospheric molecules (Joutsensaari et al. 2005). Terpenes and sesquiterpenes contribute more significant amounts of SOA as compared to isoprene (Claeys et al. 2004; Leaitch et al. 1999; Meskhidze and Nenes 2006). Atmospheric aerosols are considered to have direct effects on climate as they scatter solar radiation. Secondarily they also have indirect effects such as cloud condensation nuclei, changing cloud albedo, and the degree of cloud cover. The possibility that SOA formation from BVOC emissions cools the Earth and so a moderate temperature is dependent on BVOC emission from plants – and other similar feedbacks in Earth's system – is the focus of much current research. Hence, there is the potential for feedback between BVOC emissions, SOA, and climate.

5.4 Factors Affecting BVOC Emission

5.4.1 Taxonomic Characteristics

Trees are traditionally considered to be the key BVOC emitters. Isoprene, monoterpenes, and sesquiterpenes are emitted from many plants and they appear to have an adaptive role in protecting plants from abiotic stress. However, only some plant species emit isoprene. For example, taxonomic groups as distinct as mosses and oak trees emit isoprene, but in groups as closely allied as mosses and hornworts or oak and maple trees, both are emitters (mosses and oaks) and non-emitters (hornworts and maples) (Lerdau and Gray 2003; Hanson et al. 1999). Monson et al. (2013) studied various isoprene-emitting genera with high taxonomic diversity, such as the Fabaceae (old name: Leguminosae). Mosses (Hanson et al. 1999), ferns (Tingey et al. 1987), gymnosperms, and angiosperms (http:www.es.lancs.ac.uk/cnhgroup/iso-emissions.pdf for comprehensive list) consist of members emitting BVOCs and not emitting BVOCs. Isoprene synthases (IspSs) are the enzyme which catalyzes the conversion of DMAPP to isoprene. IspSs have been sequenced from several plant species such as the *Populus* species (Sasaki et al. 2005; Miller et al. 2001) and kudzu (*Pueraria lobata*) (Sharkey et al. 2005). According to Bohlmann et al. (1998), IspSs belong to the TSP-b family of terpene synthases, also code for monoterpene and sesquiterpene synthases in angiosperms. But these are absent in gymnosperms. Isoprene synthase genes analyzed so far have been found to have high-sequence homology with certain monoterpene synthase genes (Miller et al. 2001; Sharkey et al. 2005) and even a bifunctional acyclic monoterpene synthase/isoprene synthase gene (Sharkey et al. 2013). A small change in the gene sequences can easily alter both substrate and product specificity of IspS genes (El Tamer et al. 2003; Tholl 2006; Kampranis et al. 2007). North American oaks all emit isoprene, but many European oaks do not. Instead, among European oaks a variety of behaviors is found. Some clades emit isoprene, some emit monoterpenes in a light dependent manner, and some emit very little terpene (Loreto et al. 1998a, b; Csiky and Seufert 1999; Kesselmeier and Staudt 1999).

5.4.2 Temperature

BVOC emissions are highly temperature dependent because higher temperature increases the rate of chemical reaction and cellular diffusion and increases the vapor pressure of volatile compounds (Lerdau et al. 1994; Tingey et al. 1991; Sharkey and Yeh 2001; Fuentes et al. 1999). The emission rates of BVOCs are analyzed by their synthesis rates and physicochemical parameters like solubility, volatility, and diffusivity (Niinemets et al. 2004; Kesselmeier and Staudt 1999; Laothawornkitkul et al. 2009). BVOC emission rates are highly controlled by both external and internal factors. Although there are large uncertainties in the magnitude of emission rates of BVOCs, a recent estimate for North America (Guenther et al. 2000) suggests that of an estimated 84 $TgCyear^{-1}$ of BVOC emissions, 30 % are isoprene, 25 % are terpenoid compounds, and 40 % are non-terpenoid compounds including methanol, hexene derivatives, and 2-methyl-3-buten-2-ol.

BVOCs follow the Henry's law constant (K_H) and they partition between gas and liquid phases in the plants. Leng et al. (2013) reported that the measured K_H values for isoprene, limonene, α-pinene, and linalool at 298 K were 0.036, 0.048, 0.029, and 21.20 mol L^{-1} atm^{-1}, respectively. They also reported that diffusion of these BVOCs follow the first-order kinetics for rate of loss and diffusion coefficients of all the species as a function of temperature. Temperature affects the evaporation and release of a minimal part of the pools of BVOCs that leaks out the impermeable cell layer. Herbivory, strong winds, and forest fires are the other strong factors responsible for BVOC emissions (Litvak and Monson 1998). High humidity also plays a significant role in BVOC emission from the structures of plants containing BVOC pools.

The enzymes which are responsible to catalyze the synthesis of BVOCs can be easily controlled or influenced by temperature. Emissions of volatile terpenes have a $Q_{10} = 2–4$ at temperatures variable between 20 and 40 °C (Monson et al. 1992). Thus the elevation in atmospheric temperature is a direct effect of terpenes formed through enzymatic action.

5.4.2.1 Thermotolerance

Sharkey and Singsaas (1995) were the first researchers who reported that isoprene plays a significant role in thermotolerance function. Singsaas et al. (1997) showed that adding isoprene to an air stream (or nitrogen gas) that passed over these leaves could increase the temperature at which damage occurred from as low as 35 °C to as high as 45 °C. Thermotolerance was also mentioned with special reference to monoterpene emissions from *Quercus ilex* (oak species) by Loreto et al. (1998a, b). Several studies have reported the positive link between isoprene and photorespiration (Peñuelas and Llusià 2002). According to one study, "Thermotolerance of leaf discs from four isoprene-emitting species is not enhanced by exposure to exogenous isoprene" (Logan and Monson 1999) and was of substantial concern. They found that leaf disks' chlorophyll fluorescence held in darkness or light plus nitrogen did not increase until 45 °C, irrespective of whether isoprene was in the air stream. In their experiments the control leaf pieces did not exhibit photosynthetic damage below 45 °C. Measurements of the temperature where CO_2 uptake fell to zero were not reported in the study.

Two advancements were added in definitive experiments. First was the refinement of the thermotolerance hypothesis and the second was the use of fosmidomycin, the inhibitor that eliminates isoprene production without affecting photosynthesis. With these improvements, much stronger evidence for the thermotolerance hypothesis has been obtained (Sharkey et al. 2001). The study reported that the isoprene synthesis provides tolerance of short high-temperature episodes. The molecules similar to isoprene were also tested to find out the basic requirements for thermotolerance. One rule was apparent in the results. The compounds taken into consideration were alkenes (1,3-butadiene, 1-butene, and *cis* 2-butene) that provided thermotolerance though alkanes (n-butane, *iso*-butane, and 2-methyl-butan) did not; they even increased the damage caused by heat (Sharkey et al. 2001).

5.4.3 Soil Moisture

Soil moisture is one of the important factors responsible for BVOC emission from plants. Soil moisture is significantly correlated with BVOC (isoprene and monoterpene) emission (Gray et al. 2014). Several authors studied the global BVOC emission model with soil moisture. Guenther et al. (2006) estimated a 7 % decrease in isoprene emission due to soil moisture's effect, while Muller et al. (2008) calculated a 20 % decrease. The lower soil moisture can induce stomatal closure, which can reduce the production and emission of BVOCs coming directly from temporary storage pools inside the leaves (McBean et al. 2005).

5.5 BVOC Emission and Global Warming

Climate models suggest that, during the twenty-first century, the mean global temperature will increase by 1–6 °C (with a best estimate of 2–3 °C) (IPCC 2007). BVOC emissions are altered by varying their temperature; their emissions increase with temperature to a certain limit, beyond which enzyme degradation and physiological responses to heat stress affect the emission pattern (Guenther et al. 1993).

Emissions of BVOCs are always run along a vapor pressure gradient from the cellular level (containing relatively high concentrations) to the surrounding air of leaves (containing relatively low concentrations) due to transport. Temperature increases the emission rates of most BVOCs exponentially by enhancing the enzymatic activities of synthesis, by elevating the BVOC vapor pressure, and by decreasing the resistance of the diffusion pathway (Tingey et al. 1991). Penuelas and Llusia (2003) have suggested that increasing mean global temperatures by 2–3 °C could enhance global BVOC emissions by 25–45 %. The dependency of BVOC emission rates on temperature can be further explained with G93 algorithm, given by Guenther et al. (1993), where the emission rate of monoterpene can be expressed as:

$$E = E_s \exp\{\beta (T - T_s)\}, \tag{5.7}$$

where E and E_s are measured emission rates and basal emission rates at a standard temperature, i.e., 30 °C, respectively. T and T_s are measured and standard temperatures, respectively. β is an empirical coefficient (depends on species of vegetation and compounds emitted. With the similar model at a local temperature regime (4 °C) for Douglas fir trees, Constable et al. (1999) found a predicted increase in emission of 52 %, 46 %, and 41 % for α-pinene, β- pinene, and Δ-3 carene, respectively. Whereas isoprene emission is modeled as:

$$E = E_S \times C_T \times C_L, \tag{5.8}$$

where E is the isoprene emission rate at temperature T and PAR flux L, E_S is the emission rate measured at standard temperature T_S and PAR (1000 mmol m^{-2} s^{-1}), and the scaling factors C_T and C_L are defined by:

$$C_T = \frac{\exp\left[C_{T1} \times (T - T_s)\right] / (R \times T \times T_s)}{1 + \exp\left[C_{T2} \times (T - T_m)\right] / (R \times T \times T_s)} \tag{5.9}$$

$$C_L = \frac{\alpha \times C_{L1} \times \text{PAR}}{\sqrt{1+\alpha^2 \times \text{PAR}^2}}, \qquad (5.10)$$

where, C_{T1} (95,000 J mol^{-1}), C_{T2} (230,000 J mol^{-1}), T_m (41 °C), α (0·0027), and C_{L1} (1·066) are parameters calculated from experiments made on some BVOC-emitting plant species (Guenther et al. 1993). R is the gas constant (8·134 J K^{-1} mol^{-1}).

Applying the algorithms of emission response to temperature (Llusià and Peñuelas 2000; Tingey et al. 1991; Peñuelas and Llusia 2003) calculated global warming over the past 30 years and found that BVOCs increased global emissions by 10 % and a further 2–3 °C rise in the mean global temperature, which is predicted to occur this century (Houghton et al. 2001), could increase BVOC global emissions by an additional 30–45 %.

At a very high temperature, isoprene emission declines dramatically. It is possible that extreme temperature rises will eventually cause a decrease in isoprene emissions, irrespective of other changes to ecosystems.

Global warming can also indirectly influence global- and regional-scale BVOC emissions by altering vegetation species composition and vegetation characteristics (Wilmking et al. 2004; Starfield and Chapin 1996). Simulation model studies predict forest dieback at a lower latitude such as Amazonia (Cox et al. 2004). However, the models predicted that boreal forests will be exposed northward in a global warming scenario (Chapin et al. 2000; Kittel et al. 2000). The increase in boreal forests may increase BVOC emissions for species such as *Populus* sp. and *Picea* sp. (Lerdau and Slobodkin 2002). However, forests at lower altitudes may reduce BVOC emission, probably having no impact on global budgets.

5.6 BVOCs and Plant Signaling (Plant–Plant and Plant–Insect Herbivory)

Several authors discussed about the BVOC emissions and their role in many abiotic stress tolerance, including thermotolerance of photosynthesis and reduced oxidative stress (Sun et al. 2013; Singsaas et al. 1997; Loreto et al. 1998; Sharkey and Singsaas 1995; Possell and Loreto 2013; Llusià et al. 2005).

As per the authors of the recent twenty-first century, BVOCs are essential components for communication within a plant, between plants, and between a plant and an insect (Trowbridge and Stoy 2013; Dicke and Baldwin 2010; Duhl 2008; Baldwin et al. 2002). In plant–herbivore interactions, the volatiles can act as attractants or repellents to herbivores (Laothawornkitkul et al. 2008). For instance the α-pinene (monoterpene) released by wounded *Pinus sylvestris* L. (Scotch pine) acts as an attractant to *Hylobius abietis* (large pine weevil), and thus, previous damage of a conifer can increase herbivory damage. However, by repelling limonene attraction of *H. abietis* can be reduced (Nordlander 1991). According to Mithen (2001), heat-stressed *Brassica nigra* (black mustard) plants may become attractive to specialized feeders of Brassicaceae (older name, Cruciferae) due to the increasing emissions of allyl isothiocyanate (organosulfur compound and responsible for the pungent taste of the Cruciferae family). Arimura et al. (2001) found that BVOCs emitted from *Tetranychus* (spider mite)-infested lima bean/butter bean (*Phaseolus lunatus* L.) can activate genes encoding pathogenesis-related proteins and phenylalanine ammonia-lyase in leaves of noninfested neighboring plants, and GLVs can serve as signal compounds in plant–plant communication (Arimura et al. 2001). Later on it was found that such kind of sharing of information (between plants) depends on the diffusion and convection of the BVOC information between a sender and a receiver plant (Baldwin et al. 2002).

5.7 Conclusion

BVOCs played an important role in atmospheric chemistry and were also involved in the several mechanisms of plant physiology like plant signaling. Any biotic or abiotic stress is responsible for BVOC emission from plants. The overall effect of increasing BVOC emissions will depend on the positive and negative feedback mechanism. These emissions are released in support of defense mechanisms from plant species. This chapter signifies about the important component of atmospheric sciences as well as plant physiology, i.e., BVOCs which are still under progress; very less attention has been put toward this mechanism. The important environmental issue of global warming is also influenced by BVOCs. BVOCs also played an important role in the production of tropospheric ozone, aerosol formation, and ultimately SOA. In short, this chapter concludes with the fact that precursors play an important role in the production of secondary pollutants and these precursors like BVOCs take part in climate change issues.

References

Arigoni D, Sagner S, Latzel C, Eisenreich W, Bacher A, Zenk MH (1997) Terpenoid biosynthesis from 1-deoxy-D-xylulose in higher plants by intramolecular skeletal rearrangement. Proc Natl Acad Sci U S A 94:10600–10605

Arimura G, Ozawa R, Horiuchi J, Nishioka T, Takabayashi J (2001) Plant–plant interactions mediated by volatiles emitted from plants infested by spider mites. Biochem Syst Ecol 29(10):1049–1061

Baldwin IT, Kessler A, Halitschke R (2002) Volatile signaling in plant–plant–herbivore interactions: what is real? Curr Opin Plant Biol 5(4):351–354

Bergamaschi P, Hein R, Heimann M, Crutzen PJ (2000) Inverse modeling of the global CO cycle 1. Inversion of CO mixing ratios. J Geophys Res-Atmos 105:1909–1927

Blande JD, Tiiva P, Oksanen E, Holopainen JK (2007) Emission of herbivore-induced volatile terpenoids from two hybrid aspen (Populus tremula x tremuloides) clones under ambient and elevated ozone concentrations in the field. Glob Chang Biol 13:2538–2550

Bohlmann J, Meyer-Gauen G, Croteau R (1998) Plant terpenoid synthases: molecular biology and phylogenetic analysis. Proc Natl Acad Sci U S A 95:4126–4133

Chapin FS, McGuire AD, Randerson J, Pielke R, Baldocchi D, Hobbie SE, … Zimov SA (2000) Arctic and boreal ecosystems of western North America as components of the climate system. Glob Chang Biol 6(S1):211–223

Claeys M, Graham B, Vas G, Wang W, Vermeylen R, Pashynska V, Cafmeyer J, Guyon P, Andreae MO, Artaxo P et al (2004) Formation of secondary organic aerosols through photooxidation of isoprene. Science 303:1173–1176

Constable J, Litvak ME, Greenberg JP, Monson RK (1999) Monoterpene emission from coniferous trees in response to elevated CO2 concentration and climate warming. Glob Chang Biol 5(3):252–267

Cox PM, Betts RA, Collins M, Harris PP, Huntingford C, Jones CD (2004) Amazonian forest dieback under climate-carbon cycle projections for the 21st century. Theor Appl Climatol 78(1–3):137–156

Csiky O, Seufert G (1999) Terpenoid emissions of Mediterranean oaks and their relation to taxonomy. Ecol Appl 9:1138–1146

Daum PH, Kleinman L, Imre DG, Nunnermacker LJ, Lee Y-N, Springston SR, Newman L (2000) Analysis of the processing of Nashville urban emissions on July 3 and July 18, 1995. J Geophys Res 105:9107–9119

Dicke M, Baldwin IT (2010) The evolutionary context for herbivore-induced plant volatiles: beyond the 'cry for help'. Trends Plant Sci 15(3):167–175

Duhl AB (2008) Sesquiterpene emissions from vegetation: a review. Biogeosciences 5:761–777

Edney EO, Kleindienst TE, Jaoui M, Lewandowski M, Offenberg JH, Wang W et al (2005) Formation of 2-methyl tetrols and 2-methylglyceric acid in secondary organic aerosol from laboratory irradiated isoprene/NOX/SO2/air mixtures and their detection in ambient PM2_5 samples collected in the eastern United States. Atmos Environ 39:5281–5289

El Tamer MK, Lucker J, Bosch D, Verhoeven HA, Verstappen FWA, Schwab W et al (2003) Domain swapping of Citrus lemon monoterpene synthases: impact on enzymatic activity and product specificity. Arch Biochem Biophys 411:196–203

Fall R, Karl T, Hansel A, Jordan A, Lindinger W (1999) Volatile organic compounds emitted after leaf wounding: on-line analysis by proton-transfer- reaction mass spectrometry. J Geophys Res-Atmos 104:15963–15974

Fehsenfeld F, Calvert J, Fall R, Goldan P, Guenther AB, Hewitt CN, Lamb B, Liu S, Trainer M, Westberg H et al (1992) Emissions of volatile organic compounds from vegetation and the implications for atmospheric chemistry. Global Biogeochem Cycles 6:389–430

Fowler D, Amann M, Anderson R, Ashmore M, Depledge MH, Derwent D, Grennfelt P, Hewitt CN, Hov O, Jenkin M et al (2008) Ground-level ozone in the 21st century: future trends, impacts and policy implications. Policy Document 15/08. Royal Society, London

Franceschi VR, Krokene P, Christiansen E, Krekling T (2005) Anatomical and chemical defenses of conifer bark against bark beetles and other pests. New Phytol 162:353–375

Fuentes JD, Wang D, Gu L (1999) Seasonal variations in isoprene emissions from a boreal aspen forest. J Appl Meteorol 38:855–869

Fuentes JD, Lerdau M, Atkinson R, Baldocchi D, Bottenheim JW, Ciccioli P, Lamb B, Geron C, Gu L, Guenther A et al (2000) Biogenic hydrocarbons in the atmospheric boundary layer: a review. Bull Am Meteorol Soc 81:1537–1575

Geron C, Rasmussen R, Arnts RR, Guenther A (2000) A review and synthesis of monoterpene speciation from forests in the United States. Atmos Environ 34(11):1761–1781

Gershenzon J, McConkey M, Croteau R (2000) Regulation of monoterpene accumulation in leaves of Peppermint. Plant Physiol 122:205–213. 274

Gray CM, Monson RK, Fierer N (2014) Biotic and abiotic controls on biogenic volatile organic compound fluxes from a subalpine forest floor. J Geophys Res Biogeosci 119(4):547–556

Griffin RJ, Chen JJ, Carmody K, Vutukuru S, Dabdub D (2007) Contribution of gas phase oxidation of volatile organic compounds to atmospheric carbon monoxide levels in two areas of the United States. J Geophys Res Biogeosci 112:D10S17

Grosjean D (1995) Atmospheric chemistry of biogenic hydrocarbons—relevance to the Amazon. Quim Nova 18:184–201

Guenther A, Hewitt CN, Erickson D, Fall R, Geron C, Graedel T, Harley P, Klinger L, Lerdau M, McKay WA et al (1995) A global-model of natural volatile organic-compound emissions. J Geophys Res Atmos 100:8873–8892

Guenther AB, Geron C, Pierce T et al (2000) Natural emissions of non-methane volatile organic compounds; carbon monoxide, and oxides of nitrogen from North America. Atmos Environ 34:2205–2230

Guenther A, Karl T, Harley P, Wiedinmyer C, Palmer PI, Geron C (2006) Estimates of global terrestrial isoprene emissions using MEGAN (Model of Emissions of Gases and Aerosols from Nature). Atmos Chem Phys 6:3181–3210

Haagen-Smit AJ (1952) Chemistry and physiology of Los Angeles smog. Ind Eng Chem 44(6):1342–1346

Hanson DT, Swanson S, Graham LE, Sharkey TD (1999) Evolutionary significance of isoprene emission from mosses. Am J Bot 86:634–639

Hatakeyama S, Izumi K, Fukuyama T, Akimoto H, Washida N (1991) Reactions of OH with alpha-pinene and beta-pinene in air – estimate of global CO production from the atmospheric oxidation of terpenes. J Geophys Res Atmos 96:947–958

Houghton JT et al (2001) IPCC climate change 2001: the scientific basis. Contribution of working group I in the third assessment report of Intergovernmental Panel on Climate Change. Cambridge University Press

IPCC (2007) The physical science basis. Contribution of working group I. In: Solomon S et al (eds) Fourth assessment report of the intergovernmental panel on climate change. Cambridge University Press, pp 1–996

Joutsensaari J, Loivamaki M, Vuorinen T, Miettinen P, Nerg AM, Holopainen JK, Laaksonen A (2005) Nanoparticle formation by ozonolysis of inducible plant volatiles. Atmos Chem Phys 5:1489–1495

Kampranis SC, Ioannidis D, Purvis A, Mahrez W, Ninga E, Katerelos NA et al (2007) Rational conversion of substrate and product specificity in a Salvia monoterpene synthase: structural insights into the evolution of terpene synthase function. Plant Cell 19:1994–2005

Kesselmeier J, Staudt M (1999) Biogenic volatile organic compounds (VOC): an overview on emission, physiology and ecology. J Atmos Chem 33:23–88

Kittel TGF, Steffen WL, Chapin FS (2000) Global and regional modelling of Arctic–boreal vegetation distribution and its sensitivity to altered forcing. Glob Chang Biol 6(S1):1–18

Kroll JH, Ng NL, Murphy SL, Flagen RC, Seinfeld JH (2005) Secondary organic aerosol formation from isoprene photooxidation under high-NOx conditions. Geophys Res Lett 32:L18808

Kroll JH, Ng NL, Murphy SM, Flagan RC, Seinfeld JH (2006) Secondary organic aerosol formation from isoprene photooxidation. Environ Sci Tech 40:1869–1877

Kulmala M (2003) How particles nucleate and grow. Science 302:1000–1001

Laothawornkitkul J, Paul ND, Vickers CE, Possell M, Mullineaux PM, Hewitt CN, Taylor JE (2008) The role of isoprene in insect herbivory. Plant Signal Behav 3(12):1141–1142

Laothawornkitkul J, Taylor JE, Paul ND, Hewitt CN (2009) Biogenic volatile organic compounds in the Earth system. New Phytol 183:27–51

Leaitch WR, Bottenheim JW, Biesenthal TA, Li SM, Liu PSK, Asalian K, Dryfhout-Clark H, Hopper F, Brechtel F (1999) A case study of gas-to-particle conversion in an eastern Canadian forest. J Geophys Res Atmos 104:8095–8111

Lelieveld J, Butler TM, Crowley JN, Dillon TJ, Fischer H, Ganzeveld L, Harder H, Lawrence MG, Martinez M, Taraborrelli D et al (2008) Atmospheric oxidation capacity sustained by a tropical forest. Nature 452:737–740

Leng C, Kish JD, Kelley J, Mach M, Hiltner J, Zhang Y, Liu Y (2013) Temperature-dependent Henry's law constants of atmospheric organics of biogenic origin. J Phys Chem A 117(40):10359–10367

Lerdau M, Gray D (2003) Ecology and evolution of light-dependent and light independent phytogenic volatile organic carbon. New Phytol 157:199–211

Lerdau M, Slobodkin L (2002) Trace gas emissions and species-dependent ecosystem services. Trends Ecol Evol 17(7):309–312

Lerdau MT, Dilts SB, Westberg H, Lamb BK, Allwine EJ (1994) Monoterpene emission from Ponderosa pine. J Geophys Res 99:16609–16615

Lichtenthaler HK, Schwender J, Disch A, Rohmer M (1997) Biosynthesis of isoprenoids in higher plant chloroplasts proceeds via amevalonate-independent pathway. FEBS Lett 400:271–274

Litvak ME, Monson RK (1998) Patterns of induced and constitutive monoterpene production in conifer needles in relation to insect herbivory. Oecologia 114(4):531–540

Llusià J, Peñuelas J (2000) Seasonal patterns of terpene content and emission from seven Mediterranean woody species in field conditions. Am J Bot 87:133–140

Llusià J, Peñuelas J, Munné-Bosch S (2005) Sustained accumulation of methyl salicylate alters antioxidant protection and reduces tolerance of holm oak to heat stress. Physiol Plant 124(3):353–361

Logan BA, Monson RK (1999) Thermotolerance of leaf discs from four isoprene-emitting species is not enhanced by exposure to exogenous isoprene. Plant Physiol 120:821–825

Logan JA, Prather MJ, Wofsy SC, McElroy MB (1981) Tropospheric chemistry – a global perspective. J Geophys Res C Oceans Atmos 86:7210–7254

Loreto F, Ciccioli P, Brancaleoni E, Valentini R, De Lillis M, Csiky O et al (1998a) A hypothesis on the evolution of isoprenoid emission by oaks based on the correlation between emission type and Quercus taxonomy. Oecologia 115:302–305

Loreto F, Fo¨rster A, Du¨rr M, Csiky O, Seufert G (1998b) On them onoterpene emission under heat stress and on the increased thermotolerance of leaves of Quercus ilex L. fumigated with selected monoterpenes. Plant Cell Environ 21:101–107

McBean G, Alekseev G, Chen D, Førland E, Fyfe J, Groisman PY et al (2005) Arctic climate: past and present. Arctic climate impact assessment scientific report

Meskhidze N, Nenes A (2006) Phytoplankton and cloudiness in the Southern Ocean. Science 314:1419–1423

Mgaloblishvili MP, Khetsuriana ND, Kalandaze AN, Sanadze GA (1979) Localization of isoprene biosynthesis in poplar leaf chloroplasts. Sov Plant Physiol 26:837–842

Miller B, Oschinski C, Zimmer W (2001) First isolation of an isoprene synthase gene from poplar and successful expression of the gene in Escherichia coli. Planta 213:483–487

Mithen RF (2001) Glucosinolates and their degradation products. Adv Bot Res 35:213–232

Monson RK, Holland EA (2001) Biospheric trace gas fluxes and their control over tropospheric chemistry. Annu Rev Ecol Evol Syst 32:547–576

Monson RK, Jaeger CH, Adams WW, Driggers EM, Silver GM, Fall R (1992) Relationships among isoprene emission rate, photosynthesis, and isoprene synthase activity as influenced by temperature. Plant Physiol 98:1175–1180

Monson RK, Jones RT, Rosenstiel TN, SCHNITZLER JP (2013) Why only some plants emit isoprene. Plant Cell Environ 36(3):503–516

Müller J-F, Stavrakou T, Wallens S, De Smedt I, Van Roozendael M, Potosnak MJ, Rinne J, Munger B, Goldstein A, Guenther AB (2008) Global isoprene emissions estimated using MEGAN, ECMWF analyses and a detailed canopy environment model. Atmos Chem Phys 8:1329–1341

Ng NL, Kroll JH, Keywood MD, Bahreini R, Varutbangkul V, Flagan RC et al (2006) Contribution of first- versus second-generation products to secondary organic aerosols formed in the oxidation of biogenic hydrocarbons. Environ Sci Tech 40:2283–2297

Niinemets U et al (2004) Physiological and physicochemical controls on foliar volatile organic compound emissions. Trends Plant Sci 9:180–186

Nordlander G (1991) Host finding in the pine weevil Hylobius abietis: effects of conifer volatiles and added limonene. Entomol Exp Appl 59(3):229–237

Olcese LE, Penner JE, Sillman S (2007) Development of a secondary organic aerosol formation mechanism: comparison with smog chamber experiments and atmospheric measurements. Atmos Chem Phys Discuss 7:8361–8393

Peñuelas J, Llusià J (2001) The complexity of factors driving volatile organic compound emissions by plants. Biol Plant 44:481–487

Peñuelas J, Llusià J (2002) Linking photorespiration, monoterpenes and thermotolerance in Quercus. New Phytol 155:227–237

Penuelas J, Llusia J (2003) BVOCs: plant defense against climate warming? Trends Plant Sci 8:105–109

Penuelas J, Filella I, Gamon JA (1995) Assessment of photosynthetic radiation-use efficiency with spectral reflectance. New Phytol 131(3):291–296

Pichersky E, Gershenzon J (2002) The formation and function of plant volatiles: perfumes for pollinator attraction and defense. Curr Opin Plant Biol 5(3):237–243

Pichersky E, Noel JP, Dudareva N (2006) Biosynthesis of plant volatiles: nature's diversity and ingenuity. Science 311:808–811

Poisson N, Kanakidou M, Crutzen PJ (2000) Impact of non-methane hydrocarbons on tropospheric chemistry and the oxidizing power of the global troposphere: three-dimensional modelling results. J Atmos Chem 36:157–230

Possell M, Loreto F (2013) The role of volatile organic compounds in plant resistance to abiotic stresses: responses and mechanisms. In: Biology, controls and models of tree volatile organic compound emissions. Springer, Dordrecht, pp 209–235

Rohmer M, Knani M, Simonin P, Sutter B, Sahm H (1993) Isoprenoid biosynthesis in bacteria: a novel pathway for the early steps leading to isopentenyl diphosphate. Biochem J 295:517–524

Sasaki K, Ohara K, Yazaki K (2005) Gene expression and characterization of isoprene synthase from Populus alba. FEBS Lett 579:2514–2518

Schnitzler J-P, Arenz R, Steinbrecher R, Lehning A (1996) Characterization of an isoprene synthase from leaves of Quercus petraea (Mattuschka) Liebl. Bot Acta 109:216–221

Schwender J, Zeidler J, Gr¨oner R, M¨uller C, Focke M et al (1997) Incorporation of 1-deoxy-D-xylulose into isoprene and phytol by higher plants and algae. FEBS Lett 414:129–134

Sharkey TS (2013) Is it useful to ask why plants emit isoprene? Plant Cell Environ 36:517–520

Sharkey TD, Singsaas EL (1995) Why plants emit isoprene. Nature 374:769

Sharkey TD, Yeh S (2001) Isoprene emission from plants. Annu Rev Plant Physiol Plant Mol Biol 52:407–436

Sharkey TD, Yeh S, Wiberley AE, Falbel TG, Gong D, Fernandez DE (2005) Evolution of the isoprene bio-

synthetic pathway in kudzu. Plant Physiol 137:700–712
Sharkey TD, Gray DW, Pell HK, Breneman SR, Topper L (2013) Isoprene synthase genes form a monophyletic clade of acyclic terpene synthases in the tps-b terpene synthase family. Evolution 67(4):1026–1040
Shulaev V, Silverman P, Raskin I (1997) Airborne signalling by methyl salicylate in plant pathogen resistance. Nature 385:718–721
Silver GM, Fall R (1991) Enzymatic synthesis of isoprene from dimethylallyl diphosphate in aspen leaf extracts. Plant Physiol 97:1588–1591
Silver GM, Fall R (1995) Characterization of aspen isoprene synthase, an enzyme responsible for leaf isoprene emission to the atmosphere. J Biol Chem 270:13010–13016
Singsaas EL, Lerdau M, Winter K, Sharkey TD (1997) Isoprene increases thermotolerance of isoprene-emitting species. Plant Physiol 115:1413–1420
Sun Z, Hüve K, Vislap V, Niinemets Ü (2013) Elevated [CO_2] magnifies isoprene emissions under heat and improves thermal resistance in hybrid aspen. J Exp Bot 64:5509–5523, ert318
Tholl D (2006) Terpene synthases and the regulation, diversity and biological roles of terpene metabolism. Curr Opin Plant Biol 9:297–304
Thompson AM (1992) The oxidizing capacity of the Earth's atmosphere: probable past and future changes. Science 256:1157–1165
Thompson AM, Hogan KB, Hoffman JS (1992) Methane reductions: implications for global warming and atmospheric chemical change. Atmos Environ Part A 26(14):2665–2668
Tingey DT, Evans R, Bates EH, Gumpertz M (1987) Isoprene emissions and photosynthesis in three ferns – the influence of light and temperature. Physiol Plant 69:609–616
Tingey DT, Turner DP, Weber JA (1991) Factors controlling the emissions of monoterpenes and other volatile organic compounds. In: Sharkey TD, Holland EA, Mooney HA (eds) Trace gas emissions from plants. Academic, San Diego, pp 93–119
Trainer M, Williams EJ, Parrish DD, Buhr MP, Allwine EJ, Westberg HH, Fehsenfeld FC, Liu SC (1987) Models and observation of the impact of natural hydrocarbons on rural ozone. Nature 329:705–707
Trowbridge AM, Stoy PC (2013) BVOC-mediated plant-herbivore interactions. In: Biology, controls and models of tree volatile organic compound emissions. Springer, Dordrecht, pp 21–46
Turner GW, Gershenzon J, Croteau RB (2000) Development of peltate glandular trichomes of peppermint. Plant Physiol 124:665–679. 274
Wagner WP, Nemecek-Marshall M, Fall R (1999) Three distinct phases of isoprene formation during growth and sporulation of Bacillus subtilis. J Bacteriol 181:4700–4703
Wang YH, Jacob DJ, Logan JA (1998) Global simulation of tropospheric O_3-NOxhydrocarbon chemistry. 3. Origin of tropospheric ozone and effects of non-methane hydrocarbons. J Geophys Res 103(10):757–767
Wildermuth MC (1997) Subcellular location and biophysical regulation of foliar isoprene production (chloroplasts). PhD thesis. University of Colorado Boulder, 307 pp
Wildermuth MC, Fall R (1996) Light-dependent isoprene emission – characterization of a thylakoid-bound isoprene synthase in Salix discolor chloroplasts. Plant Physiol 112:171–182
Wildermuth MC, Fall R (1998) Biochemical characterization of stromal and thylakoid-bound isoforms of isoprene synthase in willow leaves. Plant Physiol 116:1111–1123
Wilmking M, Juday GP, Barber VA, Zald HS (2004) Recent climate warming forces contrasting growth responses of white spruce at treeline in Alaska through temperature thresholds. Glob Chang Biol 10(10): 1724–1736
Zeidler JG, Lichtenthaler HK, May HU, Lichtenthaler FW (1997) Is isoprene emitted by plants synthesized via the novel isopentenyl pyrophosphate pathway? Z Naturforsch Teil C 52:15–23

Biochemical Effects of Air Pollutants on Plants

6

Pallavi Saxena and Umesh Kulshrestha

Abstract

Urbanization and industrialization processes contribute significant amount of various air pollutants such as SO_2, NO_2, CO, particulate matter, etc. These pollutants affect plant health and emit various forms of SO_2, NOx, and O_3 which may act in combination of a variety of ways: additive, synergistic, and antagonistic. These pollutants can have a deleterious effect on a variety of biochemical and physiological processes and on the structural organization within the cells. Certain plant species are very sensitive to these pollutants resulting in well visible and measurable symptoms. Morphological damage is generally visible through lesions on the leaves, flowers, and fruits while biochemical and physiological changes which are invisible can be measured and quantified. In this chapter, biochemical effects on plants have been described. These symptoms can be used as indicators of air pollution stress for its early diagnosis and can be used as markers for a particular physiological disorder.

Keywords

Urbanization • Industrialization • Biochemical • Physiological • Morphological • Air pollution stress

6.1 Introduction

Increasing air pollution has been a matter of concern for plant health due to its adverse effects on plant physiology, biochemistry, and morphology. Some of the atmospheric gases at their supra optimum level become pollutants and evoke various types of visible and hidden plant responses which ultimately lead to reduced plant growth and productivity (Krupa et al. 1982; Srivastava 1999; Poschl 2005). The impact of such anthropogenic emission into the atmosphere and its movement into the biosphere by transformation, reaction, and modification is responsible for a variety of chronic and acute diseases at local,

P. Saxena (✉) • U. Kulshrestha
School of Environmental Sciences, Jawaharlal Nehru University, New Delhi 110067, India
e-mail: pallavienvironment@gmail.com

U. Kulshrestha, P. Saxena (eds.), *Plant Responses to Air Pollution*,
DOI 10.1007/978-981-10-1201-3_6

regional, and global scales (Rawat and Banerjee 1996). Impact on the plant community has also been studied worldwide in terms of plant–environment interactions, since the plants are much more sensitive in comparison to other organisms (Abbasi et al. 2004). The symptoms or effects in plant anatomy, physiology, or biochemistry indicate the state of the environment. Since the major system and organs of plants are exposed to the atmosphere and the leaves continuously exchange gases in and out of the systems, any change in the atmosphere is reflected in the plants' physiology.

On the other hand, plants play an important role in monitoring and maintaining the ecological balance by actively participating in the cycling of nutrients and gases like carbon dioxide and oxygen and also provide enormous leaf area for impingement, absorption, and accumulation of air pollutants to reduce the pollution level in the air environment (Escobedo et al. 2008). Biomonitoring of air pollution using plants is possible by using both native and cultivated plant species present in the studied area (Shannigrahi et al. 2004). Plants take these air pollutants from the surrounding air. Once taken, the selected pollutant may be adsorbed, absorbed, accumulated, or integrated into the plant body. If the nature of a particular air pollutant is toxic, it may injure the plant exhibiting specific symptoms. Generally, sensitive species show quicker injury symptoms than that of tolerant ones. Hence, sensitive species act as early warning indicators of pollution. On the other hand, the tolerant species help in the scavenging of air pollutants, reducing the overall pollution load (Rao 1983). Among various air pollutants, gases such as SO_2, NO_2, HF, PAN, and O_3 are highly phytotoxic. These may harm higher plants very rapidly in a drastic manner (Figs. 6.1 and 6.2). For example, the phytotoxic effect of hydrogen fluoride (HF) is well known for several plant species such as monocotyledonous ornamental plants (as tulips, gladioli), stone fruit species (as plums, peaches, and apricots), crops (like maize), and natural plants (like *Hypericum perforatum* L. and *Picea abies* L.) (Flowers et al. 2007). When plants are exposed to HF air pollution, the F^- ion accumulates in the rims and tips of the leaves and causes necrosis of leaf tissue, clearly separated from the living, green tissue by a red-brown boundary zone (Fig. 6.3) (Hogue et al. 2007).

Air pollutants cause damage to leaf cuticles and affect stomatal conductance. They can also have direct effects on photosynthetic systems, leaf longevity, and patterns of carbon allocation within plants. Pollutants interact with other environmental factors and may alter plant–environment relationships on a regional scale (Winner 1981). Air pollutants' impact on plant life are mainly of two types – directly through clear visible leaf injury (e.g., interveinal or needle chlorosis and necrosis) or indirectly on growth and reproduction. Some air pollutants cause only visible effects on plants at a much higher concentration than indirect ones. For example, chlorine (Cl_2), nitrogen dioxide (NO_2), hydrochloric acid (HCl), and ammonia (NH_3) are components that do not produce specific symptoms, but all give rise to leaf chlorosis and necrosis and growth reduction (Weinstein 1977). Thus higher plants may be used as indicators and

Fig. 6.1 Effect of PAN on milkweed leaves (Source: https://extension.umd.edu/learn/air-pollution-effects-vegetables)

Fig. 6.2 Effect of ozone on potato (Source: https://extension.umd.edu/learn/air-pollution-effects-vegetables)

Fig. 6.3 Effect of hydrogen fluoride on *Dracaena deremensis* (Source: http://mysticablog.wordpress.com)

accumulators of air pollutants for detection, recognition, and monitoring purposes. Some of the important physiological processes such as photosynthesis, respiration, carbon allocation, and stomatal functioning are known to be effected by air pollution (Darrall 1988). Certain plant species may accumulate a particular component from the air without changing it in such a way that the component can be analyzed physiochemically (qualitatively and quantitatively) after accumulation in the plants (Hung and Mackay 1997). The level of biochemical parameters such as chlorophyll, protein, soluble sugar, ascorbic acid, superoxide dismutase, and peroxidase in leaves have been found to be pollution load dependent. The effect of pollution on each of these biochemical constituents has been described below.

6.2 Effect on Pigment Content

Chlorophyll is one of the main essential parts of energy production in green plants and its amounts are significantly affected by environmental condition. Depletion in chlorophyll causes a decrease in the productivity of plants and subsequently causes plants to exhibit poor vigor. The total chlorophyll level in plants decreases under stress condition (Speeding and Thomas 1973). Bell and Mudd (1976) opined that tolerance of plants to SO_2 might be linked with the synthesis of degradation of chlorophyll. However plants maintaining their chlorophyll under polluted conditions are said to be tolerant (Singh and Verma 2007). Chlorophyll measurement is an important tool to evaluate the effects of air pollutants on the plants as it plays an important role in plant metabolism. Any reduction in chlorophyll content directly affects plant growth (Agbaire and Esiefarienhre 2009). The net photosynthetic rate is a commonly used indicator of impact of increased air pollutants on plant growth (Woo et al. 2007). Air pollution stress leads to stomatal closure, which reduces CO_2 availability in leaves and inhibits carbon fixation. Sulfur dioxides, nitrogen dioxides, and CO_2, as well as suspended particulate matter, are some of the air pollutants which are absorbed by plant leaves causing reduction in the levels of photosynthetic pigments, viz., chlorophyll and carotenoids, affecting the plant productivity directly (Joshi and Swami 2009; Honour et al. 2009). Rao and LeBlanc (1965) found that destruction of chlorophyll occurred in lichens following exposure to large doses (5 ppm for 24 h) of gaseous SOx. At this high concentration, chlorophyll molecules were degraded to pheophytin and Mg^{2+}. A similar conversion of chlorophyll to pheophytin can occur with acids or acidic substances. In this process Mg^{2+} in the chlorophyll molecule is replaced by two atoms of hydrogen, thereby changing the light-spectrum characteristic of the chlorophyll molecules. Rapid in vitro chlorophyll destruction can also be caused by free radicals produced during the oxidation of HSO_3^--catalyzed decomposition of the linoleic acid hydrogen peroxide (Peiser and Yang 1977, 1979). Shimazaki et al. (1980) presented evidence that SO_2 fumigation of leaves increases the formation of O_2^- in chloroplasts that in turn destroys chlorophylls. A superoxide radical has been shown to influence chlorophyll at very low concentrations (10–8 to 10–7 M) (Asada et al. 1977). In *Spinacia oleracea* leaves, gaseous SO_2 destroyed chlorophyll a more rapidly than chlorophyll b, but the loss of chlorophyll a was not accompanied by a corresponding increase in pheophytin a (Shimazaki et al. 1980). As scavengers of free radicals inhibited chlorophyll breakdown in *Spinacia oleracea* leaves, it was suggested that SO_2 destroys chlorophyll mainly by a free-radical oxidation. This was further supported by the observation that chlorophyll a breakdown was inhibited by superoxide dismutase. Sulfur dioxide inhibits the superoxide dismutase activity in the fumigated tissues (Shimazaki et al. 1980). Furthermore, accumulation of malondialdehyde, a lipid peroxidation product, and a decrease in chlorophyll a in SO_2-fumigated *Spinacia oleracea* leaves were related to the free-radical oxidation of chlorophyll.

Gradual disappearance of chlorophyll and concomitant yellowing of leaves is one of the most common effects on plants which may be associated with the continuously decreasing photosynthetic capacity of the plant (Joshi et al. 2009). Carotenoids which help in capturing light in the chloroplast are also affected by air pollution. Carotenoids also play a more important role in protecting the cells and live organisms as they encounter damage from free-radical oxidative cells (Fleschin et al. 2003). These pigments are more stronger than chlorophyll but much less efficient in light gathering, help the valuable but much fragile chlorophyll, and protect chlorophyll from photoxidative destruction (Joshi et al. 2009). Similar to chlorophyll, the level of carotenoids decreases with the increase in air pollution load (Joshi et al. 2009; Tripathi and Gautam 2007; Tiwari et al. 2006; Gupta et al. 2015).

6.3 Effect on Sugar Content

Soluble sugars have osmoprotectant and cryoprotectant roles and their presence is important for the plasma membrane. These are important parts in the plant structure and source of energy in all organisms. The concentration of soluble sugars is indicative of the physiological activity of a plant and it determines the sensitivity of plants to air pollution (Tripathi and Gautam 2007). Accumulation of sugars in different parts of plants is enhanced in response to the variety of environmental stresses (Prado et al. 2000). Soluble sugars have been also reported to play a protective role against stresses (Finkelstein and Gibson 2001). In this study, there was an increase in soluble sugar in polluted sites indicative of stress. Soluble sugar is an important constituent of plants which acts as a source of energy. Plants manufacture sugars during photosynthesis and breakdown during respiration (Bennett et al. 1984). The concentration of soluble sugars is indicative of the physiological activity of a plant which determines the sensitivity of plants toward air pollution. Reduction in soluble sugar content at polluted sites can be attributed to the increased rate of respiration and decreased CO_2 fixation because of chlorophyll deterioration (Wilkinson and Barnes 1973). In a polluted environment, gases such as SO_2, NO_2, and H_2S can cause more depletion of soluble sugars in the leaves under hardening conditions. In an SO_2-exposed plant, the carbohydrate content is reduced due to sulfite reaction with aldehydes and ketones of carbohydrates (Dugger and Ting 1970). Plants exposed to SO_2 exhibit increasing amounts of soluble sugars (Khan and Malhotra 1977; Koziol and Jordan 1978; Malhotra and Sarkar 1979). In *Pinus banksiana*, SO_2 fumigation (0.34 and 0.51 ppm) increased the content of the reducing sugars and reduced that of the nonreducing sugars (Malhotra and Sarkar 1979). It was suggested that the increase was due to a breakdown of polysaccharides rich in reducing sugars. Koziol and Jordan (1978) showed that SO_2 exposure of *Phaseolus vulgaris* seedlings caused a reduction in starch content. Reduction in nonstructural total carbohydrates has also been reported following SO_2

exposure of *Ulmus americana* (American elm) seedlings (Constantinidou and Kozlowski 1979).

6.4 Effect on Proline Content

Proline is a part of many proteins and enzymes and has important roles in plants as source of energy and osmoprotectant in stressed conditions (Huber 1984). Proline accumulation in abiotic stress reduces degradation of other proteins (Thomas 1991). Proline accumulation in the cells may happen because of decrease in proline degradation, increase in proline synthesis, and hydrolysis of protein (Fikriye and Omer 2005). The accumulation of proline is related to increase of tolerance against salt and drought stress in many plants (Nayar 2003). Proline acts as free-radical scavenger protecting the plants against damage due to oxidative stress. Higher exposure to air pollutants makes chloroplasts more vulnerable to generate ROS and induces oxidative stress (Woo et al. 2007). Typical environmental stress can cause excess reactive oxygen species (ROS) which are extremely reactive and cytotoxic to all organisms (Pukacha and Pukacha 2000). The deleterious effects of pollutants are caused by the production of ROS in plants, which cause peroxidative destruction of cellular constituents (Tiwari et al. 2006). Hence, higher proline in plants is considered as an indicator of higher stress like osmotic stress (Szekely 2004; Gupta et al. 2015).

6.5 Effect on Enzymatic Activities and Role of Antioxidants

Since higher plants are immobile, they experience environmental stress due to high air pollution load in the atmosphere. The ability of higher plants to scavenge the toxic effects of active oxygen seems to be a very important determinant of their tolerance to these stresses. Antioxidants are the first line of defense against free-radical damage. They are critical for maintaining the optimum health of plant cells. There are several antioxidant enzymes, peptides, and metabolites involved in the scavenging of active oxygen in plants, and their activation are known to increase upon exposure to oxidative stress (Gill and Tuteja 2010). The examples of antioxidant enzymes are superoxide dismutase (SOD), catalase (CAT), ascorbate peroxidase (APX), monodehrydroascorbate reducatse (MDHAR), dehydroascorbate reductase (DHAR) and glutathione reductase (GR) while antioxidant metabolies include phenolic and nitrogen compounds..

6.5.1 Enzymatic Activity and Peptide Defense

Information on antioxidant levels and the activity of antioxidant-regenerating enzymes are somewhat contradictory; both decreases and increases in the antioxidative capacity of the tissues have been reported (Larson 1988; Foyer and Noctor 2003; Tanou et al. 2009). Such diversification partly arises from the response specificity of a particular plant species and from different experimental conditions (stress treatment, duration of stress, assay procedure, and parameters measured).

6.5.1.1 Catalase and Peroxidase

CATs are tetrameric heme-containing enzymes with the potential to directly dismutate H_2O_2 into H_2O and O_2 and are indispensable for ROS detoxification during stressed conditions (Rao et al. 2006). CAT has one of the highest turnover rates for all enzymes: one molecule of CAT can convert approx. six million molecules of H_2O_2 to H_2O and O_2 per minute. CAT is important in the removal of H2O2 generated in peroxisomes by oxidases involved in β-oxidation of fatty acids, photorespiration, and purine catabolism. The CAT isozymes have been studied extensively in higher plants (Griffiths et al. 1989), e.g., two in *H. vulgare* (Zang et al. 2003), four in *Helianthus annuus* cotyledons (Prasad and Sharma 2004), and as many as 12 isozymes in *Brassica* (Kumar et al. 2007). Srivastava et al. (2010) reported a decrease in CAT activity in *A. doliolum* under NaCl and Cu_2O stress. Simova-Stoilova et al. (2010) reported increased CAT activity in wheat under drought stress but it was higher especially in sensitive varieties. In another study, Sharma and Dubey (2007) reported a decrease in CAT activity in rice seedlings following drought stress. It has also been reported that high light condition increased the CAT activity in *P. asperata* under drought stress (Sharma and Dubey 2007). The UV-B stress also led to a significant increase in CAT activity in *C. auriculata* seedlings (Agrawal 2007). Contrarily, Pan et al. (2006) studied the combined effect of salt and drought stress and found that it decreases the CAT activity in *Glycyrrhiza uralensis* seedlings.

Plant catalases are tetrameric homoproteins that exist as multiple isozymes encoded by nuclear genes. They are located mostly in peroxisomes and glyoxysomes, although a specific isozyme, Cat3, is present in maize mitochondria (Cho and Seo 2005). The catalase of soybean nodules is a typical homotetramer of 220 kDa (Miller et al. 2008). This enzyme may be especially abundant in the peroxisomes of determinate nodules by urease and possibly other oxidases (Tanou et al. 2009). A long-known metalloenzyme, catalase, is one of the most efficient protein catalyses known; it promotes the redox reaction.

$$2H_2O_2 \rightarrow 2H_2O + O_2$$

Hydrogen peroxide itself is not particularly reactive with the most biological precursor for more reactive oxidants such as HO. Although catalase is rather specific for H_2O_2, it reacts with a limited number of organic hydrogen peroxides such as MeOOH, using them to carry out oxidative reactions on the acceptor molecules while simultaneously reducing the peroxidic substrate. Catalase (Cat) is a high-capacity but low-affinity enzyme which destroys hydrogen peroxide. Catalase is a sink for H_2O_2 and that higher-affinity peroxidases, such as ascorbate peroxidase (APX), deal with lower concentrations (Srinivas et al. 2008). The low-catalase plants were more sensitive to stresses such as ozone and high salinity, as well hydrogen peroxide and methyl viologen.

Other important plant enzymes, the peroxidases, also function in this mode. In addition to defense against active oxygen compounds, plant peroxidases have other important cellular roles. However, in different cases endogenous auxin levels are regulated by the enzymes auxin oxidase and peroxidase (Farmer 2007). The activities of some antioxidant enzymes increase during stress treatment, and the types of enzymatic activities that increase are dependent on the form stress imposed. The enzymes whose activities increase during stress treatment may play an important role in defense against that particular stress.

6.5.1.2 Dehydroascorbate Reductase (DHAR)

DHAR is thought to play an important role in the oxidative stress tolerance of plants by regenerating ascorbate from dehydroascorbate (Foyer and Noctor 2003; Bielski et al. 1983). In some plants, DHAR activity has also been reported to increase upon exposure to high temperature, high light intensity, and water deficiency, respectively (Demirevska-Kepova et al. 2006; Zhang et al. 2003; Larson 1988). DHAR regenerates ASH from the oxidized state and regulates the cellular ASH redox state which is crucial for tolerance to various abiotic stresses leading to the production of ROS. It has also been found that DHAR overexpression also enhances plant tolerance against various abiotic stresses. In a study, under Al stress, the role of MDAR or DHAR in ASH regeneration has been studied in transgenic tobacco plants overexpressing *A. thaliana* cytosolic DHAR (DHAR-OX) or MDAR (MDAR-OX). It was found that DHAR-OX-transgenic plants showed higher levels of ASH with or without Al, whereas MDAR-OX plants only showed a higher ASH level in the absence of Al in comparison to WT. Significantly higher levels of ASH and APX in DHAR-OX plants showed better tolerance under Al stress but not MDAR-OX plants. It is clear that plants overexpressing DHAR showed tolerance to Al stress by maintaining a high ASH level (Chen and Gallie 2005). The overexpression of DHAR in tobacco protected the plants against ozone toxicity. Overexpression of DHAR increased salt tolerance in *Arabidopsis* (Ushimaru et al. 2006) and drought and ozone stress tolerance in tobacco (Eltayeb et al. 2006).

6.5.1.3 Ascorbate Reductase (APX) and Glutathione Reductase (GR)

APX is thought to play the most essential role in scavenging ROS and protecting cells in higher plants, algae, euglena, and other organisms. APX is involved in scavenging of H2O2 in water–water and ASH–GSH cycles and utilizes ASH as the electron donor. The APX family consists of at least five different isoforms including thylakoid (tAPX) and glyoxysome membrane forms (gmAPX), as well as chloroplast stromal soluble form (sAPX) and cytosolic form (cAPX) (Smith et al. 2008). APX has a higher affinity for H2O2 (μM range) than CAT and POD (mM range) and it may have a more crucial role in the management of ROS during stress. Enhanced expression of APX in plants has been demonstrated during different stress conditions. Increased leaf APX activity under Cd stress has been reported in *Ceratophyllum demersum* (George et al. 2010), *B. juncea* (Singh et al. 2008), *T. aestivum* (Li et al. 2009), and *V. mungo* (Su and Wu 2004). Noctor and Foyer 1998reported that pretreatment of *O. sativa* seedlings with H2O2 under non-heat shock conditions resulted in an increase in APX activity and protected rice seedlings from subsequent Cd stress. Enhanced activity of APX was also found in salt-stressed *A. doliolum* (Foreman et al. 2003). A significant increase in APX activity was noted under water stress in three cultivars of *P. vulgaris* (Gratao et al. 2005) and *P. asperata* (Flors and Nonell 2006). Sharma and Dubey (2007) found that mild-drought-stressed plants had higher chloroplastic APX activity than control grown plants but the activity declined at the higher level of drought stress. Pekker et al. (2002) studied the expression of cAPX in leaves of de-rooted bean plants in response to iron overload and found that cAPX expression (mRNA and protein) was rapidly induced in response to iron overload. The findings of Koussevitzky et al. (2008) suggest that cytosolic APX1 plays a key role in the protection of plants from a combination of drought and heat stress. Simonovicova et al. (2004) also reported increase in APX activity in *H. vulgare* L. cv. Alfor root tips under Al stress at 72 h.

APX and GR are the major scavengers of hydrogen peroxide in plant cells (Asada 1999) and their activities increase in response to various environmental stressors. In leaves *Arabidopsis thaliana* APX activity increased during exposure of plants to ozone, sulfur dioxide (Radotic et al. 2000) chilling, and UV-B (Koji et al. 2009). Ascorbate peroxidase (APX) and glutathione reductase (GR) activities are increased in water-stressed spinach leaves. In *Arabidopsis* leaves, the decrease in CAT activity when exposed to high temperature, high light intensity, and water deficiency preceded the increase of APX and GR activity. This decrease in CAT activity might trigger the induction of APX and GR activities by reducing the ability of cells to scavenge hydrogen peroxide (Larson 1988).

6.5.2 Metabolic Compounds' Defense

Antioxidants when added in small quantities to materials react rapidly with the free-radical intermediates of an autooxidation chain and stop it from progressing. The primary components of this antioxidant system include carotenoids, ascorbate, glutathione, vitamin E (α-tocopherols) flavonoids, phenolic acids, other phenols, alkaloids, polyamines, chlorophyll derivatives, amino acids and amines, and miscellaneous compounds. It has been recognized that naturally occurring substances too have antioxidant activity including those found in higher plants. Recently, oxygen-containing free radicals in biological systems and their role as causative agents in the etiology of a variety of chronic disorders have been the topics of interests of vegetation. It has also been reported that plants with high levels of antioxidants, whether constitutive or induced, have a greater resistance to such oxidative damage (Edwards et al. 2000; Creissen et al. 1999; Depège et al. 1998; Vierstra et al. 1982; Foyer and Halliwell 1976). A number of studies indicated that the degree of oxidative cellular damage in plants exposed to abiotic stress is controlled by the capacity of antioxidative systems (Sanchez-Rodriguez et al. 2010; Lin et al. 2008; Bartoli et al. 2004; Zhang et al. 2003; Noctor and Foyer 1998).

6.5.2.1 Phenolic Compounds

Phenolics are diverse secondary metabolites (flavonoids, tannins, hydroxycinnamate esters and lignin) abundant in plant tissues (Polidoros and Scandalios 1999). Polyphenols possess ideal structural chemistry for free-radical-scavenging activity, and they have been shown to be more effective antioxidants in vitro than tocopherols and ascorbate. Antioxidative properties of polyphenols arise from their high reactivity as hydrogen or electron donors, from the ability of the polyphenol-derived radical to stabilize and delocalize the unpaired electron (chain-breaking function), and from their ability to chelate transition metal ions (termination of the Fenton reaction) (Ferreira et al. 2002). Another mechanism underlying the antioxidative properties of phenolics is the ability of flavonoids to alter peroxidation kinetics by modification of the lipid-packing order and to decrease fluidity of the membranes (Sandalio and del Rio 1988). These changes could sterically hinder diffusion of free radicals and restrict peroxidative reactions. Moreover, it has been shown recently that phenolic compounds can be involved in the hydrogen peroxide scavenging cascade in plant cells (Harinasut et al. 2003).

6.5.2.2 Nitrogen Compounds

Alkaloids

Increasing evidence from a variety of sources have indicated that the basic nitrogen compounds of higher plants include many representatives that are potent inhibitors of various oxidatives (Gapinska et al. 2008). Caffeine, from the leaves of tea (*Thea sinensis*) and coffee (*Coffea arabica*), was shown to have antioxidative activity (in a linoleic acid oxidation test) comparable to that of butylated hydroxyanisole (BHA) and butylated hydroxytoluene (BHT). Several alkaloids of various structural types have been found to be potent inhibitors of $^1O_2$. Particularly effective are indole alkaloids such as strychnine and brucine that have a basic nitrogen atom in a rigid, cage-like structure. Such alkaloids appear to be strictly physical quenchers and are not destroyed chemically by the process of quenching. Hence, each molecule of alkaloids could inactivate many molecules of singlet oxygen (Khan et al. 2007).

Polyamines

Polyamines (spermidine and spermine) play a variety of physiological roles in plant growth and development (Singh et al. 2008; Azevedo et al. 1998). They are also potent ROS scavengers and inhibitors of lipid peroxidation (Ali and Alqurainy 2006). Furthermore, exogenous application of polyamines has been shown to protect against various stress conditions such as cold, wilting, pollution, and salinity (Leon et al. 2002). The protection of plants against ozone damage (Wang and Li 2008) by an exogenous supply of polyamines is believed to be caused by the free-radical-scavenging property of the polyamines (Azevedo et al. 1998). Also, the protection of plants against stress damage by an exogenous supply of polyamines is believed to be caused by the free-radical scavengers of the polyamines (Singh et al. 2008).

Amino Acids and Amines

Many amino acids have been tested for their antioxidant activity especially in food-based systems. Antioxidant activity has been claimed for selected amino acids such as arginine, histidine, cysteine, tryptophan, lysine, methionine, and threonine (Gapinska et al. 2008). Certain amino acids may exhibit antioxidant potential under some conditions of temperature or pH or oxygen concentration but have no effect or actually promote oxidation in others. For example, alanine and histidine were reported to inhibit the oxidation of linoleic acid at pH 9.5 and to promote it at pH 7.5 (Gapinska et al. 2008).

6.5.2.3 Other Compounds

Ascorbic Acid

Ascorbic acid (AA) has been known as a biological antioxidant. AA can directly scavenge superoxide, hydroxyl radicals, and singlet oxygen and reduce H_2O_2 to water via ascorbate peroxidase reaction (Noctor and Foyer 1998). High concentrations of ascorbic acid have been reported in many cellular environments, such as the stroma of chloroplasts where its level is 2.3×10^{-3} M. In many qualitative studies, ascorbate has been demonstrated to possess significant antioxidant activity (Ferreira et al. 2002). For example, 10^3 M ascorbate inhibited the photooxidation of a kaempferol by illuminated spinach chloroplasts. Ascorbate reduces two equivalents of $O^-{}_2$-produced H_2O_2 and the triketo derivative dehydroascorbic acid. Ascorbate also reacts with $^1O_2$ at a relatively fast rate (Noctor and Foyer 1998). AA is one of the most studied and powerful antioxidants (Wang and Li 2008; Khan et al. 2007; Larson 1988; Noctor and Foyer 1998). It has been detected in the majority of plant cell types like organelles and in the apoplast. Under physiological conditions AA exists mostly in the reduced form (90 % of the ascorbate pool) in leaves and chloroplasts (Bergmüller et al. 2003) and its intracellular concentrations can build up to a millimolar range (e.g., 20 mM in the cytosol and 20–300 mM in the chloroplast stroma (Foyer and Harbinson 1994)). The ability to donate electrons in a wide range of enzymatic and nonenzymatic reactions makes AA the main ROS-detoxifying compound in the aqueous phase. AA acts as a cofactor of violaxanthin de-epoxidase thus sustaining dissipation of excess excitation energy in chloroplasts (Khan et al. 2007). Recently, Gupta et al. (2015) have reported a higher concentration of AA as an antioxidant which has been attributed to the levels of pollution load and its stress.

6.6 Conclusion

The present chapter highlights the significant effects posed by air pollutants on plant health through biochemical parameters such as chlorophyll, proline content, and other enzymatic activities acting as bioindicators for determining the health of the plant. Moreover, among all the other air pollutants, O_3 and SO_2 affect plant metabolism mostly and can reduce the plant growth. Both morphological and physiological symptoms will be considered while analyzing the health of the plant.

References

Abbasi S, Chari A, Gajalakshmi KB, Ramesh, Ramasamy EB (2004) Approaches to greenbelt design. J Inst Public Health Engi 3:42–49

Agarwal S (2007) Increased antioxidant activity in Cassia seedlings under UV-B radiation. Biol Plant 51:157–160

Agbaire PO, Esiefarienhre E (2009) Air Pollution Tolerance Indices (APTI) of some plants around Otorogun gas plant in Delta region, Nigeria. J Appl Sci Environ Manag 13:11–14

Ali AA, Alqurainy F (2006) Activities of antioxidants in plants under environmental stress. In: Motohashi N (ed) The lutein-prevention and treatment for diseases. Transworld Research Network, Kerala, pp 187–256

Asada K (1999) The water-water cycle in chloroplasts: scavenging of active oxygens and dissipation of excess photons. Annu Rev Plant Physiol 50:601–639

Asada K, Takahashi M, Tanaka K Nakano Y (1977) Formation of active oxygen and its fate in chloroplasts. In: Hayaishi O, Asada K (eds) Biochemical and medical aspects of active oxygen. University of Tokyo Press, Tokyo/Japan, pp 45–63, 313 pp

Azevedo RA, Alas RM, Smith RJ, Lea PA (1998) Response of antioxidant enzymes to transfer from elevated carbon dioxide to air and ozone fumigation, in leaves and roots of wild-type and catalase-deficient mutant of barley. Physiol Plant 104:280–292

Bartoli CG, Gómez F, Martinez DE, Guiamet JJ (2004) Mitochondria are the main target for oxidative damage in leaves of wheat (*Triticum aestivum* L.). J Exp Bot 55:1663–1669

Bell JNB, Mudd CH (1976) Sulphur dioxide resistance in plants: a case study of Loliumperenne L. In: Mansfield TA (ed.). Effect of Air pollution on Plants. Cambridge University Press, Cambridge, pp 82–103

Bennett JH, Lee EH, Heggestad HE (1984) Biochemical aspects of plant tolerance to ozone and oxyradicals: superoxide dismutase. In: Koziol MJ, Whatley FR (eds) Gaseous air pollutants and plant metabolism. Butterworth, London, pp 413–424

Bergmüller E, Porfirova S, Dörmann P (2003) Characterization of an Arabidopsis mutant deficient in g-tocopherol methyltransferase. Plant Mol Biol 52:1181–1190

Bielski BH, Arudi RL, Sutherland MW (1983) A study of the reactivity of HO_2/O_2-with unsaturated fatty acids. J Biol Chem 258:4759–4761

Chen Z, Gallie DR (2005) Increasing tolerance to ozone by elevating folia ascorbic acid confers greater protection against ozone than increasing avoidance. Plant Physiol 138:1673–1689

Cho UH, Seo NH (2005) Oxidative stress in Arabidopsis thaliana exposed to cadmium is due to hydrogen peroxide accumulation. Plant Sci 168:113–120

Constantinidou HA, Kozlowski TT (1979) Effects of sulfur dioxide and ozone on Ulmus americana seedlings, II: carbohydrates, proteins, and lipids. Can J Bot 57:176–84

Creissen G, Firmin J, Fryer M, Kular B, Leyland N, Reynolds H, Pastori G, Wellburn F, Baker N, Wellburn A, Mullineaux P (1999) Elevated glutathione biosynthetic capacity in the chloroplasts of transgenic tobacco plants paradoxically causes increased oxidative stress. Plant Cell 11:1277–1291

Darrall NM (1988) The effect of air pollutants on physiological processes in plants. Plant Cell Environ 12(1):1–30

Demirevska-Kepova K, Simova-Stoilova L, Stoyanova ZP, Feller U (2006) Cadmium stress in barley: growth, leaf pigment, and protein composition and detoxification of reactive oxygen species. J Plant Nutr 29:451–468

Depège N, Drevet J, Boyer N (1998) Molecular cloning and characterization of tomato cDNAs encoding glutathione peroxidase-like proteins. Eur J Biochem 253:445–451

Dugger WM, Ting IP (1970) Air pollution oxidant – their effects on metabolic processes in plants. Annu Rev Plant Biol 21:215–234

Edwards R, Dixon DP, Walbot V (2000) Plant glutathione S-transferases: enzymes with multiple functions in sickness and in health. Trends Plant Sci 5:193–198

Eltayeb AE, Kawano N, Badawi GH, Kaminaka H, Sanekata T, Morishima I, Shibahara T, Inanaga S, Tanaka K (2006) Enhanced tolerance to ozone and drought stresses in transgenic tobacco overexpressing dehydroascorbate reductase in cytosol. Physiol Plant 127:57–61

Escobedo FJ, Wagner JE, Nowak DJ (2008) Analyzing the cost effectiveness of Santiago Chile's policy of using urban forests to improve air quality. J Environ Manage 86:148–157

Farmer EE (2007) Plant biology – Jasmonate perception machines. Nature 448:659–660

Ferreira RR, Fornazier RF, Vitoria AP, Lea PJ, Azevedo RA (2002) Changes in antioxidant enzyme activities in soybean under cadmium stress. J Plant Nutr 25:327–342

Fikriye K, Omer M (2005) Effects of some heavy metals on content of chlorophyll, proline and some antioxidant chemicals in beans (Phaseolusvalgaris L) seedlings. Acta Biol Cracov 47:157–164

Finkelstein RR, Gibson SI (2001) ABA and sugar interactions regulating development: cross-talk or voices in a crowd. Curr Opin Plant Biol 5:26–32

Fleschin S, Fleschin M, Nhta S, Pavel E, Mageara V (2003) Free radicals mediate protein oxidation in biochemistry. Rom Biotechnol Lett 5:479–495

Flors C, Nonell S (2006) Light and singlet oxygen in plant defense against pathogens: phototoxic phenalone phytoalexins. Acc Chem Res 39:293–300

Flowers MD, Fiscus EL, Burkey KO (2007) Photosynthesis, chlorophyll fluorescence and yield of snap bean (Phaseolus vulgaris L.) genotypes differing in sensitivity to ozone. Environ Exp Bot 61:190–198

Foreman J, Demidchik V, Bothwell JH, Mylona P, Miedema H, Torres MA, Linstead P, Costa S, Brownlee C, Jones JD, Davies JM, Dolan L (2003) Reactive oxygen species produced by NADPH oxidase regulate plant cell growth. Nature 422:442–446

Foyer CH, Halliwell B (1976) The presence of glutathione and glutathione reductase in chloroplasts: a proposed role in ascorbic acid metabolism. Planta 133:21–25

Foyer CH, Harbinson J (1994) Oxygen metabolism and the regulation of photosynthetic electron transport. In: Foyer CH, Mullineaux P (eds) Causes of photooxidative stresses and amelioration of defense systems in plants. CRC Press, Boca Raton, pp 1–42

Foyer CH, Noctor G (2003) Redox sensing and signaling associated with reactive oxygen in chloroplasts, peroxisomes and mitochondria. Physiol Plant 119:355–364

Gapinska M, Sklodowska M, Gabara B (2008) Effect of short- and long-term salinity on the activities of antioxidative enzymes and lipid peroxidation in tomato roots. Acta Physiol Plant 30:11–18

George S, Venkataraman G, Parida A (2010) A chloroplast-localized and auxin-induced glutathione S-transferase from phreatophyte Prosopis juliflora confer drought tolerance on tobacco. J Plant Physiol 167:311–318

Gill SS, Tuteja N (2010) Reactive oxygen species and antioxidant machinery in abiotic stress tolerance in crop plants. Plant Physiol Biochem 48:909–930

Gratao PL, Polle A, Lea PJ, Azevedo RA (2005) Making the life of heavy metal-stressed plants a little easier. Funct Plant Biol 32:481–494

Griffiths H, Ong BL, Avadhani PN, Ohcj G (1989) Recycling of respiratory CO2 during Crassulacean acid metabolism: alleviation of photoinhibition in Pyrrosia piloselloides. Planta 179:115–122

Gupta GP, Singh S, Kumar B, Kulshrestha UC (2015) Industrial dust sulphate and its effects on biochemical and morphological characteristics of Morus (*Morus alba*) plant in NCR Delhi. Environ Monit Assess 187:67

Harinasut P, Poonsopa D, Roengmongkol K, Charoensataporn R (2003) Salinity effects on antioxidant enzymes in mulberry cultivar. Sci Asia 29:109–113

Hogue MA, Banu MNA, Okuma E (2007) Exogenous proline and glycinebetaine increase NaCI-induced ascorbate-glutathione cycle enzyme activities, and proline improve salt tolerance more than glycinebetaine in tobacco Bright Yellow-2 suspension-cultured cells. J Plant Physiol 164:1457–1468

Honour SL, Bell JN, Ashenden TW, Cape JN, Power SA (2009) Responses of herbaceous plants to urban air pollution: effects on growth, phenology and leaf characteristics. Environ Pollut 157:1279–1286

Huber SC (1984) Biochemical Basis for effects of K-deficiency on assimilate export rate and accumulation of soluble sugars in Soyabean leaves. Plant Physiol 76:424–430

Hung H, Mackay D (1997) A novel and simple model of the uptake of organic chemicals by vegetation from air and soil. Chemosphere 35:959–977

Joshi PC, Swami A (2009) Air pollution induced changes in the photosynthetic pigments of selected plant species. J Environ Biol 35:295–298

Joshi N, Chauhan A, Joshi PC (2009) Impact of industrial air pollutants on some biochemical parameters and yield in wheat and mustard plants. Environmentalist 29:398–404

Khan AA, Malhotra SS (1977) Effects of aqueous sulphur dioxide on pine needle glycolipids. Phytochem 16:539–543

Khan NA, Samiullah, Singh S, Nazar R (2007) Activities of antioxidative enzymes, sulphur assimilation, photosynthetic activity and growth of wheat (*Triticum aestivum*) cultivars differing in yield potential under cadmium stress. J Agron Crop Sci 193:435–444

Koji Y, Shiro M, Michio K, Mitsutaka T, Hiroshi M (2009) Antioxidant capacity and damages caused by salinity stress in apical and basal regions of rice leaf. Plant Prod Sci 12:319–326

Koussevitzky S, Suzuki N, Huntington S, Armijo L, Sha W, Cortes D, Shulaev V, Mittler R (2008) Ascorbate Peroxidase 1 plays a key role in the response of Arabidopsis thaliana to stress combination. J Biol Chem 283:34197–34203

Koziol MJ, Jordan CF (1978) Changes in carbohydrate levels in red kidney bean (Phaseolus vulgaris L.) exposed to sulphur dioxide. J Exp Bot 29:1037–43

Krupa SV, Pratt GC, Teng PS (1982) Air pollution: an important issue in plant health. Plant Dis 11:429–434

Kumar P, Tewari RK, Sharma PN (2007) Antioxidant responses to enhanced generation of superoxide anion radical and hydrogen peroxide in the copper-stressed mulberry plants. Planta 223:1145–1153

Larson RA (1988) The antioxidants of higher plants. Phytochemistry 27:969–978

Leon AM, Palma JM, Corpas FJ, Gomez M, Romero-Puertas MC, Chatterjee D, Mateos RM, del Rio LA, Sandalio LM (2002) Antioxidant enzymes in cultivars of pepper plants with different sensitivity to candium. Plant Physiol Biochem 40:813–820

Li Y, Song Y, Shi G, Wang J, Hou X (2009) Response of antioxidant activity to excess copper in two cultivars of Brassica campestris ssp. chinensis Makino. Acta Physiol Plant 31:155–162

Lin A, Zhang X, Zhu YG, Zhao FJ (2008) Arsenate-induced toxicity: effects on antioxidative enzymes and DNA damage in Vicia faba. Environ Toxicol Chem 27:413–419

Malhotra SS, Sarkar SK (1979) Effects of sulphur dioxide on sugar and free amino-acid content of pine seedlings. Physiol Plant 47:223–8

Miller G, Shulaev V, Mittler R (2008) Reactive oxygen signaling and abiotic stress. Physiol Plant 133:481–489

Nayar H (2003) Accumulation of osmolytes and osmotic adjustment in water stressed wheat and maize as affected by Calcium and its antagonists. Environ Exp Bot 50:253–264

Noctor G, Foyer CH (1998) A re-evaluation of the ATP: NADPH budget during C3 photosynthesis. a contribution from nitrate assimilation and its associated respiratory activity? J Exp Bot 49:1895–1908

Pan Y, Wu LJ, Yu ZL (2006) Effect of salt and drought stress on antioxidant enzymes activities and SOD isoenzymes of liquorice (Glycyrrhiza uralensis Fisch). Plant Growth Regul 49:157–165

Peiser GD, Yang SF (1977) Chlorophyll destruction by the bisulfite-oxygen system. Plant Physiol 60:277–281

Peiser GD, Yang SF (1979) Sulfite-mediated destruction of fl-carotene. J Agric Food Chem 27:446–44

Pekker I, Telor E, Mittler R (2002) Reactive oxygen intermediates and glutathione regulate the expression of cytosolic ascorbate peroxidise during iron-mediated oxidative stress in bean. Plant Mol Biol 49:429–438

Polidoros NA, Scandalios JG (1999) Role of hydrogen peroxide and different classes of antioxidants in the regulation of catalase and glutathione S-transferase gene expression in maize (Zea mays L.). Physiol Plant 106:112–120

Poschl U (2005) Atmospheric aerosols: transformation climate and health effects. Atmos Chem 44:7520–7540

Prado FE, Boero C, Gallarodo M, Gonzalez JA (2000) Effect of NaCl on germination, growth and soluble sugar content in Chenopodium quinoa willd seeds. Bot Bull Acad Sin 41:27–3

Prasad MNV, Sharma K (2004) Zinc alleviates cadmium-induced oxidative stress in Ceratophyllum demersum L: a free-floating freshwater macrophyte. Plant Physiol Biochem 41:391–397

Pukacha S, Pukacha PM (2000) Seasonal changes in antioxidant levels of Scots Pine (*Pinus sylvestris* L.) needles exposed to air pollution – thiol and ascorbate content. Acta Physiol Plant 22:451–456

Radotic K, Ducic T, Mutavdzic D (2000) Changes in peroxidase activity and isoenzymes in spruce needles after exposure to different concentrations of cadmium. Environ Exp Bot 44:105–113

Rao DN (1983) Sulphur dioxide pollution versus plant injury with special reference to fumigation and precipitation. In: Proceedings symposium on air pollution control, vol 1. Indian Association for Air pollution Control, New Delhi, pp 91–96

Rao DN, Leblanc F (1965) Effects of SO2 on lichen algae with special reference to chlorophyll. Biologist 69:69–95

Rao KVM, Raghavendra AS, Reddy KJ (eds) (2006) Physiology and Molecular Biology of Stress Tolerance in Plants. Springer-Netherlands, Dordrecht. ISBN 10-1-4020-4224-8

Rawat JS, Banerjee SP (1996) Urban forestry for improvement of environment. J Energy Environ Monit 12:109–116

Sanchez-Rodriguez E, Rubio-Wilhelmi MM, Cervilla LM, Blasco B, Rios JJ, Rosales MA, Romero L, Ruiz JM (2010) Genotypic differences in some physiological parameters symptomatic for oxidative stress under moderate drought in tomato plants. Plant Sci 178:30–40

Sandalio M, del Rio LA (1988) Intra-organellar distribution of superoxide dismutase in plant peroxisomes (glyoxysomes and leaf peroxisomes). Plant Physiol 88:1215–1218

Shannigrahi AS, Fukushima T, Sharma RC (2004) Tolerance of some plant species considered for green belt development in and around an industrial or urban area in India: an overview. Int J Environ Stud 61(2):125–137

Sharma P, Dubey RS (2007) Modulation of nitrate reductase activity in rice seedlings under aluminium toxicity and water stress: role of osmolytes as enzyme protectant. J Plant Physiol 162:854–864

Shimazaki K, Sakaki T, Kondo N, Sugahara K (1980) Active oxygen participation in chlorophyll destruction and lipid peroxidation in S02-fumigated leaves of spinach. Plant Cell Physiol 21:1193–1204

Simonovicova M, Tamás L, Huttová J, Mistrík I (2004) Effect of aluminium on oxidative stress related enzymes activities in barley roots. Biol Plant 48:261–266

Simova-Stoilova L, Vaseva I, Grigorova B, Demirevska K, Feller U (2010) Proteolytic activity and cysteine protease expression in wheat leaves under severe soil drought and recovery. Plant Physiol Biochem 48:200–206

Singh SN, Verma A (2007) Phytoremediation of air pollutants: a review. In: Singh SN, Tripathi RD (eds) Environmental Bioremediation Technology. Springer, Berlin/Heidelberg, pp 293–314

Singh S, Anjum NA, Khan NA, Nazar R (2008) Metal-binding peptides and antioxidant defence system in plants: significance in cadmium tolerance. In: Khan NA, Singh S (eds) Abiotic stress and plant responses. IK International, New Delhi, pp 159–189

Singh S, Khan NA, Nazar R, Anjum NA (2008) Photosynthetic traits and activities of antioxidant enzymes in blackgram (Vigna mungo L. Hepper) under cadmium stress. Am J Plant Physiol 3:25–32

Smith RJ, Azevedo RA, Alas RM, Lea PA (2008) Response of antioxidant enzymes to transfer from elevated carbon dioxide to air and ozone fumigation, in leaves and roots of wild-type and catalase-deficient mutant of barley. Physiol Plant 104:280–292

Speeding DJ, Thomas WJ (1973) Effect of sulphur dioxide on the metabolism of glycolic acid by barley (Hardeumvulgare) leaves. Aust J Biol Sci 6:281–286

Srivastava HS (1999) Biochemical defense mechanisms of plants to increased levels of ozone and other atmospheric pollutants. Curr Sci 76(4):525–533

Srinivas N, Lakshmi PS, Sravanti KL (2008) Air pollution tolerance index of various plant species growing in industrial areas. Int Biannual J Environ Sci 2:203–206

Srivastava, A. K., Bhargava, P., Kumar, A., Rai, L. C. & Neilan, B. A.(2009).Molecular characterization and effect of salinity on cyanobacterial diversity in the rice fields of Eastern Uttar Pradesh, India.Saline Syst 5,4.

Su J, Wu R (2004) Stress inducible synthesis of proline in transgenic rice confers faster growth under stress conditions than with constitutive synthesis. Plant Sci 166:941–948

Szekely G (2004) The role of proline in Arabidopsis thaliana osmotic stress response. Acta Biol Szeged 48:81–81

Tanou G, Molassiotis A, Diamantidis G (2009) Induction of reactive oxygen species and necrotic death-like destruction in strawberry leaves by salinity. Environ Exp Bot 65:270–281

Thomas H (1991) Accumulation and Consumption of solutes in swards of Loliumperenne during drought and after rewatering. New Phytol 118:35–48

Tiwari S, Agarwal M, Marshall FM (2006) Evaluation of ambient air pollution impact on carrot plants at a sub urban site using open top chambers. Environ Monit Assess 119:15–30

Tripathi AK, Gautam M (2007) Biochemical parameters of plants as indicators of air pollution. J Environ Biol 28:127–132

Ushimaru T, Nakagawa T, Fujioka Y, Daicho K, Naito M, Yamauchi Y, Nonaka H, Amako K, Yamawaki K, Murata N (2006) Transgenic Arabidopsis plants expressing the rice dehydroascorbate reductase gene are resistant to salt stress. J Plant Physiol 163:1179–118

Vierstra RD, John TR, Proff KL (1982) Kaempferol 3-O-galactoside 7-O-rhamnoside is the major green fluorescing compound in the epidermis of Vicia faba. Plant Physiol 69:522–532

Wang CQ, Li RC (2008) Enhancement of superoxide dismutase activity in the leaves of white clover (*Trifolium repens* L.) in response to polyethylene glycol-induced water stress. Acta Physiol Plant 30:841–847

Weinstein LH (1977) Fluoride and plant life. J Occup Environ Med 19(1):49–78

Wilkinson TG, Barnes RL (1973) Effect of ozone on CO_2 fixation patterns in pine. Can J Bot 9:1573–1578

Winner W E (1981). The effects of SO2 on photosynthesis and stomatal behavior of Mediterranean-climate shrubs and herbs. In: Margaris NS, Mooney NS (eds) Component of productivity of Mediterranean climate region – basic and applied aspects. Dr. W Junk Publishers, pp 91–103

Woo SY, Lee DK, Lee YK (2007) Net photosynthetic rate, ascorbate peroxidase and glutathione reductase activities of Erythrina orientalis in polluted and non polluted areas. Photosynthetica 45:293–295

Zhang FQ, Shi WY, Jin ZX, Shen ZG (2003) Response of antioxidative enzymes in cucumber chloroplast to cadmium toxicity. J Plant Nutr 26:1779–1788

7 Air Pollutants and Photosynthetic Efficiency of Plants

Bhupinder Dhir

Abstract

Absorption and accumulation/integration of air pollutants by leaves induce physiological and biochemical alterations in plants. Photosynthesis is the basic physiological event affected in plants exposed to air pollutants. Reduction in leaf area, closure of stomata and the damage to the photosynthetic apparatus limit the photosynthetic capacity of plants. High concentrations of sulphur dioxide (SO_2), ozone (O_3) and nitrogen oxides (NO_x) induce stomatal closure limiting the availability of carbon dioxide (CO_2) for photosynthesis. Reactive oxygen species (ROS) generated during oxidative stress damage photosynthetic apparatus via alteration in thylakoid structure and function. The photosynthetic electron transport, carboxylation efficiency of RuBisco and chlorophyll biosynthesis are the major processes negatively affecting the photosynthetic efficiency of plants.

Keywords

Photosynthesis • Leaf area • Thylakoid structure • Stomatal closure • Reactive oxygen species • Chlorophyll

7.1 Introduction

The combustion of fuels, automobile emissions and industrial operations release toxic gases such as sulphur dioxide (SO_2), nitrogen oxides (NOx), carbon dioxide (CO_2), carbon monoxide (CO), hydrogen fluoride (HF) and particulates (PM) in the atmosphere (Agbaire and Esiefarienrhe 2009). Primary pollutants undergo further reactions such as photolysis to produce secondary pollutants viz. peroxyacetyl nitrate (PAN) in the atmosphere (Agbaire 2009). The presence of polycyclic aromatic hydrocarbons (PAHs) and heavy metals makes the particulate matter (PM) (size $\geq$100 μm) toxic (Jouraeva et al. 2002; WHO 2006; Yu et al. 2006; Uzu et al. 2010). The increase in the emission of most of the pollutants noted in the last few years with an annual increment of 0.5–2 % has adversely affected the living

B. Dhir (✉)
Department of Genetics, University of Delhi South Campus, New Delhi 110021, India
e-mail: bhupdhir@gmail.com

U. Kulshrestha, P. Saxena (eds.), *Plant Responses to Air Pollution*,
DOI 10.1007/978-981-10-1201-3_7

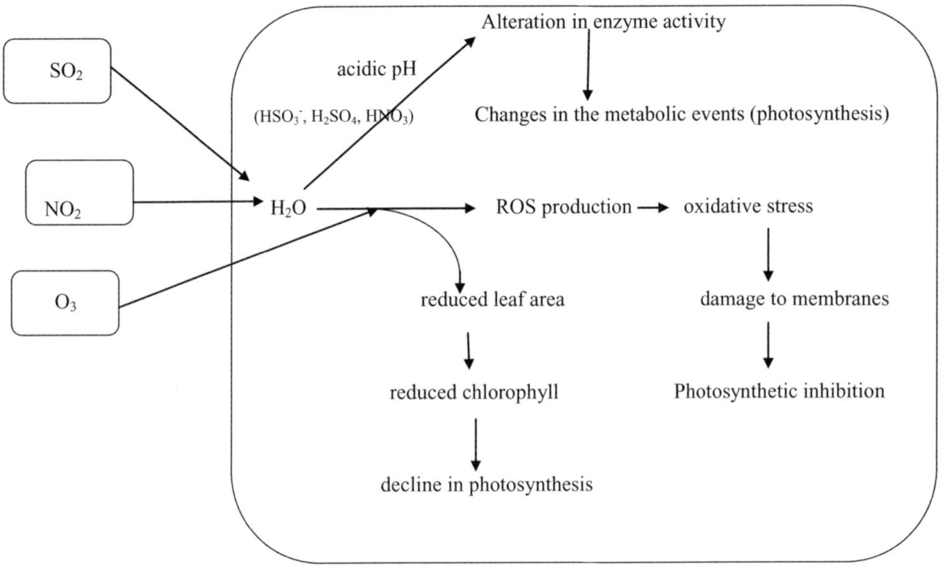

Fig. 7.1 Modes of action of various pollutants in reducing photosynthesis

organisms all over the world (Ashmore 2005). Predictions suggest the increase in the emission rate by 40–60 % by 2100 (IPPC 2007). The concentrations of gaseous pollutants are much higher in urban air in developing and developed countries (Li 2003; Tiwari et al. 2006). Among all the pollutants, carbon dioxide (CO_2) emissions have seen a drastic increase over the decades and are expected to exceed 700 μmol mol^{-1} by the end of the twenty-first century (IPCC 1997).

Exposure to high concentrations of gaseous pollutants such as NO_2, SO_2, O_3, HF, particulate matter and PAN causes detrimental effects in plants (Malhotra and Khan 1984; Guderian et al. 1985; Joshi and Swami 2009; Joshi et al. 2009; Niu et al. 2014). Absorption and accumulation of pollutants by leaves alter the physiological pH and induce oxidative stress causing metabolic and physiological disturbances in plants (Dizengremel et al. 2008). The alterations in photosynthetic efficiency adversely affect growth and productivity of plants (Maier-Maercker and Koch 1992; Tiwari et al. 2006; Ali et al. 2008; Liu and Ding 2008; Woo et al. 2007). The extent of the physiological damage in plants varies among the species depending on the pollutant exposure and sensitivity level of species (Agbaire and Esiefarienrhe 2009; Rai et al. 2009) (Fig. 7.1).

7.2 Leaf Damage

Reduction in leaf number and leaf area and premature senescence are the primary effects of air pollutant damage to foliage. The reduction in leaf blade and petiole size results in less absorption of radiation and subsequent reduction in photosynthetic rate (Fiihrer et al. 1993; Tiwari et al. 2006; Seyyednejad et al. 2009a, b; Koochak and Seyyed Nejad 2010). Foliar uptake of pollutants through stomata or cuticle or both is generally followed by accumulation in thick external walls of the epidermal cells and subsequently entry into mesophyll cells. The pollutants react with the water present in the apoplastic space and change the cellular pH which causes alterations in the major metabolic events. The structural alterations in epidermal cells, palisade and spongy parenchyma cells and stomata have been noted in response to air pollution (Robinson et al. 1998; Gostin 2009).

Chlorophyll is the principal photoreceptor pigment in photosynthesis and the index of plant productivity. Carotenoids are the accessory pigments that form an essential structural component of the photosynthetic antenna and reaction centre molecules. The reduced leaf area amounts to reduced levels of photosynthetic pigments, hence leading to lowering of photosynthetic efficiency in plants (Agbaire and Esiefarienrhe 2009; Joshi and Swami 2009; Giri et al. 2013). Decline in the production of chlorophyll and carotenoid pigments has been noted as the primary response in plants exposed to pollutants such as SO_2, NO_x, CO_2 and suspended particulate matter (Tiwari et al. 2006; Tripathi and Gautam 2007; Joshi and Swami 2009; Joshi et al. 2009; Shiragave et al. 2015). The deposition of suspended particulate matter on the leaf surface leads to closure of stomata. This affects gaseous exchange reducing the photosynthetic efficiency of plants to a greater extent. The oxidative stress resulting from generation of reactive oxygen species (ROS) causes lipid peroxidation of chloroplast membranes, and this also accounts for the loss of chlorophyll in the leaves.

The alteration in size of stomata or stomatal closure reduces the availability of CO_2 in leaves reducing the carbon fixation potential of plants (Honour et al. 2009). The decrease in the carbon fixing potential by ribulose-1,5-biphosphate carboxylase/oxygenase (RuBisCO) in plants leads to decline in production of sugars. In contrast, high accumulation of sugars and carbohydrates in plants results from altered electron transport (Britz and Robinson 2001). Reaction of compounds such as sulphite, nitrous acid (produced in the apoplast) with aldehydes and ketones of carbohydrates also correlates to their decline (Seyyednejad and Koochak 2011; Seyyednejad et al. 2011, 2013).

7.3 Effects of Gaseous Pollutants on the Photosynthetic Apparatus

In plants, photosynthetic electron flow is driven by photochemical reactions catalysed by photosystems, namely, PSII and PSI, which are linked by the electron transport chain. Both photosystems consist of reaction centres and light-harvesting complexes (LHCs), chlorophyll- and carotenoid-binding proteins, which absorb the sunlight. Light energy trapped by LHCs is transferred to antenna proteins (Yakushevska et al. 2003) and then transmitted to the reaction centre molecules. The LHCs associated with PSII and PSI contain proteins, namely, LHCII and LHCI. The photochemical activity of PSII creates a charge separation across the thylakoid membrane with a strong oxidant on the donor side capable of oxidizing water. This results in concomitant release of protons and molecular oxygen in the thylakoid lumen and also causes reduction of the primary electron acceptors of PSII. The electrons are then transported via electron transport chains in the thylakoid membrane to PSI. The cytochrome b6f complex mediates electron transport to PSI via plastocyanin (PC). PSI transfers the electrons across the membrane and reduces $NADP^+$ to form NADPH. NADPH is then used as reducing power for the biosynthetic reactions. The proton-motive force generated by linear electron flow from PSII to PSI generates ATP by F1F0-complex. Photosynthesis provides a route through which light energy generates a proton gradient across the thylakoid membrane of chloroplasts to produce ATP (Tarasingh 2013). Under normal conditions, an efficient energy dissipation cycle prevents over-reduction in the chloroplasts and creates a trans-thylakoid proton gradient in chloroplasts (Veljoviû-jovanoviû 1998; Veljovic-Jovanovic et al. 1993).

7.3.1 Ozone

High concentrations of O_3 induce stomatal closure or rupture stomatal aperture due to damage to epidermal cells making them open wide. The entry of O_3 in leaf through stomata forwards it to mesophyll cells dissolving it in the aqueous layer of the apoplast to produce reactive oxygen species (ROS) such as hydrogen peroxide (H_2O_2), hydroxyl (OH), peroxyl (OH_2) and superoxide (O_2–) radicals (Felzar et al. 2007). The stomatal closure induced by acute exposure of O_3 decreases stomatal conductance (Tiwari et al. 2006; Calatayud et al. 2007; Rai et al. 2007; Pellegrini et al. 2011a, b, 2015). Reactive oxygen species cause injury to plasma membrane via lipid peroxidation by inducing changes in membrane permeability and fluidity, potassium (K^+) exchange via ATPase reactions and calcium (Ca^{2+}) exclusion (Heath 2008; Ainsworth et al. 2014). Free radicals produced in the guard cells damage the chloroplast membranes and hence the photosynthetic apparatus. Swelling of thylakoids, increase of plastoglobuli per chloroplast and disruption of membranes lead to leakage of ions and consequent change in photosynthetic capacity (Torsethaugen et al. 1999, Plazek et al. 2000; Pellegrini 2014). Mesophyll cells are damaged directly (via oxidative stress in photoactive compartments) or indirectly by leakage of organic and inorganic solutes caused by the disorganization of the biomembranes (via change in permeability). The closure of stomata results in decreased production of energy equivalents such as ATP and NADPH, hence limiting the dark reactions of the photosynthesis.

The parameters such as quantum yield and photochemical quenching (qP) significantly decrease, while non-photochemical quenching (qN) increased in leaves of plants exposed to O_3 (Hassan 2006; Thwe et al. 2014). Increase in qN with corresponding decrease in qP indicates a non-radiative dissipation of energy as heat (Lorenzini et al. 1999; Ismail et al. 2014). The decrease in quantum yield of electron transport is responsible for downregulation of photosynthetic electron transport. The reduction of photochemical efficiency (Fv/Fm) indicates damage to PSII reaction centres (Plazek et al. 2001; Ainsworth and Long 2005; Calatayud et al. 2007; Pellegrini et al. 2011a, b; Ainsworth et al. 2012; Thwe et al. 2014). Increase in Fo suggests impairment of the transport of excitation energy from light-harvesting complexes to reaction centres. In the light reactions, the production of electrons by the water-splitting reaction in PSII (electron generation site of PSII) is impaired, and electron transport from PSII to PSI (electron donor site of PSII) is damaged. Alterations in light-reaction centres and Chl-associated proteins in photosystem I affect the carboxylation potential of the photosynthetic apparatus (Tognini et al. 1997). The reduction in the density of thylakoid proteins such as Dl, cyt f and polypeptides of OEC is noted.

The closure of stomata decreases rates of CO_2 assimilation but increases the diffusive resistance of CO_2 in the mesophyll, thereby altering allocation of carbon to different plant parts (Darrall 1989; Wolfenden and Mansfield 1990). The decline in the carboxylation efficiency affects the light and dark reactions of photosynthesis (Farage and Long 1999; Shavin et al. 1999; Castagna et al. 2001). The negative effects on the carboxylation (V_{cmax}) efficiency indicate direct oxidative damage and indirect heat-related injuries to RuBisCO. Reduced stomatal opening may be the primary mechanism for protection in plants against the damaging effects of O_3. Alternatively increased production of antioxidants has been shown to curtail oxidative stress induced by gaseous pollutants (McKee et al. 1995). A loss of RuBisCO activity primarily contributed to the decline in photosynthetic capacity of plants (Shavin et al. 1999; Castagna et al. 2001).

Loss of chlorophyll content leads to a reduction in light-harvesting and net assimilation rate. The decline in the light-harvesting capacity of chlorophyll coupled with a reduced efficiency of photosynthetic energy conversion leads to decreased net assimilation. Potassium flux changes the guard cell volume and regulates the stomatal aperture (Torsethaugen et al. 1999). Acute O_3 exposure inhibits the guard cell K^+ channels, which mediate stomatal opening, thus leading to decreased photosynthesis (Torsethaugen et al. 1999). Cytosolic Ca21 is a signalling molecule in guard cell and has been reported to inhibit the inward K1 channel due to oxidative stress. Disturbed Ca signalling involved in stomatal movement regulates oxidative stress.

Non-stomatal factors affecting the photosynthetic efficiency include (1) lower RuBP regeneration from the lower pools of Calvin cycle intermediates, (2) decreased efficiency of RuBisCO due to direct enzyme oxidation and (3) reduced CO_2 transport to the enzymes. Ozone induces reduction in level of RNA transcript for the small subunit (rbcS) and large subunits (rbcL) of RuBisCO (Feng et al. 2008). It also decreases the expression of photosynthetic genes for RuBisCO activase (Sarkar et al. 2010; Calatayud et al. 2011). Proteomic changes have been noted in RuBisCO content, photosynthetic machinery and Calvin cycle enzymes including RuBisCO activase, ATP synthase, subunit of photosystem II, aldolase, phosphoglycerate kinase and NADP-glyceraldehyde-3-phosphate dehydrogenase. The oxidation of SH groups in the Fraction 1 protein of RuBP carboxylase results in granulation of the chloroplast stroma.

Pigments such as zeaxanthin also mediate photoprotection in O_3-treated plant species. Antioxidant enzymes such as ascorbate peroxidase cause detoxification of H_2O_2 by reducing it to water through a series of reactions in the ascorbate-glutathione cycle (Noctor and Foyer 1998). The plasma membranes are protected by antioxidants such as hydrophilous ascorbate (vitamin C) and lipophilous α-tocopherol (vitamin E). Ascorbic acid (AA) plays a role in the defence against ROS generated by O_3. Apoplastic ascorbic acid protects the photosynthetic machinery. Phenylpropanoids are the compounds known to scavenge ROS in plants (Srivastava 1999; Pasqualini et al. 2003; Vollenweider et al. 2003; Fini et al. 2012; Muneer et al. 2013). The biosynthesis of phenylpropanoids increases more in sensitive species than the tolerant ones. Reduction in carbon assimilation of plants treated with O_3 increases the PAL activity which is responsible for the synthesis of phenylalanine and transcinnamic acid, precursor of phenylpropanoids. Polyamines such as putrescine (Put), spermidine (Spd) and spermine (Spm) protect plastids and thylakoid membranes against O_3 damage. Addition of exogenous polyamines reduces damage caused by O_3 in plants. Putrescine (Put), spermidine (Spd) and spermine (Spm) get associated with thylakoid membranes and various photosynthetic subcomplexes (thylakoids, PSII membranes, LHCII, PSII complex), while spermine get associated with PSII core and the reaction centre of PSII. They conjugate with hydroxycinnamic acids and protect photosynthetic machinery from O_3-triggered ROS (Langebartels et al. 1991; Navakoudisa et al. 2003). A decrease in thylakoid-bound Put significantly increases the antenna size of the LHCII, although the number of reaction centres per unit area as well as the maximal photosynthetic rate and the maximum yield of photochemistry (Fv/Fm) decreased (Reichenauer et al. 1997; Reichenauer and Bolhàr-Nordenkampf 1999). The protection by polyamines against oxidants includes: (i) scavenging of ROS, (ii) increasing the permeation of antioxidant enzyme SOD through the membranes, (iii) protecting the membranes against oxidant damage, (iv) changing the redox state of the cells or (v) regulating the expression of genes. Isoprene scavenges ROS and provides protection against oxidative stress (Sauer et al. 1999). It effectively reacts with O_3 forming hydroxymethyl hydroperoxide, thus aggravating the O_3-induced damage. It quenches O_3 by directly reacting with it in the intercellular spaces and thus counteracts the O_3 damaging effect on membranes (Sharkey 1996; Sharkey and Yeh 2001). It is suggested that isoprene produced by leaves protects the photosynthetic apparatus against oxidative stress induced by ozone. Increase in flavonoid content in plants in response to O_3 stress suggests their role in scavenging ROS including superoxide anion, hydrogen peroxide and hydroxy radical. They also play a role in peroxidase-mediated catabolism of H_2O (Loreto and Velikova 2001).

7.3.2 Sulphur Dioxide

Exposure to high SO_2 concentrations turns toxic inducing physiological and metabolic alterations in plants, hence reducing growth and productivity of plants (Agrawal and Deepak 2003; Agrawal et al. 2006). Stomata directs SO_2 in the nearby subsidiary or epidermal cells where it readily dissolves in the apoplastic water to produce sulphite (SO_3^{2}), bisulphite (HSO_3^-) and H^+ ions making the pH of the medium acidic (Liu et al. 2008, 2009). The phytotoxicity of SO_2 is due to SO_3^{2-} and HSO_3^- ions (DeKok 1990). The protons formed by the dissociation of sulphurous acid have deleterious effects on the cellular metabolism (Pfanz and Heber 1989). Photosynthesis is majorly affected by SO_2 (Darrall 1989; Agrawal et al. 2006; Chauhan and Joshi 2010). Alterations in photosynthetic capacity are ascribed to changes in permeability of plasma membrane and interference with enzymatic activities (Li et al. 2007). The response of stomata to SO_2 largely depends on leaf age, concentration and combination of pollutants (Parshina and Rygalav 1999).

The chlorophyll degradation in SO_2-exposed plants results from its strong redox properties. Chlorophyll destruction is also caused by free radicals produced during the oxidation of HSO_3^-. The chlorophyll degradation produces phaeophytin molecule and Mg^{2+}. Magnesium ions in the chlorophyll molecule are replaced by two atoms of hydrogen, thereby changing its light use efficiency. The chlorophyll b breakdown results from splitting of the phytol chain by chlorophyllase. A reduction in the Chl *a*/*b* ratio indicates that the ratio of the reaction centre pigments to light-harvesting pigments in the photosystem on the thylakoid membrane in chloroplasts is noted in SO_2-treated plants. The reaction centre is vulnerable to simulated SO_2 treatment than the light-harvesting antenna system. Decreases in Fv/Fm and ΦPSII indicate that part of the PSII photochemistry and photochemical energy conversion in PSII were inactivated by $NaHSO_3$. The reduction in electron transport from high concentrations of SO_2 is attributed to the destruction of epidermal cells adjacent to stomata and accumulation of sulphur within guard cells (Black and Unsworth 1980). The inhibition of electron transfer occurs at the site close to the reaction centre of photosystem II. Studies conducted using isolated chloroplasts indicate that oxidizing side of photosystem II is more affected by SO_2. The denaturation of the protein component in the pigment protein complex caused by destruction of disulphide bonds also account for reduction in photosynthetic efficiency. Sulphurous acid (H_2SO_3) formed by the by-product of SO_2 degradation in plant irreversibly inhibits both cyclic and non-cyclic photophosphorylation.

Short-term SO_2 fumigation causes a transient decrease in photosynthetic CO_2 uptake but increase in non-assimilatory electron transport. Inhibition in the activity of enzymes of the Calvin cycle including 3-phosphoglycerate and hexose phosphate relates to the decline in the rate of photosynthesis. The competition between CO_2 and SO_3^- for active sites of RuBisCO (Agrawal and Deepak 2003) is mainly responsible for the decline in carbon fixation by RuBisCO. Sulphur dioxide showed inhibition of CO_2 assimilation accompanied by increased reduction of the quinone acceptor, Q_A, of photosystem II, and increased oxidation of the electron donor pigment P700 of photosystem I is reported. Plants exposed to SO_2 exhibit increase in the soluble sugars. This results from the breakdown of polysaccharides. Polyhydric sugars act as scavengers of free radicals and help to cope with increasing SO_2 pollution. The decrease in content of non-structural carbohydrates and starch of damaged leaves probably corresponds with the photosynthetic inhibition or stimulation of respiration rate (Tzvetkova and Kolarov 1996).

The xanthophyll cycle (de-epoxidation) and antioxidant system (1,1-diphenyl-2-picrylhydrazyl radical-scavenging capacity) are the two protective mechanisms active under simulated SO_2 treatment (Veljovic-Jovanovic et al. 1993; Demmig-Adams and Adams 1996). The dominant component of carotenoids, lutein, participates in the non-radiative energy dissipation through the quenching of Chl fluorescence. The activation of the VAZ cycle under the influence of SO_2 is the result of a decrease in intrathylakoid pH caused by ATP consumption or as a consequence of SO_2-inhibited carbon assimilation. Plants exposed to low concentrations of $NaHSO_3$ depict the scavenging capacities against DPPH. The lutein content in leaves increases in response to SO_2 stress (Liu et al. 2006).

7.3.3 Carbon Dioxide

CO_2 concentrations above ambient consistently increase net photosynthetic rates in the short term due to improved water use efficiency. The anatomical changes such as increase in leaf thickness and alterations in cell and chloroplast development have been observed under elevated CO_2 in plants. Carbon dioxide enters the plant leaves through stomata. The closing or opening of stomata is regulated by CO_2 concentration (Joshi and Bora 2005; Ainsworth and Rogers 2007). High concentrations of CO_2 change the turgor pressure of guard cells, hence mediating the closure of stomata. Elevated atmospheric CO_2 generally reduces stomatal conductance (Darrall 1989). High concentrations of CO_2 induce closure of the stomata, thus limiting CO_2 assimilation and fixation rate (Warren et al. 2007). The stomatal closure is related to loss of membrane permeability because of oxidation of membrane channels and transport proteins or increased sensitivity of the stomata to closure signals such as internal Ca levels or abscisic acid.

Increase in atmospheric CO_2 concentrations increases photosynthetic rate because of high RuBisCO activity. Increased availability of CO_2 as the substrate leads to more carbon assimilation and hence more photosynthesis (IPCC 2007; Reddy et al. 2010). Long-term exposure to increasing CO_2 concentration leads to reduction in RuBisCO activity or concentration because the amount of RuBisCO required for maintaining the same assimilation rate decreases. Due to the diminished photorespiration rate, the energy demand per fixed carbon decreases, and consequently, the demand for compounds involved in light reactions, such as chlorophylls or carotenoids, also decreases. All these alterations result in enhanced rates of respiration. The possible changes in respiration may be due to structural changes imposed by elevated CO_2, accumulation of carbohydrates or changes in the biochemistry of respiration. Stimulation in photosynthetic efficiency is majorly reflected in C3 species (Long et al. 2004; Ainsworth and Long 2005). Species with the C4 pathway are able to concentrate CO_2 at the site of carboxylation for photosynthate production.

7.3.4 Nitrogen Oxides

The foliar uptake of NO_2 in plants occurs predominantly by stomatal openings (Darrall 1989). Plants absorb gaseous NO_2 more rapidly than nitrogen oxide (NO). It rapidly reacts with water and gets converted to HNO_2 and HNO_3 before further utilization in plant metabolism. Nitrous and nitric acid dissociates to form nitrate, nitrite and protons (Ramge et al. 1993). Nitrite is more toxic than NO_3^-. NO_2 exposure reduces chlorophyll content in plants. High concentration (100 ppb NO_2) reduces stomatal conductance. The acidic conditions produced by NO_2 influence electron flow and photophosphorylation. The damage to the photosynthetic apparatus occurs by swelling of chloroplast membranes. Biochemical and membrane injury produced from NO_2 inhibits photosynthesis by uncoupling electron transport and inducing structural alterations. Photosynthetic inhibition could also be caused by NOx because of competition for NADPH for the processes of nitrite reduction and carbon assimilation in chloroplasts. Exposure of low levels of NO_2 increases RuBP carboxylase activity. Short exposures to a relatively high concentration stimulate RuBP carboxylase as well as glycollate oxidase activity. Polyamines such as spermine and spermidine prevent NO_2-induced decline in leaf damage.

7.3.5 Pollutant Mixtures

Stomatal physiology is differently affected by pollutant mixtures in comparison to a single pollutant. Mixtures of pollutants such as O_3-NO_2, SO_2-O_3 and SO_2-NO_2 produce synergistic effects and hence cause severe damage to the photosynthetic machinery. High concentrations induce stomatal closure. Stomatal regulation of pollutant uptake is limited since O_3 fluxes to individual leaves are not reduced by elevated atmospheric CO_2. A common feature of O_3-treated leaves under ambient CO_2 was an initial stimulation of photosynthesis and stomatal conductance. The O_3-induced decline in chlorophyll content was less rapid under elevated CO_2, and photosynthesis was increased relative to the ambient CO_2 treatment. An increase in the amount of in vivo active RuBisCO may be involved in mitigating O_3-induced damage to leaves. The results obtained suggest that elevated atmospheric CO_2 has an important role in restricting the damaging effects of O_3 on photosynthetic activity during the vegetative growth (Mulholland et al. 1997). Changes in stomatal responses influence the absorption of pollutants from the pollutant mixtures. Stomatal resistance increased more in response to pollutant mixtures than single pollutants (Noormets et al. 2001, 2010). Even low concentrations of pollutant mixtures inhibit photosynthesis.

Changes in concentrations of NO_2, SO_2 and O_3 predominantly influence leaf photosynthesis rates (Teughels et al. 2005; Hassan 2010). The decline in photosynthesis predominantly occurs by reduced chlorophyll production resulting from reduced formation of precursor molecules, δ-aminolevulinic acid and protochlorophyllide. High exposure to air pollutants leads to generation of ROS, induces oxidative stress and forces chloroplasts into an excessive excitation energy level. The damage of guard cells and change in cell density occur in plants exposed to stress. The reduction in photosynthetic rates might be associated with increased stomatal conductance and decrease in components of the photosynthetic apparatus such as chlorophyll concentration, soluble protein, adenylates, RuBP regeneration and RuBisCO (ribulose-1,5-biphosphate carboxylase/oxygenase) activity (Schmidt et al. 1990). The decrease in photosynthetic parameters under CO, NOx and SO_2 might be due to the inactivation of reaction centres of photosystems which receive an initial amount of light energy and further transfer it for efficient use due to oxidative stress (Yun 2007). Ineffective energy exploitation leads to increase in energy dissipation and hence decrease in photosynthesis. Long-term treatments to low-concentration mixtures of O_3, NO_2 and SO_2 are known to increase the D-1 protein content in the reaction centre of photosystem II; however, exposure to high concentrations reduces D-1 protein content. Inhibition of photosynthesis also occurs due to loss of carboxylation efficiency resulting from reduced RuBisCO activity (Farage et al. 1991). The loss of RuBisCO protein might be due to the progressive depletion of biochemical pathway associated with signal transduction and gene regulation, and excessive production of ROS which leads to incorrect folding or assembly of proteins, and consequent protein degradation.

The partial stomatal closure induced by elevated atmospheric CO_2 has been suggested to decrease the impact of air pollutants such as O_3 by restricting their uptake by plant (Allen 1990). Studies suggest that elevated CO_2 increases the amount of active RuBisCO, thereby providing strong evidence that this may be a mechanism contributing to the limitation of O_3-induced damage. The responses of stomatal conductance to pollutant mixtures are cultivar and species specific (Darrall 1989). Mixtures of SO_2 and NO_2 decrease photosynthesis. Studies suggest that SO_2 and NO_2 exposure under ambient CO_2 conditions prevents stomatal closure, while similar treatments applied under elevated CO_2 promoted closure (Atkinson et al. 1991). Both SO_2 and NO affect chloroplast ultrastructure by causing swelling of the lumen within the thylakoids of chloroplasts. Swelling of thylakoid membranes has been caused by ionic disturbances and acidification caused directly by products of the pollutants (Pukacki et al. 2000). Active transport of pollutant products such as sulphite and sulphate into the chloroplast takes place by means of the phosphate translocator of plastid envelopes, and additional orthophosphate appears to enhance the influx of sulphur anions. Indirect proton uptake via a shuttle involving both nitrite and HNO_2 may consequently interfere with such events by causing a breakdown of the trans-envelope pH gradient inhibitory effects of nitrite upon CO_2 fixation. Significant reductions of cyclic photophosphorylation have been noted in plants exposed to SO_2 + NO_2 treatments. Elevated CO_2 and O_3 affect apparent quantum yield, photosynthesis, carboxylation efficiency and electron transport capacity. The negative impact of elevated O_3 gets exaggerated by elevated CO_2.

Pollutants like SO_2, NO_2 and H_2S cause depletion of soluble sugars in the leaves. The reaction of sulphite with aldehydes and ketones of carbohydrates causes reduction in carbohydrate content (Tripathi and Gautam 2007). The negative effect of hazardous gases on carbon metabolism is a result of their possible interaction with the reactive centre of ribulose bisphosphate carboxylase. The decrease in total carbohydrates and sucrose content of damaged leaves probably corresponds with photosynthetic inhibition. Higher starch accumulation in damaged leaves results from export from the mesophyll.

Singlet oxygen and hydrogen peroxide (H_2O_2) content increased in CO-, NOx- and SO_2-treated leaves. Increased ROS act as a signal to induce defence responses such as antioxidant enzymes of ascorbate-glutathione cycle to CO, NOx and SO_2 gas stress. High exposure to CO, NOx and SO_2 concentrations causes disturbance in ascorbate-glutathione pathways. Increased APX activity and CAT activity in stressed plants indicate their role in detoxification of ROS and hence resistance in plants. Proton gradients across thylakoid membranes are harnessed by chloroplast coupling factors.

7.4 Effect of Other Pollutants on the Photosynthesis

7.4.1 Peroxyacyl Nitrates

Peroxyacetyl nitrate (PAN), a phytotoxic pollutant, is absorbed mainly through stomata. Low concentrations of PAN inhibit electron transport, photophosphorylation and CO_2 fixation in chloroplasts. The inhibitory effect of PAN on enzymes has been attributed to its ability to oxidize SH groups in proteins and metabolites such as cysteine, reduced glutathione, CoA, lipoic acid and methionine.

7.4.2 Fluorides

The gaseous form, i.e. hydrogen fluoride (HF), is absorbed through the leaves. Gaseous fluoride absorption through stomata influences stomatal responses. The significant reduction in epidermal cell size and stomata results from inhibition in cell elongation. Studies suggest that after passing through the cell wall, fluoride attacks cytoplasmic membranes and is partially retained there and thereafter transferred to vacuoles. Chloroplasts have also been suggested as the site of fluoride accumulation in leaves. Exposure to HF causes a decline in photosynthetic CO_2 fixation. Low concentrations of fluoride (1.3–12 ppb) cause depression in the amounts of chlorophyll. The quantum yield (Y (II)) of PSII, the relative electron transport rate (ETR), non-regulatory energy dissipation quantum yield (Y(NO)), non-photochemical quenching coefficient (NPQ) and the relative physiological characteristics show reduction in response to HF. The significant changes in the chlorophyll levels influence the photosynthetic capacity under chronic fluoride exposure conditions (Doley 1988; Kumar and Rao 2008; Yang et al. 2015).

7.4.3 Acidic Rain and Fog

Exposure to highly acidic fog (pH 1.68) causes significant depression in net photosynthesis due to reduction in leaf buffering capacity and extensive leaf injury. This accounted for decrease in growth and yield (Trumble and Walker 1991). Defoliation caused by acidic conditions stimulates the chlorophyll destruction and reduces fruit yield significantly due to the imbalanced source to sink ratio. The depression in photosynthesis results from reduction in photosynthetically active leaf area. Carbon dioxide assimilation rates are significantly decreased (20 %) by acidic fog with pH values of 2.5 and 3.0, while rates of stomatal resistance increased. Epidermal, palisade and spongy mesophyll cells are shrunken or obliterated. Damage to cells results from acidic conditions in the leaf (Odiyi and Bamidele 2014). Stomatal closure followed by increased stomatal resistance causes reduction in assimilation rate.

Simulated acid rain treatment affects mainly the epidermal cells causing erosion of the cuticle and altering the leaf permeability (Takemoto et al. 1988). Plants polluted with simulated acid rain show symptoms such as chlorosis, necrosis, stunted growth, lesion, suppression of leaf production, leaf curling, withering of leaves, leaf abscission and leaf death (Silva et al. 2006). The greater foliar injury is associated with the decreased chlorophyll content and the damage to the photosynthetic apparatus (Liu et al. 2010). Chlorophyll content was significantly reduced by simulated acid rain treatment at pH 2.0 and pH 3.0. Reduction was due to the removal of Mg^+ from the tetrapyrrole ring of the chlorophyll molecules by H^+ or due to the increase of transpiration by acid rain (Evans et al. 1997).

7.4.4 Particulate Matter

Accumulation of particulate matter (PM) has negative impact on plants (Nawrot et al. 2011). The deposition of particulate matter on leaves clogs stomata, hence reducing the absorption of photosynthetically active radiation (PAR). The clogging of stomata affects the gaseous exchange. Accumulation of PM on foliage decreases the chlorophyll production. The reduction in the availability of light and CO_2 (stomatal conductance), low chlorophyll content and increased stomatal resistance negatively affect photosynthesis (Vardaka et al. 1995; Beckett et al. 1998;

Farmer 2002; Heerden et al. 2007; Rai et al. 2010). High stomatal resistance makes the CO_2 flow to chloroplast more difficult and results in lower light access. Fluorescence studies showed significant decrease in photosynthetic efficiency of PSII (Fv/Fm) with increasing dust deposition on leaves in plants such as *Quercus coccifera* and *Z. prismatocarpum* (Vardaka et al. 1995; Heerden et al. 2007; Przybysz et al. 2014).

7.5 Conclusions

The photosynthetic apparatus is an essential part of plant cell as it assimilates and fixes atmospheric CO_2 during photosynthesis. Light energy initiates the excitation of the light-harvesting complexes I and II (LHCI and LHCII) and is further transferred into reaction centres (photosystems I and II, PSI and PSII). The electron transport reactions transfer electrons from PSII to PSI, where they are reenergized and used for the production of NADPH and ATP in an enzymatically regulated reaction used in CO_2 fixation in stroma. During the course of light reactions occurring in the thylakoid membrane, protons are accumulating into the thylakoid lumen creating a pH difference between the chloroplast stroma and lumen.

High concentrations of air pollutants damage leaves and reduce leaf area, thereby affecting the photosynthetic efficiency of plants to a significant extent. The damage to the photosynthetic apparatus accounts for the decline in photosynthetic efficiency, but each pollutant differs in its mode of action for damaging the photosynthetic apparatus of leaves. SO_2 damages the leaf by acidifying action, weakening the PS II donor side and inhibiting Calvin cycle activation, while O_3 reacts with cell membranes, primarily the plasma membrane. In general, pollutants enter the leaf through stomata, react with the water in the mesophyll cells and trigger the synthesis of ROS. Oxidative stress caused by ROS damages cell membranes and denatures critical enzymes. Reactive oxygen species react with cell membranes of the photosynthetic apparatus, thereby inhibiting PS II donor side and Calvin cycle. Alterations in the stomatal conductance affect rate of carboxylation and electron transport. Loss of chlorophyll content reduces light-harvesting capacity. This affects the efficiency of photosynthetic energy conversion leading to decreased net assimilation and dissipation of excess energy as heat. Exposure of pollutant concentrations makes the stomata wide open decreasing the efficiency of photosystem II for CO_2 assimilation and hence lowering carbon sequestration. Acclimatory decreases in photosynthesis have also been associated with reductions in RuBisCO activity, content and oxidative damage-induced injuries. Increase in atmospheric CO_2 concentration initially increases its diffusion into the leaf supporting photosynthesis. Increase in CO_2 concentration promote carboxylation efficiency of RuBisco thereby affecting photorespiration. This adversely affects the stomatal control which decreases the transpiration rate but increases water use efficiency (WUE) (the ratio of assimilated carbon to water loss). In a nutshell, direct oxidative damage, as well as indirect heat-related injuries to the photochemical apparatus, induces downregulation of photosynthesis.

References

Agbaire PO (2009) Air pollution tolerance indices (APTI) of some plants around Erhoike-Kokori oil exploration site of Delta state. Niger Int J Phys Sci 4:366–368

Agbaire PO, Esiefarienrhe EJ (2009) Air pollution tolerance indices (apti) of some plants around Otorogun gas plant in Delta state. Niger Appl Sci Environ Manag 13(1):11–14

Agrawal M, Deepak SS (2003) Physiological and biochemical responses of two cultivars of wheat to elevated levels of CO_2 and SO_2, singly and in combination. Environ Pollut 121:189–197

Agrawal M, Singh B, Agrawal SB, Bell JNB, Marshall F (2006) The effect of air pollution on yield and quality of mungbean grown in periurban areas of Varanasi. Water Air Soil Pollut 169:239–254

Ainsworth EA, Long SP (2005) What have we learned from 15 years of free-air CO_2 enrichment (FACE)? A meta-analytic review of the responses of photosynthesis, canopy properties and plant production to rising CO_2. New Phytol 165:351–372

Ainsworth EA, Rogers A (2007) The response of photosynthesis and stomatal conductance to rising [CO_2]: mechanisms and environmental interactions. Plant Cell Environ 30:258–270

Ainsworth EA, Yendrek CR, Sitch S, Collins WJ, Emberson LD (2012) The effects of tropospheric ozone on net primary productivity and implications for climate change. Annu Rev Plant Biol 63:637–661

Ainsworth EA, Serbin SP, Skoneczka JA, Philip A (2014) Using leaf optical properties to detect ozone effects on foliar biochemistry Townsend. Photosynth Res 119:65–76

Ali A, Alfarhan A, Aldjain I, Bokhari N, Al-Taisan W, Al-Rasheid K, Al-Quraishi S (2008) Photosynthetic responses of pea plants (*Pisum sativum* L. cv. Little marvel) exposed to climate change in Riyadh city, KSA. Afr J Biotechnol 7(15):2630–2636

Allen LH (1990) Plant responses to rising carbon dioxide and potential interactions with air pollutants. J Environ Qual 19:15–34

Ashmore MR (2005) Assessing the future global impacts of ozone on vegetation. Plant Cell Environ 28:949–964

Atkinson CJ, Wookey PA, Mansfield TA (1991) Atmospheric pollution and the sensitivity of stomata on barley leaves to abscisic acid and carbon dioxide. New Phytol 117:535–541

Beckett KP, Freer-Smith P, Taylor G (1998) Urban woodlands: their role in reducing the effects of particulate pollution. Environ Pollut 99(3):347–360

Black VJ, Unsworth MH (1980) Stomatal responses to sulphur dioxide and vapor pressure deficit. J Exp 31:667–677

Britz SJ, Robinson JM (2001) Chronic ozone exposure and photosynthate partitioning into starch in soybean leaves. Int J Plant Sci 162:111–117

Calatayud V, Cervero J, Sanz MJ (2007) Foliar, physiological and growth responses of four maple species exposed to ozone. Water Air Soil Pollut 185:239–254

Calatayud V, García-Breijo FJ, Cervero J, Reig-Armiñana J, Sanz MJ (2011) Physiological, anatomical and biomass partitioning responses to ozone in the Mediterranean endemic plant *Lamottea dianae*. Ecotoxicol Environ Saf 74:1131–1138

Castagna A, Nali C, Ciompi S, Lorenzini G, Soldatini GF, Ranieri A (2001) Ozone exposure affects photosynthesis of pumpkin (*Cucurbita pepo*) plants. New Phytol 152:223–229

Chauhan A, Joshi PC (2010) Effect of ambient air pollutants on wheat and mustard crops growing in the vicinity of urban and industrial areas. NY Sci J 3:52–60

Darrall NM (1989) The effect of air pollutants on physiological processes in plants. Plant Cell Environ 12:1–30

DeKok LJ (1990) Sulphur metabolism in plants exposed to atmospheric sulphur. In: Rennenberg H, Brunold C, DeKok LJ, Stulen I (eds) Sulphur nutrition and sulphur assimilation in higher plants, Fundamental, environment and agricultural aspects. SPB Academic Publishing, The Hague, pp 125–138

Demmig-Adams B, Adams WW III (1996) The role of xanthophyll cycle carotenoids in the protection of photosynthesis. Trends Plant Sci 1:21–26

Dizengremel P, Le Thiec D, Bagard M, Jolivet Y (2008) Ozone risk assessment for plants: central role of metabolism-dependent changes in reducing power. Environ Pollut 156:11–15

Dole Y (1988) Fluoride-induced enhancement and inhibition of photosynthesis in four taxa of *Pinus*. New Phytol 110:21–31

Evans LS, Gmor NF, Dacosta F (1997) Leaf surface and histological perturbations of leaves of *Phaseolus vulgaris* and *Helianthus annuus* after exposure to simulated acid rain. Am J Bot 4:304–313

Farage PK, Long SP (1999) The effects of O_3 fumigation during leaf development on photosynthesis of wheat and pea: an in vivo analysis. Photosynth Res 59:1–7

Farage PK, Long SP, Lechner EG, Baker NR (1991) The sequence of the change within the photosynthetic apparatus of wheat following short term exposure to ozone. Plant Physiol 95:529–535

Farmer A (2002) Effects of particulates. In: Bell JNB, Treshow M (eds) Air pollution and plant life. Wiley, Hoboken, pp 187–199

Felzer BS, Cronin T, Reilly JM, Melillo JM, Wang Z (2007) Impacts of ozone on trees and crops. Compt Rendus Geosci 339:784–798

Feng ZZ, Kobayashi K, Ainsworth EA (2008) Impact of elevated ozone concentration on growth, physiology and yield of wheat (*Triticum aestivum* L.): a meta-analysis. Glob Chang Biol 14:2696–2708

Fiihrer G, Payer HD, Pfanz H (1993) Effects of air pollutants on the photosynthetic capacity of young Norway spruce trees. Response of single needle age classes during and after different treatments with O_3, SO_2, or NO_2. Trees 8:85–92

Fini A, Guidi L, Ferrini F, Brunetti C, Di Fernando M, Biricolti S, Pollastri S, Calamai L, Tattini M (2012) Drought stress has contrasting effects on antioxidant enzymes activity and phenylpropanoid biosynthesis in *Fraxinus ornus* leaves: an excess light stress affair? J Plant Physiol 169:929–939

Giri S, Srivastava D, Deshmukh K, Dubey P (2013) Effect of air pollution on chlorophyll content of leaves. Curr Agric Res 1, doi: http://dx.doi.org/10.12944/CARJ.1.2.04

Gostin IN (2009) Air pollution effects on the leaf structure of some Fabaceae species. Not Bot Hort Agrobot Cluj 37(2):57–63

Guderian R, Tingey DT, Rabe R (1985) Effects of photochemical oxidants on plants. In: Guderian R (ed) Air pollution by photochemical oxidants. Springer, Berlin, pp 129–334

Hassan IA (2006) Effects of water stress and high temperature on gas exchange and chlorophyll fluorescence in *Triticum aestivum* L. Photosynthetica 44(2):312–315

Hassan IA (2010) Interactive effects of O_3 and CO_2 on growth, physiology of potato (*Solanum tuberosum* L.). World J Environ Sustain Dev 7:1–12

Heath RL (2008) Modification of the biochemical pathways of plants induced by ozone: what are the varied route to changes? Environ Pollut 55:453–463

Honour SL, Bell J, Nigel B, Ashenden TA, Cape J, Power N, Sally A (2009) Responses of herbaceous plants to urban air pollution: effects on growth, phenology and leaf surface characteristics. Environ Pollut 157:1279–1286

IPCC (2007) Climate change 2007, synthesis report. Contribution of Working Groups I, II and III to the Fourth Assessment Report of the Intergovernmental Panel on Climate Change, Geneva

Ismail IM, Basahi JM, Hassan IA (2014) Gas exchange and chlorophyll fluorescence of pea (*Pisum sativum* L.) plants in response to ambient ozone at a rural site in Egypt. Sci Total Environ 497–498:585–593

Joshi N, Bora M (2005) What have we learned from 15 years of free-air CO_2 enrichment (FACE)? meta-analytic review of the responses of photosynthesis, canopy properties and plant production to rising CO_2. New Phytol 165:351–372

Joshi PC, Swami A (2009) Air pollution induced changes in the photosynthetic pigments of selected plant species. J Environ Biol 30:295–298

Joshi N, Chauhan A, Joshi PC (2009) Impact of industrial air pollutants on some biochemical parameters and yield in wheat and mustard plants. Environmentalist 29:398–404

Jouraeva VA, Johnson DL, Hassett JP, Nowak DJ (2002) Differences in accumulation of PAHs and metals on the leaves of *Tilia* × *euchlora* and *Pyrus calleryana*. Environ Pollut 120(2):331–338

Koocha H, Seyyed Nejad SM (2010) Some morphological changes and physiological responses due to air pollution in *Prosopis juliflora* plant. In: Proceedings of the 16th national and 4th international conference of biology, 4–6 November 2010, Mashhad, p 1414

Kumar KA, Bhaskara Rao AV (2008) Physiological responses to fluoride in two cultivars of mulberry. World J Agric Sci 4(4):463–466

Langebartels C, Kerner K, Leonardi S, Scharaudner M, Trost M, Heller W, Sandermann H Jr (1991) Biochemical plant response to ozone: I. Differential induction of polyamine and ethylene biosynthesis in tobacco. Plant Physiol 95:882–889

Li MH (2003) Peroxidase and superoxide dismutase activities in fig leaves in response to ambient air pollution in a subtropical city. Arch Environ Contam Toxicol 45:168–176

Li B, Xing D, Zhang L (2007) Involvement of NADPH oxidase in sulfur oxide- induced oxidative stress in plant cells. Natl Prod Rep 6:628–534

Liu YJ, Ding H (2008) Variation in air pollution tolerance index of plants near a steel factory: implications for landscape-plant species selection for industrial areas. WSEAS Trans Environ Dev 4:24–32

Liu N, Peng CL, Lin ZF, Zhang LL, Pan XP (2006) Changes in photosystem II activity and leaf reflectance features of several subtropical woody plants under simulated SO_2 treatment. J Integr Plant Biol 48:1274–1286

Liu J, Zhang XL, Xu XF, Xu HH (2008) Comparison analysis of variation characteristics of SO2, NOx, O3 and PM2.5 between rural and urban areas, Beijing. Environ Sci 29:1059–1065

Liu N, Lin ZF, Guan LL, Lin GZ, Peng CL (2009) Light acclimation and HSO3 – damage on photosynthetic apparatus of three subtropical forest species. Ecotoxicology 18(7):929–938

Liu KH, Mansell RS, Rhue RD (2010) Cation removal during application of acid solution into air dry soil columns. Soil Sci Soc J 4:1747–1753

Long SP, Ainsworth EA, Rogers A, Ort DR (2004) Rising atmospheric carbon dioxide: plants FACE the future. Annu Rev Plant Biol 55:591–628

Lorenzini G, Nali C, Ligasacchi G, Ambrogi R (1999) Effects of ozone on photosynthesis of Mediterranean urban ornamental plants. Acta Hortic 496:335–338

Loreto F, Velikova V (2001) Isoprene produced by leaves protects the photosynthetic apparatus against ozone damage, quenches ozone products, and reduces lipid peroxidation of cellular membranes. Plant Physiol 127(4):1781–1787

Maier-Maercker U, Koch W (1992) The effect of air pollution on the mechanism of stomatal control. Trees 7:12–25

Malhotra SS, Khan A (1984) In: Treshow M (ed) Biochemical and physiological impact of major pollutants air pollution and plant life. John Wiley and Sons, London, England

McKee IF, Farage PK, Long SP (1995) The interactive effects of elevated CO2 and O3 concentration on photosynthesis in spring wheat. Photosynth Res 45:111–119

Mulholland BJ, Craigon J, Black CR, Colls JJ, Atherton J, Landon G (1997) Impact of elevated atmospheric CO_2 and O_3 on gas exchange and chlorophyll content in spring wheat (*Triticum aestivum* L.). J Exp Bot 48:1853–1863

Muneer S, Kim T, Choi B, Lee B, Lee JH (2013) Effect of CO, NOx and SO_2 on ROS production, photosynthesis and ascorbate–glutathione pathway to induce *Fragaria annasa* as a hyperaccumulator. Redox Biol 17:91–98

Navakoudisa E, Lu¨tzb C, Langebartelsc C, Lu¨tz-Meindld U, Kotzabasisa K (2003) Ozone impact on the photosynthetic apparatus and the protective role of polyamines. Biochim Biophys Acta 1621:160–169

Nawrot B, Dzieranowski K, Gawroski SW (2011) Accumulation of particulate matter, PAHs and heavy metals in canopy of small-leaved lime. Environ Prot Nat Res 49:52–60

Niu J, Feng Z, Zhang W, Zhao P, Wang X (2014) Non-stomatal limitation to photosynthesis in *Cinnamomum camphora* seedlings exposed to elevated O_3. PLoS ONE 9(6):e98572

Noctor G, Foyer CH (1998) Ascorbate and glutathione: keeping active oxygen under control. Annu Rev Plant Physiol Plant Mol Biol 49:249–279

Noormets A, McDonald EP, Dickson RE, Kruger EL, Sober A, Isebrands JG, Karnosky DF (2001) The effect of elevated carbon dioxide and ozone on leaf- and branch-level photosynthesis and potential plant-level carbon gain in aspen. Trees-Struct Funct 15:262–270

Noormets A, Kull O, So˜ber A, Kubiske ME, Karnosky DF (2010) Elevated CO_2 response of photosynthesis depends on ozone concentration in aspen. Environ Pollut 158:992–999

Odiyi O, Bamidele JF (2014) Effects of simulated acid rain on growth and yield of Cassava *Manihot esculenta* (Crantz). J Agric Sci 6:96–101

Parshina OV, Rygalov VY (1999) Structural and functional changes in photosynthetic apparatus of wheat under exposure to sulfur dioxide fumes. Life Support Biosphys Sci 6(3):199–207

Pasqualini S, Piccioni C, Reale L, Ederli L, Della Torre G, Ferranti F (2003) Ozone induced cell death in tobacco cultivar Bel W3 plants. The role of programmed cell death in lesion formation. Plant Physiol 133:1122–1134

Pellegrini E (2014) PSII photochemistry is the primary target of oxidative stress imposed by ozone in *Tilia americana*. Urban For Urban Green 13:94–102

Pellegrini E, Carucci MG, Campanella A, Lorenzini G, Nali C (2011a) Ozone stress in Melissa officinalis plants assessed by photosynthetic function. Environ Exp Bot 73:94–101

Pellegrini E, Francini A, Lorenzini G, Nali C (2011b) PSII photochemistry and carboxylation efficiency in *Liriodendron tulipifera* under ozone exposure. Environ Exp Bot 70:217–226

Pellegrini E, Campanella A, Paolocci M, Trivellini A, Gennai C, Muganu M et al (2015) Functional leaf traits and diurnal dynamics of photosynthetic parameters predict the behavior of grapevine varieties towards ozone. PLoS ONE 10(8):e0135056

Pfanz H, Heber U (1989) Determination of extra- and intracellular pH values in relation to the action of acidic gases on cells. In: Linkens HF, Jackson JF (eds) Modern methods of plant analysis NS Vol 9. Gases in plant and microbial cells. Springer Verlag, Berlin/Heidelberg, pp 322–343

Plazek A, Rapacz M, Skoczowski A (2000) Effects of ozone fumigation on photosynthesis and membrane permeability in leaves of spring barley, meadow fescue, and winter rape. Photosynthetica 38:409–413

Plazek A, Hura K, Rapacz H, Zur I (2001) The influence of ozone fumigation on metabolic efficiency and plant resistance to fungal pathogens. J Appl Bot 75:8–13

Przybysz A, Popek R, Gawroska H, Katarzyna G, Karolina O, Wrochna M, Gawroski SW (2014) Efficiency of photosynthetic apparatus of plants grown in sites differing in level of particulate matter. Acta Sci Pol Hortorum Cultus 13(1):17–30

Pukacki PM (2000) Effects of sulphur, fluoride and heavy metals pollution on the chlorophyll fluorescence of Scots pine (*Pinus sylvestris* L.) needles. Dendrobiology 45:83–88

Rai R, Agrawal M, Agrawal SB (2007) Assessment of yield losses in tropical wheat using open top chambers. Atmospheric 41:9543–9554

Rai A, Kulshrestha K, Srivastava PK, Mohanty CS (2009) Leaf surface structure alterations due to particulate pollution in some common plants. Environmentalist 30:18–23

Rai A, Kulshrestha EK, Srivastava EPK, Mohanty (2010) Leaf surface structure alterations due to particulate pollution in some common plants. Environmentalist 30:18–23

Ramge P, Badeck FW, Plochl M, Kohlmaier GH (1993) Apoplastic antioxidants as decisive elimination factors within the uptake process of nitrogen dioxide into leaf tissues. New Phytol 125:771–785

Reddy AR, Rasineni GK, Raghavendra AS (2010) The impact of global elevated CO_2 concentration on photosynthesis and plant productivity. Curr Sci 99:46–57

Reichenauer TG, Bolhàr-Nordenkampf HR (1999) Mechanisms of impairment of the photosynthetic apparatus in intact leaves by ozone. Verlag der Zeitschrift für Naturforschung, Tübingen

Reichenauer T, Bolhar-Nordenkampf HR, Soja G (1997) Chronology of changes within the photosynthetic apparatus of *Populus nigra* under ozone stress. Phyton 37:245–250

Robinson MF, Heath J, Mansfield TA (1998) Disturbances in stomatal behaviour caused by air pollutants. J Exp Bot 49:461–469

Sarkar A, Rakwal R, Agrawal SB, Shibato J, Ogawa Y, Yoshida Y, Agrawal GK, Agrawal M (2010) Investigating the impact of elevated levels of O_3 on tropical wheat using integrated phenotypical, physiological, biochemical and proteomics approaches. J Proteome Res 9:4565–4584

Sauer F, Schafer C, Neeb P, Horie O, Moortgat GK (1999) Formation of hydrogen peroxide in the ozonolysis of isoprene and simple alkenes under humid conditions. Atmos Environ 33:229–241

Schmidt W, Neubauer C, Kolbowski J, Schreiber U, Urbach W (1990) Comparison of effects of air pollutants (SO_2, O_3, NO_2) on intact leaves by measurements of chlorophyll fluorescence and P700 absorbance changes. Photosynth Res 25(3):241–248

Seyyednejad SM, Koochak H (2011) A study on air pollution-induced biochemical alterations in *Eucalyptus camaldulensis* Aus. J Basic Appl Sci 5(3):601–606

Seyyednejad SM, Niknejad M, Yusefi M (2009a) Study of air pollution effects on some physiology and morphology factors of *Albizia lebbeck* in high temperature condition in Khuzestan. J Plant Sci 4:122–126

Seyyednejad SM, Niknejad M, Yusefi M (2009b) The effect of air pollution on some morphological and biochemical factors of *Callistemon citrinus* in petrochemical zone in South of Iran. Asian J Plant Sci 8:562–565

Seyyednejad SM, Niknejad M, Koochak H (2011) A review of some different effects of air pollution on plants. Res J Environ Sci 5:302–309

Seyyednejad SM, Koochak H, Vaezi J (2013) Some biochemical responses due to industrial air pollution in *Prosopis juliflora* plant. J Biol Today's World 2:471–481

Sharkey TD (1996) Isoprene synthesis by plants and animals. Endeavor 20:74–78

Sharkey TD, Yeh S (2001) Isoprene emission from plants. Annu Rev Plant Physiol Plant Mol Biol 52:407–436
Sharkey TD, Chen X, Yeh S (2001) Isoprene increases thermotolerance of fosmidomycin-fed leaves. Plant Physiol 125:2001–2006
Shavin S, Maurer S, Matyssek R, Bilger W, Scheidegger C (1999) The impact of ozone fumigation and fertilization on chlorophyll fluorescence of birch leaves (*Betula pendula*). Trees 14:10–16
Shiragave PD, Ramteke AA, Patil SD (2015) Plant responses to vehicular pollution: specific effect on photosynthetic pigments of plants at divider of NH-4 highway. Cent Eur J Exp Biol 4(2):1–4
Silva LC, Oliva MA, Azevedo AA, Araujo JM (2006) Responses of resting plant species to pollution from an iron pelletization factory. Water Air Soil Pollut 175(1–4):241–256
Srivastava HS (1999) Biochemical defence mechanisms of plants in response to increased levels of ozone and other atmospheric pollutants. Curr Sci 76:525–533
Takemoto BK, Bytnerowicz A, Olszyk DM (1988) Depression of photosynthesis, growth, and yield in field-grown green pepper (*Capsicum annuum* L.) exposed to acidic fog and ambient ozone. Plant Physiol 88:477–482
Tarasing CS (2013) Role of protein phosphorylation in adaptive responses of the photosynthetic apparatus. Int J Res Sci Technol 2: 92–96
Teughels H, Nijs I, Van Hecke P, Impens I (2005) Competition in a global change environment: the importance of different plant traits for competitive success. J Biogeogr 22:297–305
Thwe AA, Vercambre G, Gautier GHF, Phattaralerphong J, Kasemsap P (2014) Response of photosynthesis and chlorophyll fluorescence to acute ozone stress in tomato (Solanum *lycopersicum* Mill.). Photosynthetica 52(1):105–116
Tiwari S, Agrawal M, Marshall FM (2006) Evaluation of ambient air pollution impact on carrot plants at a sub urban site using open top chambers. Environ Monit Assess 119:15–30
Tognini M, Ranieri A, Castagna A, Nali C, Lorenzini G, Soldatini GF (1997) Ozone-induced alterations in thylakoid protein patterns in pumpkin leaves of different age. Phyton 37(3):277–282
Torsethaugen G, Pell EJ, Assmann SM (1999) Ozone inhibits guard cell K^+ channels implicated in stomatal opening. Proc Natl Acad Sci U S A 96:13577–13582
Tripathi AK, Gautam M (2007) Biochemical parameters of plants as indicators of air pollution. J Environ Biol 28:127–132
Trumble JT, Walker GP (1991) Acute effects of acidic fog on photosynthetic activity and morphology of *Phaseolus lunatus*. Hortscience 26(12):1531–1534
Tzvetkova N, Kolarov D (1996) Effect of air pollution on carbohydrate and nutrient concentrations in some deciduous tree species. Bulg J Plant Physiol 22:53–63
Uzu G, Sobanska S, Sarret G, Munoz M, Dumat C (2010) Foliar lead uptake by lettuce exposed to atmospheric fallouts. Environ Sci Technol 44:1036–1042
van Heerden PDR, Krüger GHJ, Kilbourn LM (2007) Dynamic responses of photosystem II in the Namib Desert shrub, *Zygophyllum prismatocarpum*, during and after foliar deposition of limestone dust. Environ Pollut 146:34–45
Vardaka E, Cook CM, Lanaras T, Sgardelis SP, Pantis JD (1995) Effect of dust from a limestone quarry on the photosynthesis of *Quercus coccifera*, an evergreen schlerophyllous shrub. Bull Environ Contam Toxicol 54:414–419
Veljovic-Jovanovic S, Bilger W, Heber U (1993) Inhibition of photosynthesis, acidification and stimulation of zeaxanthin formation in leaves by sulfur dioxide and reversal of these effects. Planta 191(3):365–376
Veljoviû-jovanoviû S (1998) Active oxygen species and photosynthesis: mehler and ascorbate peroxidase reactions. Iugoslav Physiol Pharmacol Acta 34:503–522
Vollenweider P, Ottiger M, Günthardt-Goerg MS (2003) Validation of leaf ozone symptoms in natural vegetation using microscopical methods. Environ Pollut 124:101–118
Warren CR, Low M, Matyssek R, Tausz M (2007) Internal conductance to CO_2 transfer of adult Fagus sylvatica: variation between sun and shade leaves and due to free-air ozone fumigation. Environ Exp Bot 59:130–138
Wolfenden J, Mansfield TA (1990) Physiological disturbances in plants caused by air pollutants. Proc R Soc Edinb 97B:117–138
Woo SY, Lee DK, Lee YK (2007) Net photosynthetic rate, ascorbate peroxidase and glutathione reductase activities of *Erythrina orientalis* in polluted and non-polluted areas. Photosynthetica 45(2):293–295
World Health Organization (WHO) (2006) Health risks of particulate matter from long-range transboundary air pollution. Joint WHO/Convention Task Force on the Health Aspects of Air Pollution
Yakushevska AE, Keegstra W, Horton P (2003) The structure of photosystem II in Arabidopsis: localization of the CP26 and CP29 antenna complexes. Biochemistry 42:608–613
Yang B, Sun T, Chen H, Wang Z, Yang H, Guo X, Chen X (2015) Effects of hydrogen fluoride-stress on physiological characteristics of Theaceae tree seedlings. International conference on education, management and computing technology
Yu L, Mai B, Meng X, Bi X, Sheng G, Fu J, Peng P (2006) Particle-bound polychlorinated dibenzo-p-dioxins and dibenzofurans in the atmosphere of Guangzhou. China Atmos Environ 40(1):96–108
Yun MH (2007) Effect of ozone on CO_2 assimilation and PSII function in plants with contrasting pollutant sensitivities. Diss Abstr Int 68:10

8 Effect of Air Pollutants on Plant Gaseous Exchange Process: Effect on Stomata and Respiration

Anshu Gupta

Abstract

Air pollution has become an extremely serious problem. Air pollutants affect both plants and animals. Under polluted conditions, plants develop different physiological, morphological and anatomical changes. Pollutants cause damage to cuticular waxes by which then they enter the leaves through stomata. This further leads to injury to plants which can be either acute or chronic. Changes in stomata due to air pollutants which seem to be small can be of great consequence with respect to survival of the plant during stress. These effects can further lead to disturbing the water balance of leaf or whole plant. Respiration also gets affected because of the exposure of plants to air pollutants. The present paper deals with the effect of air pollutants on stomata as well as on respiration leading to affect gaseous exchange.

Keywords

Air pollutants • Stomata • Stress • Respiration

8.1 Introduction

Air pollution has become an extremely serious problem for the modern industrialised world. The prime concern for today's world is changes in the gaseous composition of earth's atmosphere. Fossil fuel consumption has accelerated due to increase in human population, industrial revolution, technological advancement and urbanisation (Watson et al. 1990). The atmospheric concentration of CO_2 has increased from about 275 ppm prior to industrial revolution to a present value of 365 ppm, and it is increasing at the rate of 1–1.5 ppm/year (Conway et al. 1994). Its concentration is expected to be doubled by the middle of the next century (IPCC 1990).

Uncontrolled use of fossil fuels in industries and transport sectors has led to the increase in concentrations of gaseous pollutants such as SO_2, NO_x, etc. (Rai et al. 2011). The general state of the environment, including air quality, is

A. Gupta (✉)
School of Environmental Sciences (SES), Jawaharlal Nehru University (JNU), New Delhi 110067, India
e-mail: anshu.guptaevs@gmail.com

U. Kulshrestha, P. Saxena (eds.), *Plant Responses to Air Pollution*,
DOI 10.1007/978-981-10-1201-3_8

deteriorating in many cities of the developing countries. World Bank studies in selected cities of developing countries have shown that swelling urban populations and the growth of industrial activities and automotive traffic in Asia have caused serious air pollution (World Bank 2009). The adverse effects of air pollution have been associated with three major sources: sulphur dioxide and solid particulates from fossil fuels; photochemical oxidants and carbon monoxide from motor vehicles and miscellaneous pollutants such as hydrogen sulphide, lead and cadmium emitted by smelters, refineries, manufacturing plants and vehicles (Birley and Lock 1999).

It is a known fact that 60 % of air pollution in city is caused by automobiles only. On sensitive species of both plants and animals, the effect of these pollutants is observed at acute level. Plants are considered for investigation of effect of auto exhaust pollutants. Response of plants towards air is being assessed by the air pollution tolerance index (APTI). Some plant species and varieties are so sensitive that they can be conveniently employed as biological indicators or monitors of specific pollutants. They can further assist the planner in managing the urban cities (Horaginamani and Ravichandran 2010). Agarwal and Bhatnagar (1991) studied APTI of some selected plants and described *Mangifera indica* as reliable bioaccumulator plant. Air pollution affects plants mainly through the uptake of pollutants through stomata. Sulphur dioxide and ozone are the two most important pollutants that affect the plants (Emberson 2004). SO_2 is a widespread phytotoxic air pollutant in the environment with ambient concentration of about 0.001 ppm in the air (Allen 1990).

8.2 Plant Responses to Air Pollutants

Air pollution may or will have harmful effects on living things and materials. It may interfere with biochemical and physiological processes of plants to an extent, which ultimately leads to yield losses (Heck et al. 1988). Studies have shown that under polluted conditions, plants develop different morphological, physiological and anatomical changes (Inamdar and Chaudhari 1984; Iqbal 1985; Gravano et al. 2003; Dineva 2004). Sulphur dioxide, one of the most prominent phytotoxic by-products of fossil fuel burning, is also rising progressively in large areas around the world, especially in developing countries. Both elevated CO_2 and SO_2 are anthropogenic stress factors and have potential influence on biological systems including agricultural crops (Aggarwal and Deepak 2003).

Sulphur dioxide is a widespread toxic air pollutant which can cause positive effects on physiological and growth characteristics of plants at low concentrations, especially in plants growing in sulphur- deficient soil (Darrall 1989) when the sulphate might be metabolised to fulfil the demand for sulphur as a nutrient (De Kok 1990).

Increased uptake of SO_2 can cause toxicity and reduce growth and productivity of plants due to accumulation of sulphite or sulphate, by interacting with different physiological processes, and also it damages tissues and pigments (Darrall 1989; Agrawal and Verma 1997). In certain cases, SO_2-induced reduction in plant growth and alteration of physiological and biochemical processes are not accompanied with visible foliar symptoms (Crittendem and Read 1978). Reduction in yield is also reported without visible symptoms when plants are treated with low concentration of SO_2 for long duration (Godzik and Krupa 1982).

Sulphur is necessary for the general metabolism of plants because it is a major component of amino acids, proteins and some vitamins. In healthy leaves, sulphur content ranges from 500 to 14,000 ppm by dry weight (0.5–14 mg/g dry weight) depending upon species. Concentrations below 250 ppm are considered critical, giving rise to deficiency symptoms and to the substitution of selenium (when available) for Sulphur in amino acids and proteins (Treshow 1970). Part, or all, of the sulphur requirements of plants may be met by direct uptake of SO_2 from the atmosphere if it is present at very low concentrations. On the other hand, if the concentration of SO_2 increases beyond a certain critical level that may vary with species (biochemical threshold level), it can result in the general disruption of photosynthesis, respiration and other fundamental cellular processes. Injury becomes irreversible, leading to death, as concentration and time of exposure increase further. Tolerance varies with many factors of the plant and of its environment (Malhotra and Hocking 1976).

8.3 Entry and Effects of Pollutants on Plants

The following are the effects and route of the pollutants entering the plant leaf through stomata, affecting respiration and other gas exchange processes.

8.3.1 Uptake of Pollutants

The most susceptible part of a plant to injury is the leaf due to the presence of abundant stomata which permit the penetration of pollutants into the tissues of the leaves. Boundary layer resistance is the first barrier of gaseous air pollutants which varies with a number of factor including wind speed, size, shape and orientation of leaves (Heath et al. 2009). More pollutants enter the leaves at higher wind speed as boundary layer resistance declines. Waxy cuticle is a potential barrier to most of the pollutants but the cells most exposed to air pollution action are epidermal cells. However, cuticular waxes can be dissociated by acidic gases and these gases can enter the leaves by penetrating the cuticle (Rai et al. 2011).

8.3.2 Effect on Cuticle and Stomata

Cuticle and stomata are the first receptors or targets where the pollutants encounter. Stomata provide the direct path through which the gases enter the leaf, but the direct impact on cuticle must also be considered. The response of stomata to air pollutants is varying and varies from species to species. It also varies with concentration, age of the plants as well as environmental conditions (Abeyrante and Illeperuma 2006). Plant species differ in their ability to mitigate traffic pollution due to differences in their leaf surface characteristics which include epicuticular wax, cuticle, epidermis, stomata and trichomes (Neinhuis and Barthlatt 1998).

Pollutants absorbed by guard cells and subsidiary cells may initially affect the stomatal aperture. Sulphur dioxide has a notable effect in stimulating stomatal opening (Mansfield and Majernick 1970), interacting with CO_2 and atmospheric moisture.

Different plant species can respond differently when exposed to same concentrations of SO_2 (Biggs and Davis 1980). It can cause opening of stomata in one species and closing in another (Mudd 1975). It has also been reported that short-term exposure to SO_2 causes stomatal opening, whereas long-term exposure can lead to partial closure (Abeyrante and Illeperuma 2006). The effects of SO_2 and acid deposition are well seen on cuticular waxes and are well documented (Fowler et al. 1980). Degradation of cuticular waxes due to air pollution has been seen in species such as Scots Pine. Due to air pollution and acid deposition, the weathering of needle cuticle is many times faster in unpolluted forest areas. Similar observations have been described in lichens and mosses (Huttunen and Lane 1983). Due to this evapotranspiration would be greater which would be critical in arid environments. SO_2 had been found to show decrease in photosynthesis and respiration in cultured lichen symbionts (Showman and Rudolph 1971). Air pollutants and oxidative stresses can also have a marked effect on the Ca^{2+} homeostasis of guard cells and the intracellular machinery responsible for stomatal movement (McAnish et al. 2002). Pollutants like SO_2 enter the leaves mainly through the stomata, resultant injury is classified as either acute or chronic. Abeyrante and Illeperuma (2006) have given a plot showing the average values calculated for stomatal pore width versus SO_2 concentration at the three sampling sites (Fig. 8.1). Sampling site 1 recorded high SO_2 conc. as compared to other two sites. Site 1 had 50 % of the pore size of the stomatal of leaves as compared to other two sites.

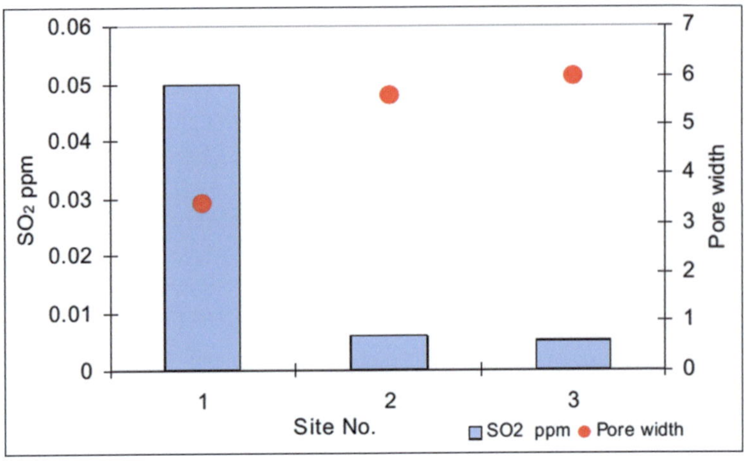

Fig. 8.1 Correlation between average SO2 concentration and pore width (Source: Abeyrante and Illeperuma 2006)

Acute injury results in the appearance of symptoms like two-sided (bifacial) lesions that usually occur between veins and along the margins of the leaves occasionally. Rai and Kulshrestha (2006) have suggested that due to air pollutants the inhibited cell elongation, leaf area and consequently the increase in cell frequency resulted in reduction in the size of stomata and epidermal cells. In order to avoid entry of harmful constituents of exhaust which can otherwise cause adverse effects, the reduction in the size of stomata could be considered as an adaptive response (Satyanarayana et al. 1990; Salgare and Thorat 1990).

Distorted shapes of stomata observed in *Pongamia pinnata* populations exposed to exhaust pollution might have resulted due to lowering of pH in cytoplasm of guard cells and thus change in the turgor relations of the stomata complex (Kondo et al. 1980) due to physiological injury within the leaf (Ashenden and Mansfield 1978). Further, Rai and Mishra (2013) have illustrated that the plants growing along the roadsides have modified leaf surface characters including stomata and epidermal cells due to the stress of automobile exhaust emission with high traffic density in urban areas.

Rahul and Jain (2014) have reported that dust particles of a range less than 5 mm in diameter can interfere with the mechanism of stomatal pores. These small openings are largely responsible for the basic respiration and transpiration function of plants.

Most of the air pollutants which are known to show effect on stomata, are natural components of the atmosphere, but they are present now in higher concentrations in the atmosphere than their natural concentration. The changes in the stomata due to air pollutants which seem to be small can be of great consequence with respect to survival of a plant during stress (Robinson et al. 1998).

Stomatal resistance should be considered as the main obstacle to Ozone flux (Kollist et al. 2000), the direct reaction of the pollutant with cell wall ascorbate is frequently involved (Plochl et al. 2000). The first detoxifying layer which represents the antioxidant system found in the cell (apoplasm + symplasm) at the time of Ozone attack will scavenge ozone and its derivatives (Fig. 8.2). This system is highly linked to the level of ascorbate and especially apoplastic ascorbate, which was primarily proposed as a good indicator for ozone tolerance (Turcsanyi et al. 2000; Tausz et al. 2007).

8.3.3 Effect on Plant Water Balance

Many atmospheric pollutants interfere with the control of stomatal aperture even when present at low concentrations. Therefore they have potential to upset the water balance of the leaf or the whole plant. Pollutants such as SO_2 and CO_2 cause stomatal closure at higher concentrations, whereas at low concentrations the stomatal conductance is often increased (Robinson et al. 1998).

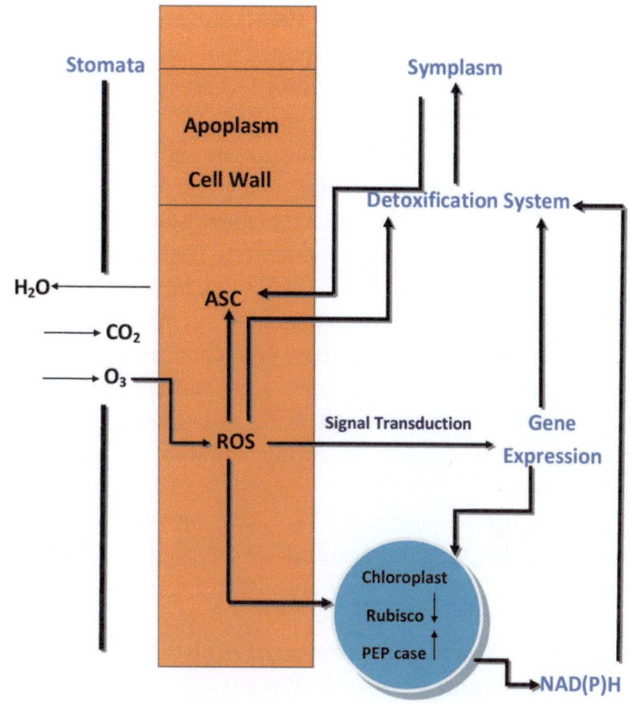

Fig. 8.2 Summary of the relationships between stomatal uptake, metabolic changes and detoxification system under chronic ozone attack in plant cells. *ASC* ascorbate, *PEPcase* phosphoenolpyruvate carboxylase, *ROS* reactive oxygen species, *Rubisco* ribulose-1,5-bisphosphate carboxylase (Source: Dizengremel et al. 2008)

8.3.4 Effect on Respiration

Exposure to air pollutants may result not only in damage of leaf, reduction in growth and yield of crops but it also interferes with physiological processes (Unsworth and Ormrod 1982). Exposure of plants to air pollutants at high concentration for a long period of time results in the development of symptoms of visible injury and associated physiological disturbances. These responses are generally irreversible and may lead to reduction in plant growth and yield. Many factors including plants, pollutants and environment will affect the sensitivity of plants to a range of pollutants. These include toxicity of the pollutant, concentration, frequency and duration of exposure to pollutant, stomatal behaviour, pollutant uptake by plants and prevailing environmental conditions like sunlight, humidity and temperature.

Although the process of respiration includes dark respiration and photorespiration which are important components of carbon budget, the evidences for pollutant-induced modification of respiration are less well documented than for photosynthesis. Since these processes of respiration occur in several sites in the cell, including mitochondria, peroxisomes, cytoplasm, these processes are very much vulnerable to pollutant attack (Koziol and Whatley 2013).

In a study by Aggarwal and Deepak (2003), investigating the long-term influence of elevated concentration of CO_2 and SO_2, singly and in combination on the physiological and biochemical characters of two cultivars of wheat (*Triticum aestivum*), showed that the respiration rate, total phenolics and total soluble sugars increased in response to SO_2. Dark respiration (Rs) increased in response to SO_2- and $CO_2 + SO_2$-treated plants as compared to control. In contrast, elevated CO_2 caused decline in Rs insignificantly. Rs increased at individual treatment of SO_2 because the series of reactions leading to detoxification of SO_2 are ATP mediated which is provided by respiration. In contrast to this there was an insignificant decline in Rs due to CO_2 enrichment (Aggarwal and Deepak 2003). Respiration rates were found to be function of both leaf nitrogen and carbohydrate concentration (Tjoelker et al. 1999). Aggarwal and Deepak (2003) have also reported that rate of respiration was affected by declining leaf nitrogen and increasing TNC in response to CO_2.

A number of studies have been done on the effect of SO_2 on respiration and oxidative phosphorylation. In contrast to the above study, sulphur dioxide has been reported to reduce respiration in plants (Gilbert 1968). Ballantyne (1973) showed sodium sulphite inhibited ATP formation in both bean and corn mitochondria. This inhibition due to sulphite was partially reversed by the addition of oxidised glutathione to the reaction mixture following addition of mitochondria.

Although most of the workers have investigated photosynthetic response, effects on respiratory processes have also been observed.

8.3.4.1 Respiratory Response to High Concentration of Pollutants

When plants get exposed to high concentration of pollutants, plants develop visible injury to the tissues. Depending upon the degree of injury, the respiration is either inhibited or stimulated. As the repair processes utilise energy because of this, the rate of respiration in the non-damaged tissues adjacent to these necrotic areas is increased. A wasteful loss of carbohydrate and energy which is normally used in growth occurs due to the enhancement of respiration in response to high concentration of pollutants. If exposures are not extreme, physiological processes are altered but no visible damage occurs. If stress periods are prolonged, these effects may lead to reductions in growth in the long run (Koziol and Whatley 2013). Reduced respiration in plants grown at elevated CO_2 has common response but not universal (Ziska and Bunce 1993). Carbon dioxide serves as the substrate for photosynthesis. Results of several experiments at elevated CO_2 have indicated stimulation of photosynthesis and reduction in photorespiration, thereby increasing the growth and productivity of plants (Allen 1990). Under CO_2 enrichment, the amount of carbon fixed is greater than the amount of carbon lost and therefore growth and productivity are enhanced (Ryan 1991).

8.3.4.2 Respiratory Response to Low Concentration of Pollutants

On exposure to a low concentration of pollutants, stimulation of respiration is usually exhibited by plants, which may be due to the operation of detoxification and repair mechanism. In response to pollutants, shift from the glycolytic pathway to the pentose phosphate pathway is often observed. If the pollutant exposure periods are short, this enhanced use of energy by the plant is likely to be of benefit. So it prevents the pollutants to reach the sensitive metabolic sites like photosynthetic pathways within the cell. Prior to any observed depression of photosynthesis, indeed respiration can be affected (Koziol and Whatley 2013).

8.3.4.3 Effect of Pollutants on Photorespiration

Evidences generally are not available to allow an assessment of effect of pollutants on photorespiration. This is due to some reasons like difficulties involved in measuring rates of respiration in the light and also due to the fact that early investigations were unaware of the existence of this process. Indeed, photorespiration is a wasteful process; pollutant-induced effects may be beneficial to the growth of the plants because in the short term, the rates of net photosynthesis will increase (Koziol and Whatley 2013).

8.3.4.4 Changes in Respiration in Association with Photosynthesis

If the rate of photosynthesis in plants is very high, a small change in the rate of respiration will not effect significantly on the carbon balance of the plant. On the other hand, if the rate of photosynthesis is very low, change in respiration can lead to change in growth and yield of the plant (Koziol and Whatley 2013).

If the environmental conditions like light and temperature are limiting for the plant photosynthesis and plant is exposed to very high concentration of pollutants, under these conditions, photosynthesis will be severely reduced. Under such conditions change in respiration rate could alter significantly the carbon balance of the plant. This may lead to premature leaf drop, senescence. Jones and Mansfield (1982) have reported that greater reduction in photosynthetic rates has been seen in plants exposed to higher level of pollutants under light-limited conditions than the plants under light higher irradiance.

Evidences are there to show the response to pollutants from the non-photosynthetic portion of the plant such as roots. A reduction in the activity of root will have consequences not only upon root growth but also for the whole plant, if the plant is growing in a stressful environment.

8.4 Conclusion

The study shows that leaf characters including cuticle, stomata, epidermal cells, and guard cells get affected due to stress induced by the air pollutants. This further affects the gaseous exchange as well as respiration in plants. This is an indicator of environmental stress. The effects of individual pollutants are quite variable because they vary from species to species. Changes in leaf in characters induced due to the effect of air pollutants seem to be small, but during the survival of the plant in stress, they can be of great consequence.

References

Abeyratne VDK, Ileperuma OA (2006) Impact of ambient air pollutants on the stomatal aperture of *Argyreia populifolia*. Ceylon J Sci 35(1):9–15

Agarwal SK, Bhatnagar DC (1991) Auto vehicular air pollution induce pigment and ascorbic acid changes in avenue plants. Acta Ecol 13(1):1–4

Agrawal M, Deepak SS (2003) Physiological and biochemical responses of two cultivars of wheat to elevated levels of CO_2 and SO_2, singly and in combination. Environ Pollut 121(2):189–197

Agrawal M, Verma M (1997) Amelioration of sulphur dioxide phytotoxicity in wheat cultivars by modifying NPK nutrients. J Environ Manag 49(2):231–244

Allen LH (1990) Plant responses to rising carbon dioxide and potential interactions with air pollutants. J Environ Qual 19(1):15–34

Ashenden TW, Mansfield TA (1978) Extreme pollution sensitivity of grasses when SO_2 and NO_2 are present in the atmosphere together. Nature 273:142–143

Ballantyne DJ (1973) Sulphite inhibition of ATP formation in plant mitochondria. Phytochemistry 12(6):1207–1209

Biggs AR, Davis DD (1980) Stomatal response of three birch species exposed to varying doses of SO_2. J Am Soc Hortic Sci 100:514–516

Birley MH, Lock K (1999) The health impacts of peri-urban natural resource development. Liverpool School of Tropical Medicine, Liverpool

Intergovermental Panel on Climate Change (1990) In: Houghton JT, Callander BA (eds) (1992) Climate change 1992: the supplementary report to the IPCC scientific assessment. Cambridge University Press, Cambridge

Conway TJ, Tans PP, Waterman LS (1994) Atmospheric CO_2 records from sites in the NOAA/CMDL air sampling network. Trends 93:41–119

Crittenden PD, Read DJ (1978) The effects of air pollution on plant growth with special reference to sulphur dioxide. New Phytol 80(1):49–62

Darrall NM (1989) The effect of air pollutants on physiological processes in plants. Plant Cell Environ 12(1):1–30

DeKok LJ (1990) Sulphur metabolism in plants exposed to atmospheric sulphur. In: Rennenberg H, Brunold C, DeKok LJ, Stulen I (eds) Fundamental, environment and agricultural aspects. SPB Academic Publishing, The Hague, pp 125–138

Dineva SB (2004) Comparative studies of the leaf morphology and structure of white ash Fraxinus americana L. and London plane tree *Platanus acerifolia Willd* growing in polluted area. Dendrobiology 52:3–8

Dizengremel P, Le Thiec D, Bagard M, Jolivet Y (2008) Ozone risk assessment for plants: central role of metabolism-dependent changes in reducing power. Environ Pollut 156(1):11–15

Emberson LD (2004) Air pollution and crops, RAPIDC workshop report, SEI-Y, University of York, UK. http://www.york.ac.uk/inst/sei/rapidc2.html

Fowler D, Cape JN, Nicholson IA, Kinnaird JW, Paterson IS (1980) The influence of a polluted atmosphere on cuticle degradation in Scots pine (*Pinus sylvestris*). In: International conference on the Ecological impact of acid precipitation. Sandefjord (Norway), 11–14 Mar 1980

Gilbert OL (1968) Biological indicators of air pollution. Ph.D thesis, University Newcastle upon Tyne

Godzik S, Krupa SV (1982) Effects of sulfur dioxide on growth and productivity of crop plants. In: Unsworth MH, Ormrod DP (eds) Air pollution in agriculture and horticulture. Butterworths, London, pp 247–265

Gravano E, Giulietti V, Desotgiu R, Bussotti F, Grossoni P, Gerosa G, Tani C (2003) Foliar response of an Ailanthus altissima clone in two sites with different levels of ozone-pollution. Environ Pollut 121(1):137–146

Heath RL, Lefohn AS, Musselman RC (2009) Temporal processes that contribute to nonlinearity in vegetation responses to ozone exposure and dose. Atmos Environ 43:2919–2928

Heck WW, Taylor OC, Tingey DT (1988) Assessment of crop loss from air pollutants. Elsevier Applied Science, London

Horaginamani SM, Ravichandran M (2010) Ambient air quality in an urban area and its effects on plants and human beings: a case study of Tiruchirappalli, India. Kathmandu Univ J Sci Eng Technol 6(2):13–19

Huttunen S, Laine K (1983) Effects of air-borne pollutants on the surface wax structure of Pinus sylvestris needles. In: Annales Botanici Fennici, JSTOR. Finnish Botanical Publishing Board, pp 79–86

Inamdar JA, Chaudhari GS (1984) Effects of environmental pollution on leaf epidermis and leaf architecture. J Plant Anat Morphol 1:1–8

Iqbal MZ (1985) Cuticular and anatomical studies of white clover leaves from clean and air-polluted areas. Pollut Res 4:59–61

Jones T, Mansfield TA (1982) The effect of SO_2 on growth and development of seedlings of *Phleum pratense* under different light and temperature environments. Environ Pollut (Ser A) 27:57–71

Kollist H, Moldau H, Mortensen L, Rasmussen SK, Jørgensen LB (2000) Ozone flux to plasmalemma in barley and wheat is controlled by stomata rather than by direct reaction of ozone with cell wall ascorbate. J Plant Physiol 156(5):645–651

Kondo N, Maruta I, Sugahara K (1980) Research report from the National Institute for Environmental Studies, Yatabe, Japan 11:127–136

Kozioł MJ, Whatley FR (2013) Gaseous air pollutants and plant metabolism. Butterworth-Heinemann

Malhotra SS, Hocking D (1976) Biochemical and cytological effects of sulphur dioxide on plant metabolism. New Phytol 76:227–237

Mansfield TA, Majernik O (1970) Can stomata play a part in protecting plants against air pollutants? Environ Pollut 1(2):149–154

McAinsh MR, Evans NH, Montgomery LT, North KA (2002) Calcium signalling in stomatal responses to pollutants. New Phytol 153(3):441–447

Mudd JB (1975) Sulfur dioxide. In: Responses of plants to air pollution, Academic Press New York, pp 9–22

Neinhuis C, Barthlott W (1998) Seasonal changes of leaf surface contamination in beech, oak, and ginkgo in relation to leaf micromorphology and wettability. New Phytol 138(1):91–98

Plochl M, Lyons T, Ollerenshaw J, Barnes J (2000) Simulating ozone detoxification in the leaf apoplast through the direct reaction with ascorbate. Planta 210(3):454–467

Rahul J, Jain MK (2014) An investigation in to the impact of particulate matter on vegetation along the national highway: a review. Res J Environ Sci 8(7):356

Rai A, Kulshrestha K (2006) Effect of particulates generated from automobile emission on some common plants. J Food Agric Environ 4(1):253

Rai P, Mishra RM (2013) Effect of urban air pollution on epidermal traits of road side tree species, Pongamia pinnata (L.) Merr. J Environ Sci Toxicol Food Technol 2(6):2319–2402

Rai R, Rajput M, Agrawal M, Agrawal SB (2011) Gaseous air pollutants: a review on current and future trends of emissions and impact on agriculture. J Sci Res 55:77–102

Robinson MF, Heath J, Mansfield TA (1998) Disturbances in stomatal behaviour caused by air pollutants. J Exp Bot 49:461–469

Ryan MG (1991) Effects of climate change on plant respiration. Ecol Appl 1(2):157–167

Salgare SA, Thorat VB (1990) Effect of auto-exhaust pollution at Andheri (West), Bombay on the micromorphology of some trees. J Ecobiol 2(4):267–272

Satyanarayana G, Pushpalatha K, Acharya UH (1990) Dust loading and leaf morphological trait changes of plants growing in automobile polluted area. Adv Plant Sci 3(1):125–130

Showman RE, Rudolph ED (1971) Water relations in living, dead, and cellulose models of the lichen *Umbilicaria papulosa*. Bryologist 74:444–450

Tausz M, Grulke NE, Wieser G (2007) Defense and avoidance of ozone under global change. Environ Pollut 147(3):525–531

Tjoelker MG, Reich PB, Oleksyn J (1999) Changes in leaf nitrogen and carbohydrates underlie temperature and CO_2 acclimation of dark respiration of five boreal tree species. Plant Cell Environ 22:767–778

Treshow M (1970) Environment and plant response. McGraw-Hill, New York

Turcsányi E, Lyons T, Plöchl M, Barnes J (2000) Does ascorbate in the mesophyll cell walls form the first line of defence against ozone? Testing the concept using broad bean (*Vicia faba L.*). J Exp Bot 51(346):901–910

Unsworth MH, Ormrod DP (1982) Effects of gaseous air pollution in agriculture and horticulture. Butterworth-Heinemann.

Watson RH, Rodhe H, Oeschger H, Siegenthaler V (1990) Greenhouse gases and aerosols. In: Houghton JT, Jenskins GJ, Ephramus JJ (eds) Climate change: the IPCC scientific assessment. Cambridge University Press, Cambridge, pp 1–40

World Bank (2009) The world Bank annual report 2009. Year in review

Ziska LH, Bunce JA (1993) Inhibition of whole plant respiration by elevated CO2 as modified by growth temperature. Physiol Plant 87(4):459–466

Tropospheric Ozone: Impacts on Respiratory and Photosynthetic Processes

9

Harpreet Kaur

Abstract

Ozone is an important oxidant of the post-industrialised era. Numerous detrimental effects have been attributed to ozone on human health; plants too are gravely affected by its increasing concentration. Of about 15–16% of global temperature changes can be attributed to increase in tropospheric ozone levels in the present time. This chapter presents a review of effects on net primary productivity, photosynthesis and respiration of plants as a response to ozone exposure. In general, exposure to ozone decreases photosynthesis, increases dark respiration and decreases net primary productivity. These variations are however affected by other factors like level of exposure, age of plant and type of plant among many others.

Keywords

Ozone • Photosynthesis • Respiration • Net primary productivity • Plants

9.1 Introduction

Tropospheric ozone is an important greenhouse gas and, above background concentrations, acts as an important air pollutant which is dangerous to human (Lippmann 1993; Burnett et al. 1997) and plant health (Fiscus et al. 2005; Felzer et al. 2005; Ainsworth et al. 2012). Fifteen to 16% of the total global change in temperature post-industrialization has been estimated to be contributed by tropospheric ozone (Baier et al 2005). After particulate matter, ozone is a leading cause of human morbidity and mortality. Globally 0.7 million deaths/year have been attributed to tropospheric ozone pollution (Anenberg et al. 2010).

In plant kingdom, human-induced tropospheric ozone not only interferes with ecosystem functioning (like carbon storage) (Sitch et al. 2007; Nikolova et al. 2010; Galant et al. 2012) and forest productivity (Karnosky et al. 2007) but also poses a great threat to agriculture by affecting crop yields (Feng et al. 2008). Various studies indicate its effect on net primary productivity which results from interference of ozone with

H. Kaur (✉)
School of Environmental Sciences, Jawaharlal Nehru University, New Delhi 110067, India
e-mail: richaaj5@gmail.com

U. Kulshrestha, P. Saxena (eds.), *Plant Responses to Air Pollution*,
DOI 10.1007/978-981-10-1201-3_9

respiration and photosynthetic pathways. Due to its detrimental effects on plants, it is phytotoxic at near concentrations of 0.20 ul I' and above (Reich 1983).

Tropospheric O_3 is a product of photochemical reactions whose main precursors are nitrogen oxides (NOx), carbon monoxide (CO), methane (CH_4), and volatile organic compounds (VOCs) (Seinfeld and Pandis 2012). One of the main steps in ozone formation is photolysis of oxygen molecule which is more rapid at higher temperatures. Therefore, high O_3 production occurs in conditions of strong sunlight and high temperatures. It is worth noticing here that these high temperatures can also favor maximum plant photosynthesis and growth in temperate ecosystems. However, extremes of sunlight and temperature can lead to plant stress, in which case high [O_3] and maximum stomatal conductance and O_3 uptake are no longer coincident. The sensitivity of O_3 production depends on the NOx emission levels, as mentioned above. In rural areas of industrialized countries with moderate NOx levels, O_3 formation reactions dominate. In these regions, which include many of the major crop-growing areas of the world, the rate of O_3 formation increases with increasing [NOx], and O_3 formation is referred to as NOx limited. Among the plethora of air pollutants affecting plant growth and function, ozone emerges as an important one (Ainsworth et al. 2012).

9.2 Ozone Uptake and Effects

In general, pollutant uptake is reduced when stomata are closed and essentially no plant injury takes place. Opening and closing of stomata is affected by the turgor changes in the guard cells and the adjacent subsidiary cells. As a result, the factors which indirectly or directly influence the size of stomatal aperture influence uptake of pollutants like ozone. Studies have revealed that the cell membranes are the primary site of ozone action. In other words ozone affects those molecular configurations which are vital to cellular function. Primarily ozone increases the permeability of the plasmalemma which causes an ionic imbalance. This further creates metabolic imbalance. For example, Keitel and Arndt (1983) showed that tobacco plants exhibited a loss of turgor within few minutes of exposure to 0.18 ppm of ozone, indicating the effect of ozone on permeability of membrane. Ozone exposure effects on permeability by Perchorowicz and Ting (1974), Evans and Ting (1973), etc. further support this theory and suggest that continued ozonation causes irreparable injury to membrane components.

9.3 Effect of Ozone on Carbon Metabolism

Acute exposure to ozone is capable of increasing the reactive oxygen species in the guard cells. These ROSs are the main entry points for changes in net primary productivity of plants on ozone exposure (38). It also has the potential to generate more ROSs, including hydrogen peroxide, superoxide radicals, hydroxyl (OH^-) radicals, and NO (Ainsworth et al. 2012). It is well established that plant growth in chronic O_3 is characterized by decreased rates of CO_2 assimilation at the leaf level (Fiscus et al. 2005).

9.3.1 Case of Woody Plants

Study on trees like poplar, pine (*Pinus elliottii*), pond pine (*P. serotina*), white pine (*P. strobus*), and loblolly pine (*P. taeda*) (seedlings of the latter) indicated that there was a general pattern of photosynthesis reduction and stimulation of respiration. However, these changes depend upon age of the leaves and level and duration of exposure besides other factors.

Chronic exposure of hybrid poplar (*Populus deltoides* × trichocarpa) plants to low concentrations of ozone had negative impact upon net photosynthetic capacity, dark respiration, and leaf chlorophyll contents. 0.20 µl/l O_3 exposure had no immediate effect on net photosynthesis, but chronic exposure to 0.125 or 0.085 µl per liter displayed gradual effects on CO_2 exchange. These included greater dark respiration rates in developing and young mature leaves and lower net photosynthesis in fully expanded leaves of all ages. The leaf maturation and aging processes in themselves have correlation with changes in CO_2 exchange. Age-related changes in net photosynthesis, dark respiration, light-saturation, chlorophyll contents of leaf, and apparent quantum yield were relatively typical. They followed the pattern of increasing activity during maturation and gradual decline with subsequent increase in age. It was observed that hybrid poplar leaves were capable of sustaining high rates of photosynthesis for extended periods of time; over 50 days at >20 mg dm2 h-' and over 75 days at >10 mg dm 2 h-'. Effects of exposure to low concentrations of O_3 were indicated by differences in age-related patterns of CO_2 exchange between control and ozone-treated plants (Barnes 1972; Reich 1983).

9.3.2 Case of Herbaceous Plants

Sun et al. (2014) studied effect of ozone on various aspects of plant growth in soybean. The study showed that light-saturated photosynthesis observed a decrease with increasing O_3 levels and developmental stages. This decrease was higher in young leaves compared to fall in the old leaves. The overall photosynthesis decreased by 7 % per 10 ppb increase in O_3 levels. The overall average leaf starch decreased 8 % per 10 ppb increase in O_3. The overall average leaf starch content was 60 % lower in the old leaves than in the young leaves. No significant level of change was observed in leaf sucrose, glucose, and fructose with O_3 levels during early reproductive stages. However, there was a decrease with increasing O_3 levels during late reproductive stages, and it was lower in the old leaves than in the young leaves.

The critical O_3 levels that cause 10 % decrease varied among those measured parameters with average around 50 ppb for [O_3]. The estimated critical O_3 levels are between 48 and 49 ppb for photosynthesis such as light-saturated photosynthesis and RUBISCO carboxylation; between 45 and 47 ppb for photosynthetic pigments such as chlorophyll; between 46 and 48 ppb for leaf major carbohydrates such as starch, sucrose, fructose, and glucose; and 53 ppb for electron transport rates. The critical O_3 levels that cause 10 % decrease in seed yield compared to the ambient control are 49 ppb [Bassin et al. 2007]. Photosynthetic pigments were the most sensitive parameter to O_3 among all variables measured. Ozone affects RUBISCO-limited photosynthesis more than RuBP-limited photosynthesis, indicating RUBISCO is the early target of O_3 damage.

9.4 Ozone and Primary Metabolism

It is well established that decreased rates of carbon dioxide assimilation describe growth of a plant in chronic O_3 levels at the leaf level (Ashmore 2005; Fiscus et al. 2005). This can therefore be attributed to the basis of O_3-mediated reductions in ecosystem net primary productivity. An analysis of O_3 impact on crops as well as tree species has emphasized its role in altering light-saturated photosynthesis and revealed that angiosperm trees, soybean (*Glycine max*), wheat (*Triticum aestivum*), and rice (*Oryza sativa*) were significantly affected by ambient or near-ambient [O_3] (Morgan et al. 2003; Flowers et al. 2007; Ainsworth 2008; Feng et al. 2008). O_3-induced reductions in primary metabolism are well correlated with the capacity at the cellular level for CO_2 fixation, based on studies of RUBISCO transcript levels, protein level, and enzyme activity.

Already ozone leads to decreased carbon availability from O_3-mediated changes in primary metabolic processes, and plant carbon balance is further impacted by indirect losses. Cost of detoxification is needed to counter the reactive oxygen species increase generated by O_3. Although the role of apoplastic ascorbate in countering the damage against ozone has been documented, the dissolution chemistry of ozone in apoplast is not completely understood (Luwe et al. 1993; Fuhrer and Booker 2003; Conklin and Barth 2004).

9.5 Conclusion

Sensitivity to ozone is associated with leaf characteristics related to the ability of ozone to diffuse into the leaves and further diffusion of O_3 through intercellular space into mesophyll cells (Dixon et al. 1994; Dermondy et al. 2008). Photosynthesis, total nonstructural carbohydrate (TNC) levels, and many metabolites and amino acids as well as seed yield are highly correlated to each other and exhibit a linear decrease with increasing ozone levels. Loss of seed yield is mainly a result of the loss of photosynthetic capacity which is non-stomatal in origin and to the length of growing season and onset of canopy senescence. Ozone interacts with developmental stages and leaf ages with higher damage at later reproductive stages and in older leaves. This has been supported by extensive experimental and modeling studies. Further it can be said that the effects of O_3 on vegetation can feed back to the climate system through alterations to carbon sequestration. It should be noted that climate change itself can alter natural emissions of O_3 precursors, some of which are also radiative forcing agents. The complex set of interactions and feedbacks emphasizes the need to take O_3 pollution seriously at local, regional, and hemispheric scales. More efforts are required to improve our understanding of O_3 pollution biology such that appropriate emissions control measures can be introduced to limit O_3 impacts on ecosystem services.

References

Ainsworth EA (2008) Rice production in a changing climate: a meta-analysis of responses to elevated carbon dioxide and elevated ozone concentration. Glob Chang Biol 14:1642–1650

Ainsworth EA, Yendrek CR, Sitch S, Collins WJ, Emberson LD (2012) The effects of tropospheric ozone on net primary productivity and implications for climate change*. Annu Rev Plant Biol 63:637–661

Anenberg SC, Horowitz LW, Tong DQ, West JJ (2010) An estimate of the global burden of anthropogenic ozone and fine particulate matter on premature human mortality using atmospheric modeling. Environ Health Perspect 118(9):1189

Ashmore MR (2005) Assessing the future global impacts of ozone on vegetation. Plant Cell Environ 28:949–964

Baier M, Kandlbinder A, Golldack D, Dietz KJ (2005) Oxidative stress and ozone: perception, signalling and response. Plant Cell Environ 28(8):1012–1020

Barnes RL (1972) Effects of chronic exposure to ozone on photosynthesis and respiration of pines. Environ Pollut (1970) 3(2):133–138

Bassin S, Volk M, Suter M, Buchmann N, Fuhrer J (2007) Nitrogen deposition but not ozone affects productivity and community composition of alpine grassland after 3 yr of treatment. New Phytol 175:523–534

Burnett RT, Brook JR, Yung WT, Dales RE, Krewski D (1997) Association between ozone and hospitalization for respiratory diseases in 16 Canadian cities. Environ Res 72(1):24–31

Conklin PL, Barth C (2004) Ascorbic acid, a familiar small molecule intertwined in the response of plants to ozone, pathogens, and the onset of senescence. Plant Cell Environ 27:959–970

Dermody O, Long SP, McConnaughay K, DeLucia EH (2008) How do elevated CO_2 and O_3 affect the interception and utilization of radiation by a soybean canopy? Glob Chang Biol 14:556–564

Dixon RK, Brown S, Houghton RA, Solomon AM, Trexler MC, Wisniewski J (1994) Carbon pools and flux of global forest ecosystems. Science 263:185–190

Evans LS, Ting IP (1973) Ozone-induced membrane permeability changes. Am J Bot 60:155–162

Felzer B, Reilly J, Melillo J, Kicklighter D, Sarofim M et al (2005) Future effects of ozone on carbon sequestration and climate change policy using a global biogeochemical model. Clim Change 73:345–373

Feng ZZ, Kobayashi K, Ainsworth EA (2008) Impact of elevated ozone concentration on growth, physiology and yield of wheat (Triticum aestivum L.): a meta-analysis. Glob. Chang Biol 14:2696–2708

Fiscus EL, Booker FL, Burkey KO (2005) Crop responses to ozone: uptake, modes of action, carbon assimilation and partitioning. Plant Cell Environ 28:997–1011

Flowers MD, Fiscus EL, Burkey KO, Booker FL, Dubois J-JB (2007) Photosynthesis, chlorophyll fluorescence, and yield of snap bean (Phaseolus vulgaris L.) genotypes differing in sensitivity to ozone. Environ Exp Bot 61:190–198

Fuhrer J, Booker F (2003) Ecological issues related to ozone: agricultural issues. Environ Int 29:141–154

Galant A, Koester RP, Ainsworth EA, Hicks LM, Jez JM (2012) From climate change to molecular response: redox proteomics of zone-induced responses in soybean. New Phytol 194:220e229

Karnosky D, Skelly JM, Percy KE, Chappelka AH (2007) Perspectives regarding 50 years of research on effects of tropospheric ozone air pollution on US forests. Environ Pollut 147:489e506

Keitel, A., & Arndt, U. (1983). Ozone-induced turgidity losses of tobacco (Nicotiana tabacum var. Bel W3)-an indication to rapid alterations of membrane permeability. Angewandte Botanik (Germany, FR)

Lippmann M (1993) Use of human lung tissue for studies of structural changes associated with chronic ozone exposure: opportunities and critical issues. Environ Health Perspect 101(Suppl 4):209

Luwe MWF, Takahama U, Heber U (1993) Role of ascorbate in detoxifying ozone in the apoplast of spinach (Spinacia oleracea L.) leaves. Plant Physiol 101:969–976

Morgan PB, Ainsworth EA, Long SP (2003) How does elevated ozone impact soybean? A meta-analysis of photosynthesis, growth and yield. Plant Cell Environ 26:1317–1328

Nikolova PS, Andersen CP, Blaschke H, Matyssek R, Haberle KH (2010) Belowground effects of enhanced tropospheric ozone and drought in a beech/ spruce forest (Fagus sylvatica L./Picea abies [L.] Karst). Environ Pollut 158:1071–e1078

Perchorowicz JT, Ting IP (1974) Ozone effects on plant cell permeability. Am J Bot 61:787–793

Reich PB (1983) Effects of low concentrations of O_3 on net photosynthesis, dark respiration, and chlorophyll contents in aging hybrid poplar leaves. Plant Physiol 73(2):291–296

Seinfeld JH, Pandis SN (2012) Atmospheric chemistry and physics: from air pollution to climate change. Wiley, New York

Sitch S, Cox PM, Collins WJ, Huntingford C (2007) Indirect radiative forcing of climate change through ozone effects on the land-carbon sink. Nature 448:791–795

Sun J, Feng Z, Ort DR (2014) Impacts of rising tropospheric ozone on photosynthesis and metabolite levels on field grown soybean. Plant Sci 226:147–161

Air Pollution Stress and Plant Response

10

Irina Gostin

Abstract

Air pollution is a major problem in modern society. During the last century, the interactions between plants and different types of air pollutants were investigated: many studies on the influence of environmental pollution were focused on physiological, biochemical and ultrastructural aspects. The cuticle covers the epidermis external walls from all aerial organs of plants, and it is the main barrier between the plant body and the environment. Therefore, there was a permanent contact between cuticular surface and various pollutants from the atmosphere. If the interaction between air pollutants and the leaves affects the cuticle, the changes in its structure can be considered as biomarkers of air pollution and it can serve in diagnosis. Urban pollution is responsible for the damage of trichomes, cuticle and stomatal guard cells, significantly affecting foliar morphology. Scanning electron microscopy investigations of the leaves from plants growing in polluted sites revealed a remarkable difference in size of the stomatal pores, ruptured of the guard cells, damage of cuticle and epicuticular wax. Some authors investigated structural modifications which occurred in the vegetative organs of different species of plants under the effect of air pollution. The reaction of different species to the modified environmental conditions is strongly correlated with their structural and functional features. In our case studies, the plants originating from extensively polluted areas shows substantial changes in their anatomy; assimilatory tissues contain elevated amounts of tannin or polyphenolic compounds; frequency of the calcium oxalate crystals is increased; and transfusion parenchyma shows the highest degree of alterations. Fluctuating asymmetry (FA) is expected to increase with increasing stress. Our results show that higher asymmetry levels were observed in unpolluted sites than in polluted sites. These data indicating that plants living in the stressful

I. Gostin (✉)
Alexandru Ioan Cuza University, Iasi, Romania
e-mail: irinagostin@yahoo.com

U. Kulshrestha, P. Saxena (eds.), *Plant Responses to Air Pollution*,
DOI 10.1007/978-981-10-1201-3_10

habitats are more symmetrical and consequently these three cosmopolite species could be used as an 'index of habitat quality.

Keywords
Environmental pollution • Stomatal guard cells • Foliar morphology • Anatomy • Epicuticular wax

Air pollution is a major problem in modern society. Even though air pollution is usually a greater problem in cities, pollutants contaminate air everywhere on the planet. These substances include various gases and tiny particles or particulates that can affect human health and cause significant environmental damage. Air pollution was earlier considered as a local problem around large point sources. But due to acid rain, smog and long-range transport of pollutants, it has become rather a regional problem. The transboundary nature of pollutants was clearly evident when areas far away from sources of air pollution also showed higher concentrations of air pollutants (Rai et al. 2011). The main air pollutants are sulphur dioxide, ozone, fluorides, nitrates, oxides of nitrogen and particulate matters (Mudd and Kozlowski 1975).

According Kozioł and Whatley (1984), the first mention of air pollution damage to the vegetation was made by Evelyn (1661) in *Fumifugium* or *The Inconvenience of the Air and Smoke or London Dissipated*, in which he noticed that polluted air kills "our bees and flowers".

First investigations concerning the effect of air pollutants on plants were carried out by Stockhardt (1871) who investigated the effect of smoke on *Picea* and *Abies* trees in Germany.

In the last century, the interactions between plants and different types of air pollutants were investigated by many authors: most studies on the influence of environmental pollution were focused on physiological and ultrastructural aspects (Heumann 2002; Psaras and Christodoulakis 1987; Velikova et al. 2000). Studies concerning the anatomy of the vegetative organs under conditions of pollution have been also carried out (Bermadinger et al. 1988; Dineva 2004; DaSilva et al. 2005).

The emission of air pollutants (resulting from industrial, domestic or social activities) can create problems for plants, because of toxic gases with which they come into contact. At the ultrastructural level, there are many changes in the vegetative and reproductive organs (Ahmad et al. 2005; Patel and Devi 1984; Psaras and Christodoulakis 1987; Soikkeli and Karenlampi 1984; Da Silva et al. 2005). These modifications accompany morphological changes such as chlorosis and adjacent necrosis and consist in destruction of chlorophyll pigments, thylakoid degeneration and plastid lysis (Lendzian and Unsworth 1983; Psaras and Christodoulakis 1987).

10.1 The Cuticle and Its Interaction with Atmospheric Pollutants

The cuticle covering the epidermis external walls from all aerial organs of plants is the main barrier between the plant body and the environment. Therefore, there was a permanent contact between the cuticular surface and various pollutants from the atmosphere. If the interaction between air pollutants and the leaves affects the cuticle, the changes in its structure can be considered as biomarkers of air pollution, and it can serve in diagnosis.

The inner part of the cuticle is characterized by its interaction of their lipid components with the external part of the cell walls, containing pectin and cellulose. The next layer of the cuticle can be easily removed (enzymatically) from the cell wall; the outer layer is predominantly composed of cutin and epicuticular waxes. Cutin is responsible for the mechanical integrity of the epidermis

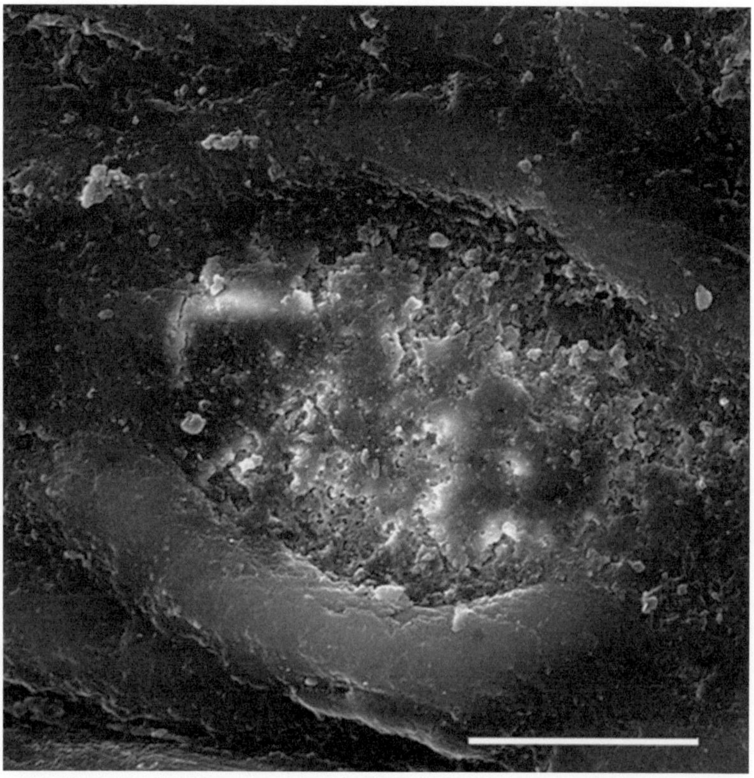

Fig. 10.1 SEM microphotograph of *Abies alba* 2-year-old leaf; damaged wax structure in the stomatal antechamber (original)

and consists predominantly of esterified fatty acids; there are also a small amount of free fatty acids in the cuticle matrix, and they are responsible for transcuticular transport (Kerstiens 1994).

Epicuticular wax (quite heterogeneous, containing a wide range of long-chain saturated hydrocarbons and their derivatives) is impregnated into the outer part of the cuticular matrix, forming a continuous layer above it; it is the primary barrier between external factors and the internal part of the plant. Above epicuticular wax may exist discontinuous deposits of crystalline wax; it plays an important role in the interaction between leaves with acid rain and can play an important role in retaining solid pollutants by increasing the active surface of the cuticle.

Epicuticular wax structure changes with age of the plant; initially, it has a fine, crystalline structure, but in time, it can be altered by contact with acid rain or the precipitation, by mechanical abrasion due to contact with microparticles carried by the wind or by contact with leaves or other adjacent surfaces (Berg 1989). In consequence, epicuticular wax crystals slowly degrade over time, and this leads to changing in the appearance of leaves and their affinity for water. There are various studies on the differences in structure and water affinity of leaf cuticle from polluted areas compared to those in unpolluted areas (Grill et al. 1987; Bermadinger et al. 1988; Bermadinger-Stabentheier 1995). Typical symptoms are observed with leafage: merging of the wax crystals and occurrence of a compact, uninterrupted layer of wax above the normal one, covering the cuticle (Fig. 10.1); this layer of wax can completely occlude or close the stomatal pores or the stomatal chamber, when the stomata are located under the epidermis level.

Grill et al. (1987) investigated the morphology of epicuticular wax from the leaves of healthy spruce trees compared with leaves of affected ones due to air pollution, in different parts of Germany. This has a normal aspect in healthy trees, from unpolluted areas, while in those affected changes were found; the leaf micromorphological appearance shows an older age than of real analysed leaves. Following these findings, the authors concluded that the state of the cuticle is rather an indicator of the overall vigour of the plant and less of the degree of its exposure to different pollutants.

Bermadinger-Stabentheiner (1995) investigated the alterations to epicuticular waxes independent of pollution effects, recorded in various field studies, in order to distinguish changes induced by air pollutants from the artefacts caused by mechanical or chemical injury to the leaf surface. Experimentally, the air-permeable bags isolated portions of leafy branches, thus avoiding the influence of air pollutants. On smoothed and squashed wax layers, a significant regrowth of wax tubes was occasionally observed, unrelated to needle age.

Urban dust was responsible for the damage of trichome, epidermis, cuticle and stomatal guard cells significantly affecting foliar morphology. Gupta et al. (2015c) investigate the effect of dustfall on morphological characteristics of *Morus* (*Morus alba*). Plants from a polluted site (Sahibabad, India – located near an important juncture of two national highways) exhibited more damage to these morphological parts, suggesting that industrial dust is injurious to the plants. Scanning electron microscopy (SEM) investigations revealed a remarkable difference in size of the stomatal pores, rupture of the guard cells and damage of cuticle and epidermis cell at both sites at the abaxial surface (Gupta et al. 2015c). Dust particles deposited in and around stomata on abaxial leaf surface obstructing the stomatal pores were observed (Gupta et al. 2015b). Similar results were obtained in the case of a medicinal plant arjun (*Terminalia arjuna*) from two sites with different characteristics in the National Capital Region (NCR) of Delhi (Gupta et al. 2015a). Cuticle and epicuticular wax changes were observed as an effect of the particulate deposition on the leaf surface. The obstructed stomata and smaller size of the stomatal pores were observed at the Sahibabad site where pollution is higher as compared to the control site.

Our investigations were carried out on the leaves of *Populus nigra* (poplars) and *Pinus sylvestris* (Scots pine) in polluted urban areas (Iasi City, Romania (N, 47009′27,996″; E, 27036′45,763″), elevation 79 m above sea level) with high traffic values (>10,000 vehicles/day); the control samples were collected from Iasi Botanical Garden Anastasie Fatu. The poplar leaves collected from the polluted area showed the presence of solid massive deposits (Fig. 10.2b–d). They are present only on the leaves from urban areas, are affected by heavy traffic, and are almost absent on the leaves from the clean area (botanical garden), used as a control (Fig. 10.2a).

The surface of Scots pine needles is covered with a smooth layer of epicuticular wax, with few crystals (Fig. 10.3a). In the suprastomatal chambers, waxy filigree tubes could be observed. On the smooth cuticle of fully expanded needles, the wax tubules fused into an amorphous surface with time. On the polluted needles (Fig. 10.3b), the needle surface shows numerous particulate debris; the fine structure of the epicuticular wax is not longer visible. At the same time, on the cuticle surface, fungus mycelia are very frequent; they are currently associated with the solid deposits from the needles.

In another study, structural and micromorphological changes from *Abies alba* leaves collected from polluted sites (from the adjacent area of Ceahlău National Park, NE Romania) and from the park area were investigated (Gostin 2010). The control sample (M) was collected from the protected area of the park, whereas the test samples were taken from the Tasca railway station (V1) and from the neighbouring carrying station of a cement factory (V2).

The area of silver fir forests decreased significantly during the last 200 years in most European countries. Reasons for this decline are related to the human impact, through deforestation, overexploitation and promotion of faster-growing tree species, clear-cut forestry, improper management and air pollution (Wolf 2003). Symptoms are

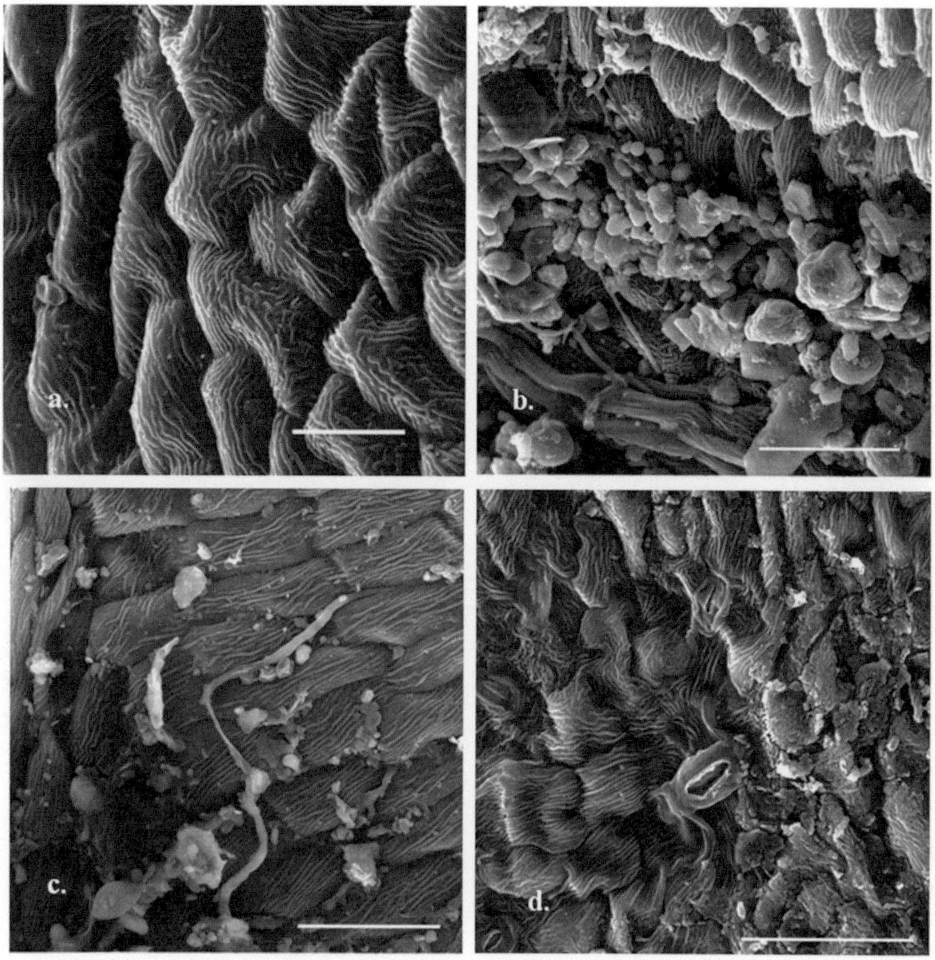

Fig. 10.2 SEM microphotographs illustrating the aspect of the *Populus nigra* leaf surfaces. Epidermis: (**a**) upper epidermis, control sample (ribbed cuticle, without solid deposits, can be observed); (**b**) upper epidermis, polluted sample (massive foreign deposits can be observed); (**c**, **d**) lower epidermis, polluted sample (**a** *scale bar* = 20 μm, **b–d** *scale bar* = 50 μm) (original)

manifested by needle browning, early defoliation and stand mortality (Krause et al. 1986).

The SEM observations on the silver fir tree stomata (on 2-year-old needles from control area) are almost completely occluded by anastomosed wax tubes, covered by small and uniformly disposed granules (Fig. 10.4a–c). On the needles from polluted sites, the majority of stomata have the stomatal antechamber covered with a compact crust of amorphous wax (Fig. 10.4d–f).

Scanning electron microscopy (SEM) research made by Ivanescu and Gostin (2007) on the surfaces of leaves from different species of gymnosperms accomplishing more or less pronounced defoliations and in individuals with foliar chlorosis and/or necrosis revealed the role of deposits in causing these phenomena.

Massive deposits of lime and cement dust on the surfaces of gymnosperm leaves from Bicaz area cover the stomatal pores, alter the cuticular relief changing the pattern of the cuticular striations and adjust the proportion between crystallized and amorphous waxes, in favour of the latter one. This may help to the development of microflora (fungi and algae) which, once installed, cover portions that are photosynthetically active and issue a series of toxic substances that affect the overall condition of the leaf.

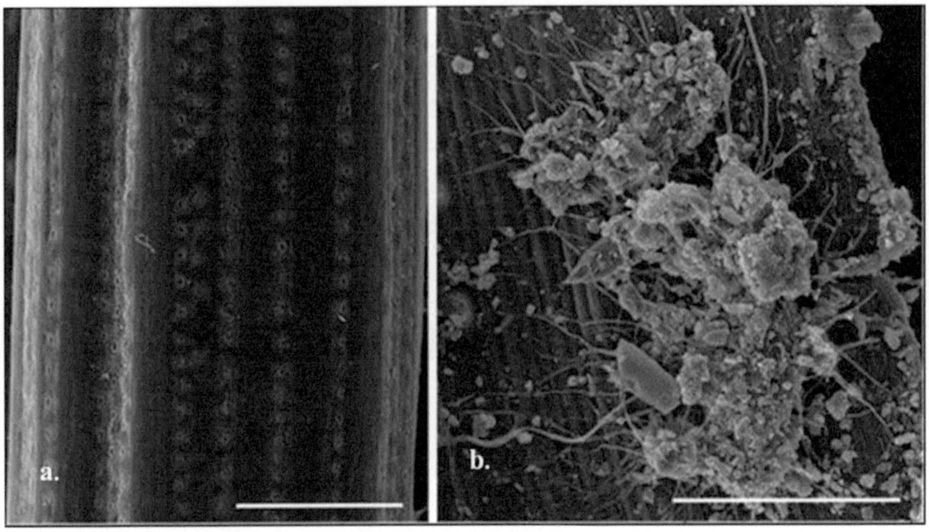

Fig. 10.3 SEM microphotographs from the abaxial surface of the Scots pine needles: (**a**) sample from witness area (*scale bar* = 500 μm), (**b**) damaged sample from urban polluted area (*scale bar* = 200 μm) (original)

Some authors consider the presence of a microflora on the leaf surfaces to be normal; also, there are opinions according to which its presence is a sign of important physiological disorders, which are not yet phenotypically manifest; therefore, this is another possible answer to the controversial problem of the massive defoliations of trees (Ivanescu and Gostin 2008).

The exposure to the air pollutants caused changes in the cuticular wax. This is represented especially by an accelerate fusion and degradation of tubular wax structures which cover the stomatal pores. This reaction was observed also by Grill et al. (1987) and Viskari (2000) in Norway spruce needles and by Bacic and Popovic (1998) and Bacic et al. (2005) in silver fir needles. The transformation of tubular wax into an amorphous one is normally caused by ageing, but in polluted areas this process is more rapid. In the analysed samples, the amorphous crust covers only partially the stomatal pores; Bacic et al. (2005) find completely obstructed stomata in polluted areas from Croatia. The environmental conditions (such as temperature) and different components of the air pollutants (organic compounds, acid rain) could influence the degree of injuries that occur at this level (Günthardt-Goerg and Vollenweider 2007; Vollenweider et al. 2008).

10.2 Structural Changes Under the Effect of Air Pollution on Plants

Structural modifications which occurred in the vegetative organs of different species of plants under the effect of air pollution were investigated from many authors. The reaction of different species to the altered environmental conditions is strongly correlated with their structural and functional features. Christodoulakis and Fasseas (1990) show no significant changes in *Laurus nobilis* (a resistant xerophytic plant) leaf structure exposed to air pollutants in Athens. Histo-anatomical studies regarding the modifications that occurred in a tree's leaves under the effect of air pollutants are numerous (Alvarez et al. 1998; Vollenweider et al. 2003; Dineva 2004; Da Silva et al. 2005; Maranho et al. 2006; Gostin and Ivanescu 2007; Vollenweider et al. 2008).

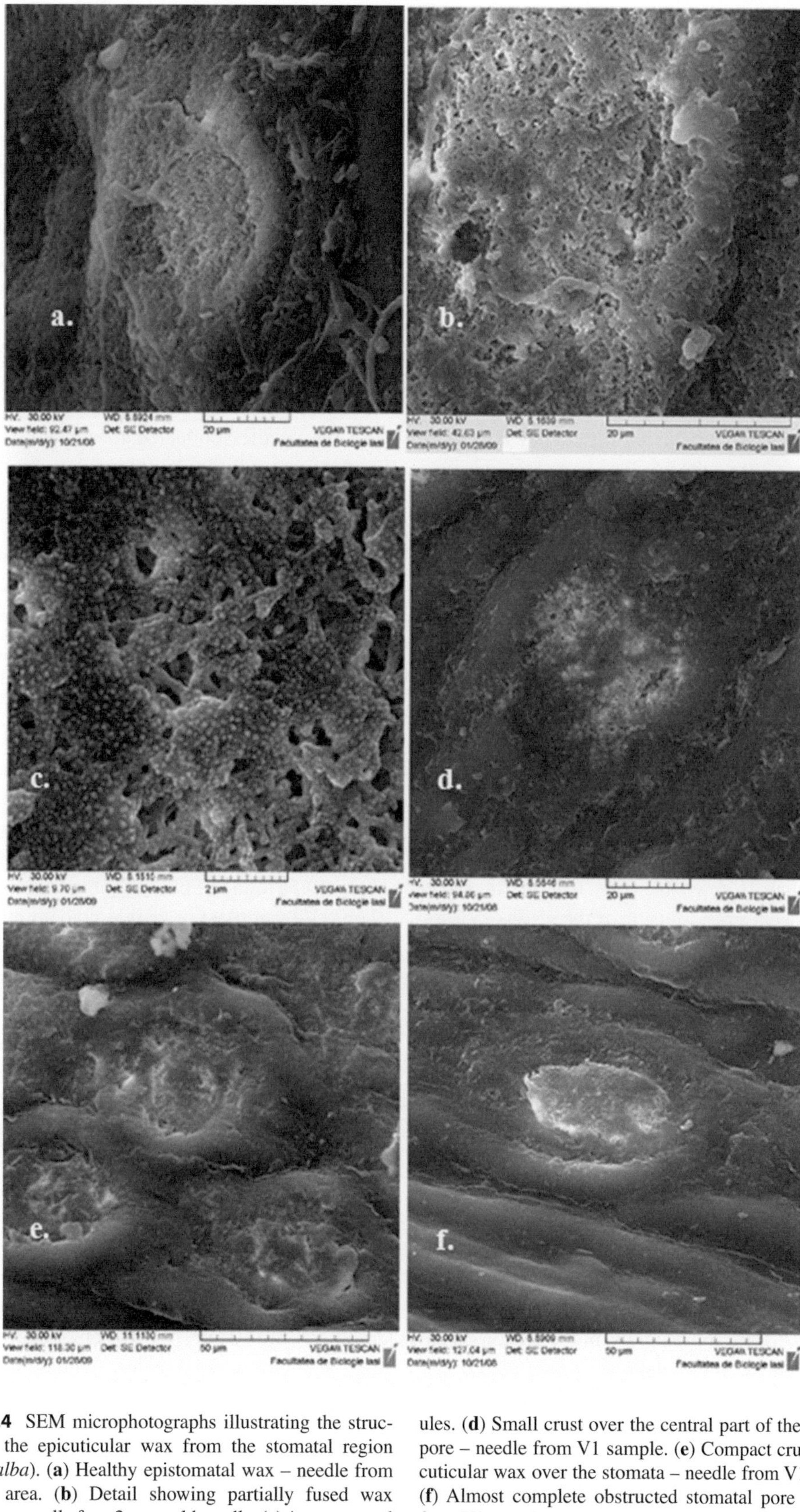

Fig. 10.4 SEM microphotographs illustrating the structure of the epicuticular wax from the stomatal region (*Abies alba*). (**a**) Healthy epistomatal wax – needle from control area. (**b**) Detail showing partially fused wax tubules, normally for a 2-year-old needle. (**c**) Anastomosed wax tubes, covered by small and uniformly disposed granules. (**d**) Small crust over the central part of the stomatal pore – needle from V1 sample. (**e**) Compact crust of epicuticular wax over the stomata – needle from V1 sample. (**f**) Almost complete obstructed stomatal pore – needle from V2 sample (*scale bar* **a**, **b** = 20 µm; **c** = 2 µm; **d** = 20 µm; **e**, **f** = 50 µm) (Gostin 2010)

10.2.1 Case Study 1

The vegetal material consists of 2-year-old leaves *of Abies alba*, which have been collected from Ceahlău National Park (Romania, Neamt County) and its adjacent area. The control sample (M) was collected from the protected area of the park, and the test samples were taken from the Tasca railway station (V1) and from the main road, in the neighbouring carrying station of a cement factory (V2) (Gostin 2010).

The leaves of silver fir tree from polluted sites show moderately visible injuries. They appear as brown spots along the needles; the affected leaves are no more than 20 % from the total (Gostin 2010).

The structural characteristics of the leaves collected from unpolluted sites are similar with those described in the literature (Fig. 10.5a). The epidermis is unilayered; all the cell walls are thick and lignified. The hypodermis is discontinuous especially in the vicinity of the stomata. The mesophyll consists in bilayered palisade parenchyma (under the upper epidermis) and multilayered spongy parenchyma (under the lower epidermis). In all assimilatory cells, numerous crystals of calcium oxalate located in the exterior of the cell wall are visible (Fig. 10.5b). The extracellular occurrence (outside of the walls of mesophyll cells, which face the intercellular spaces) of the calcium oxalate crystals in Pinaceae leaves was noticed by Fink (1991) in *Picea abies*.

The histo-anatomical analysis in the case of the needles collected from polluted areas was made through visibly affected areas. Over the epidermis cells (including the stomata), solid deposits are visible (Fig. 10.7d). Five to ten leaves were investigated for each sample; in the majority of them, epidermis cells, as well as the stomatal cells, are full of tannin (especially from V2); in other samples epidermis cells exhibit thicker external walls and a wide lumen. Crushed cells are visible sporadically in the region of the midvein of the lower epidermis (top of the leaf from V2) (Fig. 10.6e). Usually, the hypodermis is not affected, but sometimes, the lumen is also filled with tannin (Fig. 10.5e, f). The damages induced by local pollutants affect both palisade and spongy parenchyma (Fig. 10.5c, e, f). The structural modifications begin with the protoplast alteration and the increase in thickness of the cellular walls (Fig. 10.6c, d). In the affected areas from the assimilatory tissues, the cells contain elevated amounts of tannin (in order to establish the chemical nature of the brown deposits, histological identification with alcoholic solution of vanillin was performed – Fig. 10.7a, b). In other areas, variable deposits of polyphenolic compounds were observed (Figs. 10.5e and 10.6a) (Gostin 2010).

In the needles collected from polluted sites, the density of the calcium oxalate crystals is higher than those originated from the control zone (Fig. 10.5d). Crystals are visible within the cells, along their walls, and precipitated outside the assimilatory cells, connected to the external parts of the walls or scattered in the intercellular space (Gostin 2010).

In the central cylinder, especially the transfusion parenchyma is the most intensely affected (Fig. 10.5e). Many cells from the abaxial part are crushed and their place was taken by large lacunas. Occasionally, in the phloem (V1 and V2 samples), the cell walls become sinuous and slightly thicker, while the lumen becomes narrow (Figs. 10.5e and 10.6a). In other leaves (Fig. 10.6b) (V2 samples), a visible phloem hypertrophy could be noticed. These are the results of an intense and unilateral cambial activity, followed by an increase of the number of sieve cell layers and the modification in the phloem shape (Gostin 2010). Sometimes, in V2 samples, strongly damaged leaf, probably consecutively of insect attack, with mesophyll cells fulfilled with tannin and large areas of necrosis in assimilatory tissues could be observed (Fig. 10.6f).

Histological changes in the silver fir tree leaves affected by the pollutants occur especially in the central cylinder and in the assimilatory tissues. The transfusion tissue is frequently collapsed in the vicinity of the phloem from the vascular bundles. In a review, Gunthardt-Goerg and Vollenweider (2007) notice that the toxic metals enter in leaves through the stomata and show an accumulation in gradient connected to the vein system.

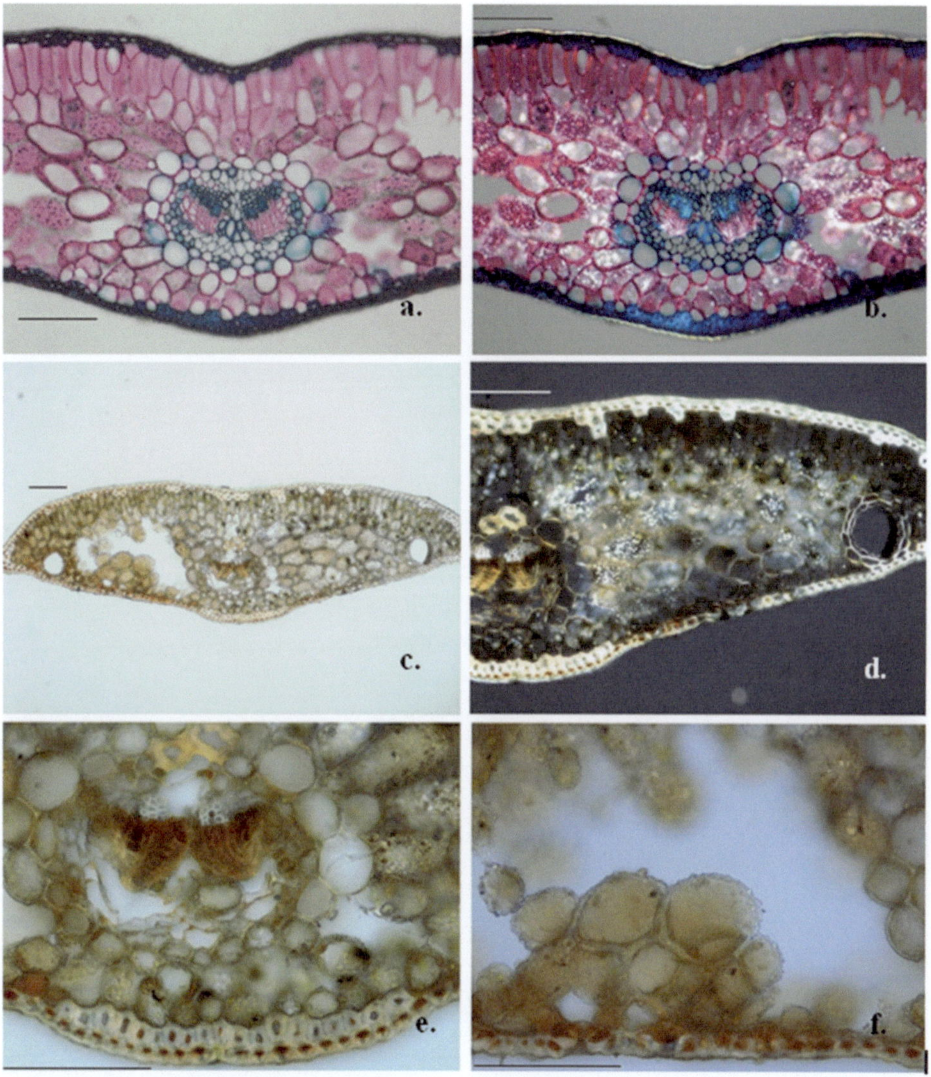

Fig. 10.5 **a** Cross section through the median part of an *Abies alba* needle from control area; the hypodermis is discontinuous both under the lower and the upper epidermis; the palisade parenchyma is unilayered. (**b**) The same section view in polarized light. Calcium oxalate crystals could be observed linked by the external walls of the cells, especially in spongy parenchyma. (**c**) Cross section through the median part of an injured needle from V1 samples. Tannin accumulation and polyphenolic compounds both in palisade and spongy parenchyma are indicated. (**d**) The same section (higher magnification and polarized light). Numerous calcium oxalate crystals could be observed inside and outside of the cells. (**e**) Cross section through the median part of an injured needle from V1 samples. Lower epidermis and hypodermis cells are fulfilled with tannin; in spongy parenchyma cells, dark spots – polyphenolic compounds – could be observed; the central cylinder structure is altered; the transfusion tissue is collapsed; and the phloem from vascular bundles is hypertrophied. (**f**) Detail from the lateral inferior part of the needle – all epidermis cells (including stomata) and hypodermis cells are fulfilled with tannin; the cell walls of the necrotic spongy parenchyma cells are very thick (*scale bar* = 100 μm) (Gostin 2010)

The increase of cell wall thickness from the mesophyll is a common response of the plant leaves to the stress caused by different air pollutants, including heavy metals. Necrosis involving thickening and pigmentations of the cell walls of palisade parenchyma was described by Alvarez et al. (1998) for *Abies religiosa* needles exposed to ozone. In *Abies alba* needles, wall thickening

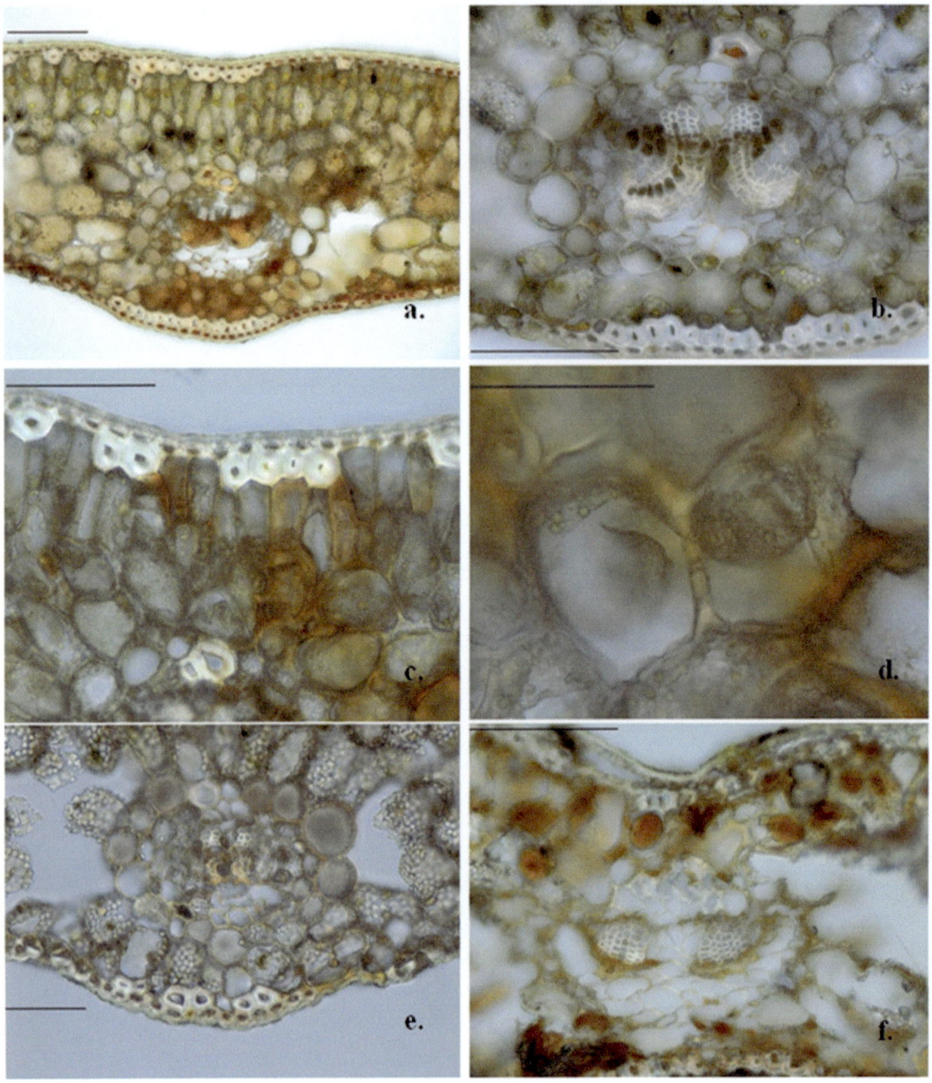

Fig. 10.6 Cross sections through the median part of a needle from V2 area. (**a**) Epidermis and hypodermis cells contain tannin; the spongy parenchyma cells located under the central cylinder are necrotic, with tannin in the walls; and in some palisade parenchyma cells, polyphenolic compounds could be observed. (**b**) Particular modification in the phloem shape from a vascular bundle. (**c**, **d**) Very thick cell walls in assimilatory tissue. (**e**) Cross section through the *top* of a needle. Crushed cells could be observed in lower epidermis. (**f**) Strongly damaged leaf, probably consecutively of insect attack – mesophyll cells fulfilled with tannin and large areas of necrosis in assimilatory tissues (**a**–**c**, **e**, **f** *scale bar* = 100 μm, **d** *scale bar* = 50 μm) (Gostin 2010)

involves both palisade and spongy parenchyma. This symptom is correlated with important tannin accumulation in the affected cells (identified by vanillin test) (Gostin 2010).

Plants can synthesize and accumulate a comprehensive spectrum of phenolics in response to physiological stimuli and stress (Dixon and Paiva 1995); in analysed samples, dark deposits of polyphenolic compounds could be observed into the cells of the palisade and spongy parenchyma. Usually, on the same section, both phenolic compounds and tannin deposit could be observed. The distribution pattern of these zones is random. The intense damages that occur in conducting and in assimilatory tissues are related with the pollution sensitivity of silver fir. The capacity of

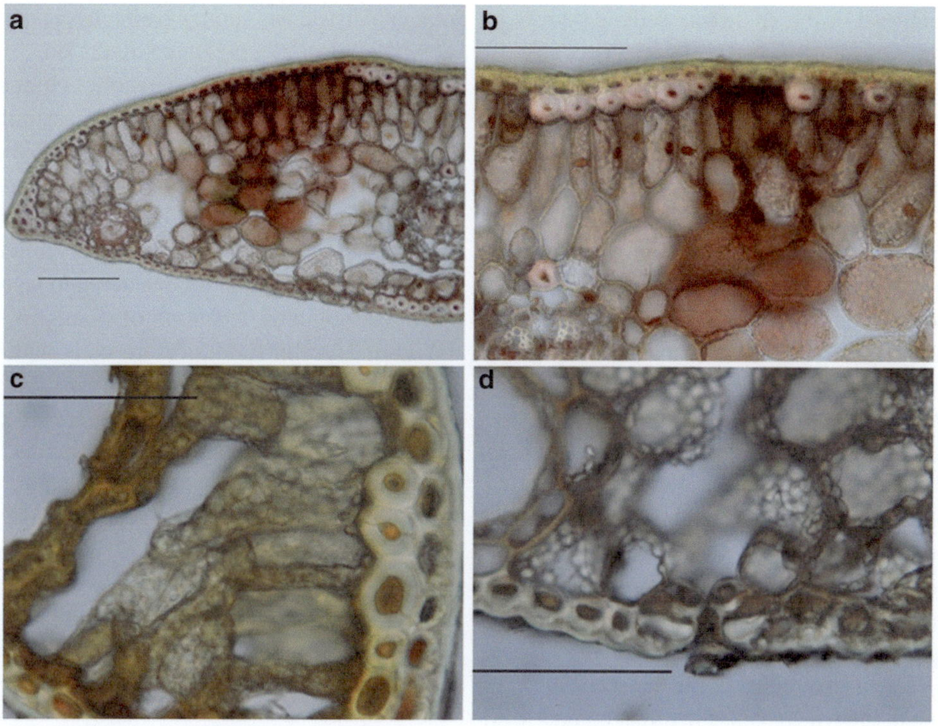

Fig. 10.7 Cross sections through the median part of a needle from V2 area: (**a**, **b**) tannin identification with alcoholic solution of vanillin. (**c**) Crushed and necrotic cells from assimilatory tissue. (**d**) Stomata covered by solid deposits (**a**, **b** *scale bar* = 100 μm, **c**, **d** *scale bar* = 50 μm) (Gostin 2010)

some species to isolate and store the pollutants in relatively inactive tissues, away from more active and vital leaf tissues, represents a detoxification response and implicitly a protective mechanism (Choi et al. 2001; Gunthardt-Goerg and Vollenweider 2007). This kind of reaction was not observed in silver fir needles.

10.2.2 Case Study 2

Four Fabaceae species (*Lotus corniculatus*, *Trifolium montanum*, *T. pratense* and *T. repens*) from Ceahlău National Park (Romania, Neamt County) and its adjacent area were investigated from a histological point of view, in order to establish the influence of the roadside pollution on the leaves' structure; plant species growing along roadsides are considerably modified due to the stress of automobile exhaust emission (Sukumaran 2014). These plant species were chosen because they are widespread and can be found in different polluted sites. The control sample (M) was collected from the protected area of the park and the variants from Tasca railway station (V1) and near the main road of the zone (V2). For each site (M, V1, V2), five plants were collected. Four leaves from the central part of each plant were investigated (Gostin 2009).

In all analysed species, the leaves show dorsiventral structure. The upper epidermis consists of large cells covered by a thin cuticle. The mesophyll consists in a uni- or bilayered palisade parenchyma, with short cells, and a plurilayered spongy parenchyma, with round cells. The lower epidermis structure is similar to the upper one. The leaves are amphistomatic (with stomata in both epidermises). All *Trifolium* species leaves show some differences between the plants collected from control and from polluted areas. The thickness of the foliar lamina slightly decreases under the influence of the air pollutants; the height of the palisade cells and the diameter of the spongy cells decrease too (Gostin 2009). The external wall of the epidermis cells is thicker in V1 and V2 samples compared with the control in *Trifolium repens* and *Lotus corniculatus* leaves.

The pollution stress had a negative impact on the leaves of the investigated species. Nevertheless, these species are quite resistant to the air pollutant actions, and despite the observed modifications, they continue to grow and reach flowering stage. The reduction of the plant's growth as consequence of the pollution stress was underlined by different authors (Gupta and Iqbal 2005; Maruthi Sridhar et al. 2005, 2007).

At the level of the epidermis cells, the increase of the external wall thickness serves as a more efficient barrier against the pollutants penetrating into the leaf. Gaseous pollutants, which still enter into the mesophyll via stomata, may strongly interact with the surface of assimilatory cells, assuming that the larger proportion of intercellular space is positively correlated with a larger inner surface area of that tissue (Gostin 2009). As the first tissue which comes in direct contact with the air pollutants, the epidermis shows crushed cells in *Trifolium montanum* and *T. repens* (Figs. 10.8c and 10.10a–c). Dark deposits are visible over the external cell walls in *Trifolium montanum* V1 samples (Fig. 10.8a); they are absent in the control variant (Fig. 10.8b).

Leaf anatomy of the investigated species also showed reduction in epidermis and palisade parenchyma cells in polluted leaves as compared with those collected from nonpolluted sites. Significant injuries were particularly observed in spongy parenchyma in *T. montanum* and *T. pratense* leaves (Figs. 10.8f and 10.9d, f). The vascular tissues did not show visible changes; only the accumulation of polyphenolic compounds could be noticed at this level (Fig. 10.8e). Sometimes, the substomatal chamber from *Lotus corniculatus* leaves (V2 area) was fulfilled with dark deposits (Fig. 10.9b). Reduction in stomatal closure under metal stress is not uncommon. Decrease in size of the stomata resulting from an inhibitory action of a pollutant may, in fact, represent an avoidance mechanism (Iqbal et al. 1996).

In this study, the spongy parenchyma cells show no significant decrease in the polluted areas. Iqbal (1985) has shown significant reduction in palisade and spongy parenchyma in leaves of white clover of a polluted population.

The presence of the phenolic compounds (dark deposits from the epidermis, assimilatory and vascular tissues) (Figs. 10.8e, f; 10.9b, c, f; and 10.10a, f) indicates that long-term exposure to air pollutants leads to enhanced accumulation of these compounds. The enhanced accumulation of phenolics and lignin is considered to be one of the most common reactions of plants to stress (Wild and Schmitt 1995). Rare cells with large amounts of tannin are visible in *Trifolium pratense* and *Lotus corniculatus* leaves (Figs. 10.9e and 10.10e). Condensed tannins were apparently involved in Cd detoxification. They can complex more or less efficient different heavy metals (Vollenweider et al. 2006).

All investigated species were proven to have some anatomical adaptations indicating a significant potential for resistance to air pollutants.

10.3 Relationship Between Pollution and Fluctuating Asymmetry

Developmental stability measures are potential tools in monitoring environmental stress in plants (Tracy et al. 1995). It is based on the organism's ability to minimize random perturbations during development and is often used as a measurement of the effects of environmental perturbations on organisms (Graham et al. 1993; Moller and Swaddle 1997).

Fluctuating asymmetry (FA) is expected to increase with increasing stress (Palmer and Strobeck 1986; McKenzie and Clarke 1988; Leary and Allendorf 1989). Sometimes, in the field investigations (regarding different types of stress), the results are quite opposite (Graham et al. 2010).

The degree of fluctuating asymmetry of a specific feature could depend on its functional importance, because a stabilized development should be more strongly selected in the characters performing critical functions for an organism, in which FA might be detrimental, than in those characters functionally less important (Palmer and Strobeck 1986).

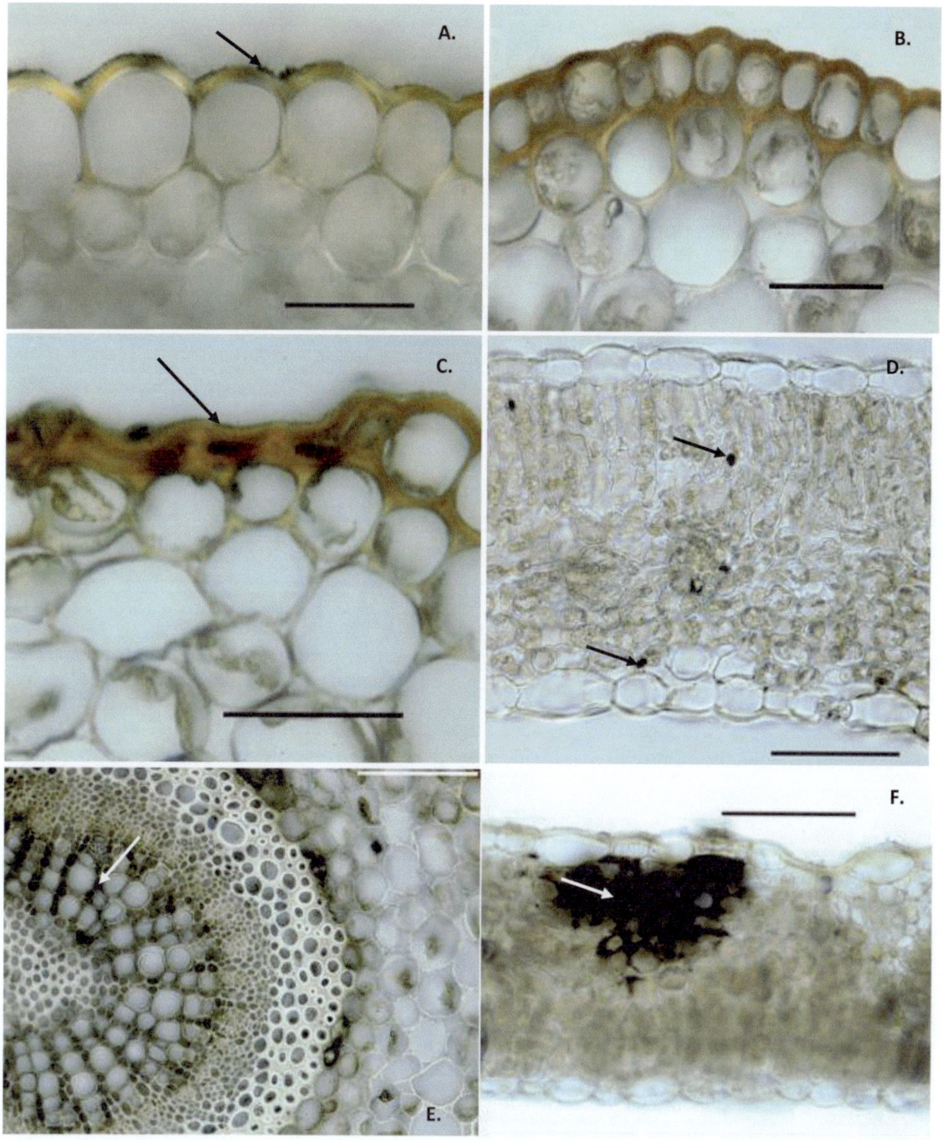

Fig. 10.8 Cross sections through the leaf of *Trifolium montanum*: (**a**) epidermis from the midvein (V1) (dark deposits over the external cell wall could be observed); (**b**) epidermis from the midvein (**c**); (**c**) epidermis from the midvein (V2) (crushed epidermis cells) (*scale bar* = 50 μm); (**d**) cross section through the leaf from V1 sample (*small dark* spots in the mesophyll); (**e**) cross section through the midvein (V2); (**f**) cross section through the leaf from V2 sample (*large dark* deposits in the mesophyll) (*scale bar* = 100 μm) (Gostin 2009)

Hódar (2002) suggests that plants living in the more stressful sites are more symmetrical, more symmetrical plants respond less to yearly variations in drought stress, and the response to yearly variations in drought depends on the climatic conditions in which a tree is living. Habitat quality refers to the ability of the environment to provide conditions appropriate for individual and population persistence (Krausman 1999). The quality of population habitat is, therefore, expected to negatively relate to environmental perturbations (Velikovik and Perisic 2006). However, the main drawback of the use of FA as a diagnostic tool is the difficulty in discriminating the genetic from the environmental components producing FA in the field (Rettig et al. 1997).

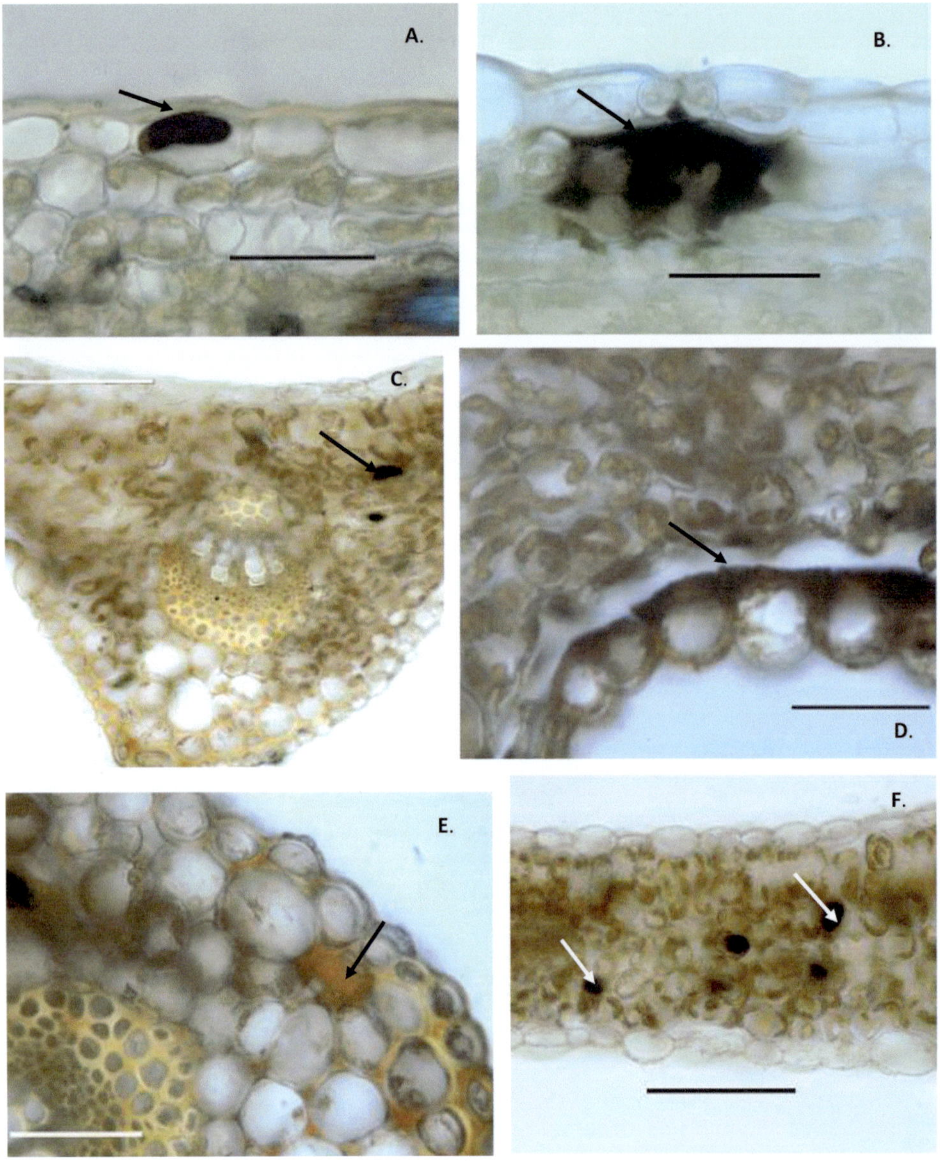

Fig. 10.9 Cross sections through the leaf of *Trifolium montanum* and *Trifolium pratense*. (**a**) *T. montanum* – upper epidermis from V1 leaf (*dark* deposit into an epidermis cell); (**b**) lower epidermis from V2 leaf (*dark* deposit in a substomatal chamber) (*scale bar* = 50 μm). (**c**) *T. pratense* – cross section through the midvein of V1 leaf (*scale bar* = 100 μm); (**d**) cross section through the midvein of the V2 leaf (detachment of lower epidermis could be observed); (**e**) cross section through the midvein of V2 leaf (cell with tannin) (*scale bar* = 50 μm); (**f**) cross section through the lamina of V1 leaf (dark deposits in the mesophyll) (*scale bar* = 100 μm) (Gostin 2009)

Parsons (1992) has mentioned that relatively severe stress is needed to increase FA under field conditions. Increasing asymmetry tends, therefore, to occur in stressed marginal habitats. Genetic perturbations implying genomic stress include certain specific genes, directional selection, inbreeding and chromosome balance alterations (Gostin 2008).

We investigated developmental instability, measured by fluctuating asymmetry (FA), of three Fabaceae species (*Lotus corniculatus*, *Trifolium pratense* and *T. repens*) growing in pol-

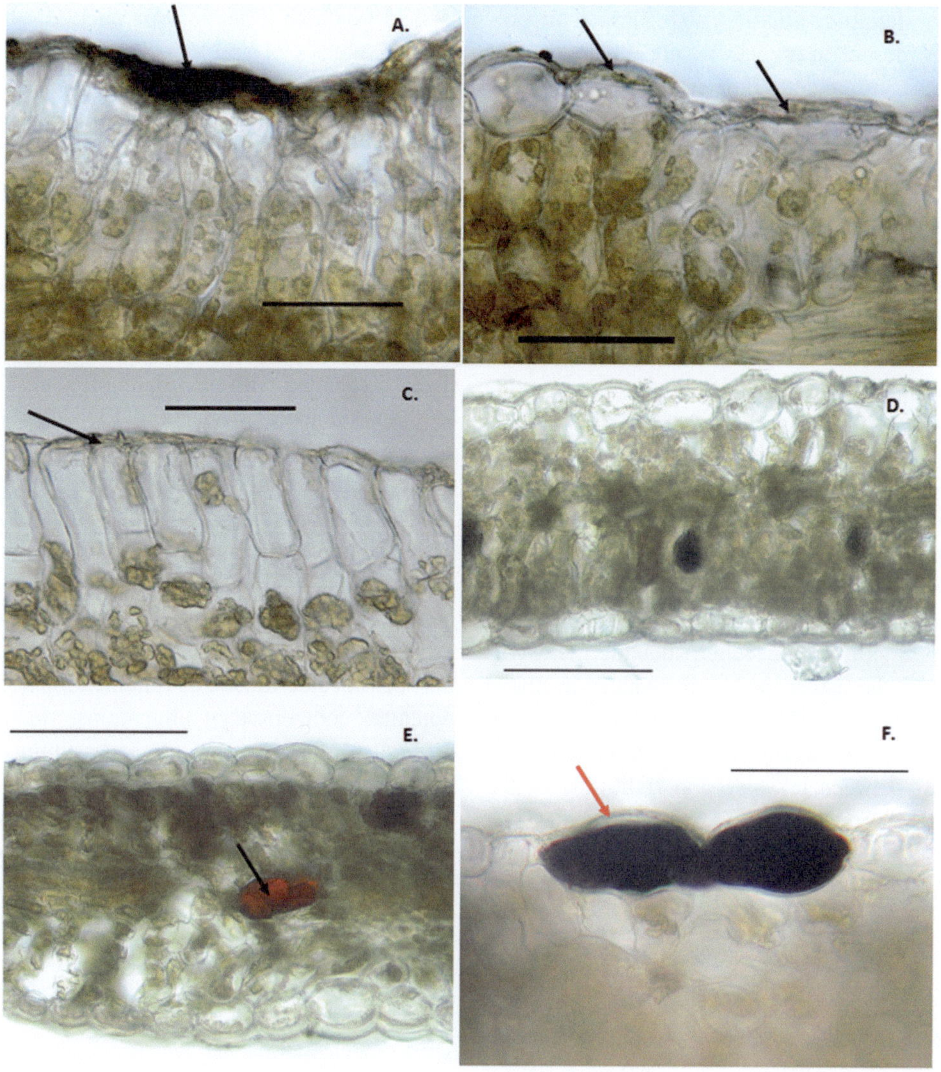

Fig. 10.10 Cross sections through the leaf of *Trifolium repens* and *Lotus corniculatus*. (**a**) *Trifolium repens* – cross section through the leaf of V1 sample (completed crushed epidermis with dark deposit); (**b**, **c**) cross section through the leaf of V2 samples. (**b**) Normal epidermis cells near crushed epidermis cells, **c** destroyed upper epidermis) (*scale bar* = 50 μm). (**d**) *Lotus corniculatus* – cross section through the V1 leaf; (**e**) cross section through the V2 leaf (cell with tannin) (*scale bar* = 1000 μm); (**f**) upper epidermis from V1 leaf (dark deposit into an epidermis cell) (*scale bar* = 50 μm) (Gostin 2009)

luted and unpolluted sites. This parameter was estimated and compared between three populations of each investigated species, two from areas with different pollution gradients and one from a clean area; leaf width (LW and RW, which is the distance from the midrib to the right and left margins) was measured from 30 leaves of individuals from each area, and the rapport of LW/RW was considered for FA determinations.

The vegetal material consists in whole plants of *Lotus corniculatus*, *T. pratense* and *T. repens* collected from Ceahlău National Park and its adjacent area. The control sample (M) was collected from the protected area of the park and the variants from Tasca railway station (V1) and near the main road of the zone (V2) (Gostin 2015). Different morphological parameters were measured with a digital calliper (0.01 mm accuracy).

The stationary conditions (regarding air pollutants present in the area) were obtained from Neamt Environmental Protection Agency (Tables 10.1 and 10.2).

The results of the measurement of different morphological parameters from investigated species are presented in Tables 10.3, 10.4 and 10.5.

The same type of variation was noticed at the leaflet from all investigated species (larger dimensions in control plants and a visible decrease in the V1 and V2 samples). The differences in size are more evident in *Lotus corniculatus* leaflets (1.45 × 1.12 cm in control site versus 0.67 × 0.23 cm in V2 area). In the case of *Trifolium pratense* and *T. repens* leaflets, the dimensional differences are lower (2.85 × 1.72 cm in control versus 2.24 × 1.23 cm in V2 area and 1.43 × 0.89 cm in control versus 1.1 × 0.63 cm in V2 area, respectively). The rapport between left and right sides of the leaflet, showing the asymmetry of the leaflet, has higher values in plants from control area (M) for *Lotus corniculatus* and *Trifolium pratense*. In *L. corniculatus* leaflets, the rapport between left side and right side of the leaflets was higher in control area 1.035 than in polluted areas (V1 and V2). The same situation was observed in the case of *T. pratense* (0.890 in control area versus 0.966 and 1.013 in V1 and V2 sites). This means that the asymmetry degree is more evident in the polluted sites. In the case of *Trifolium repens*, the most symmetrical plants were found in V1 and control areas (1 and 1.022, respectively); the asymmetry degree increases in V2 (polluted areas) (0.968).

Leaf asymmetry has been shown to be positively associated with environmental stress such as pollution load; Moller and Swaddle (1997) suggested that adverse environmental conditions may affect the evolutionary process at different levels. The possibility of local adaptation of plants to stressful conditions might recommend the use of fluctuating asymmetry as a good diagnostic character to assess stress tolerance. In principle, genetic or environmental influences may disrupt side-to-side communication and thus potential compensatory allocations, making it difficult for structures to preserve symmetry or recover from stress (Freeman et al. 2003).

In a recent review Graham and co-workers (2010) reported a failure to detect the expected increase in bilateral FA under the impacts of natural environmental stressors.

Our results (different values of FA in different species from the same genus – *Trifolium*) suggest the necessity for particular studies for each analysed species; this is because each species responds differently to different environmental conditions (Gostin 2015). It seems that only the exposure to an episode of acute pollution of some sensitive species leads to increases in FA. Chronic pollution, especially in the case of resistant species, is not accompanied by increases in FA.

Table 10.1 Pollutant concentrations in the investigated areas

	Pollutant type		
Collection point	Dust suspensions (mg/m^3)	NO_2 (μg/m^3)	SO_2 (μg/m^3)
Control (protected area) (M)	0.021	7.651	4.719
Tasca railway station (V1)	0.121	17.152	5.164
Main road (V2)	0.281	20.974	6.833

Gostin (2015)

Table 10.2 Pollutant concentrations in the investigated areas

	Pollutant type					
Collection point	Pb	Cd	Zn	Cr	Cu	Co
Control ((M)	45.31	0.214	210.05	4.05	27.02	3.59
Tasca railway station (V1)	104.11	0.798	887.06	21.31	122.90	20.85
Main road (V2)	138.30	2.817	1673.72	78.84	193.07	41.91

Gostin (2015)

Table 10.3 Results of the measurement of different morphological parameters of *Lotus corniculatus* vegetative aerial organs (from different polluted sites and CNP)

Morphological parameter (cm)	M	V1	V2
Leaflet length (mean)	1.45	0.75	0.67
Leaflet width (mean)	1.12	0.39	0.23
Leaflet left side	0.56	0.195	0.115
Leaflet right side	0.58	0.193	0.117
Rapport between left side and right side	1.035	0.989	1.017

Gostin (2015)

Table 10.4 Results of the measurement of different morphological parameters of *Trifolium pratense* leaflets (from different polluted sites and CNP)

Morphological parameter (cm)	M	V1	V2
Leaflet length (mean)	2.85	2.58	2.24
Leaflet width (mean)	1.72	1.39	1.23
Leaflet left side	0.810	0.683	0.619
Leaflet right side	0.910	0.707	0.611
Rapport between left side and right side	0.890	0.966	1.013

Gostin (2015)

Table 10.5 Results of the measurement of different morphological parameters of *Trifolium repens* leaflets (from different polluted sites and CNP)

Morphological parameter (cm)	M	V1	V2
Leaflet length (mean)	1.43	1.33	1.1
Leaflet width (mean)	0.89	0.76	0.63
Leaflet left side	0.45	0.38	0.31
Leaflet right side	0.44	0.38	0.32
Rapport between left side and right side	1.022	1	0.968

Gostin (2015)

References

Ahmad SH, Reshi Z, Ahmad J, Iqbal MZ (2005) Morpho-anatomical responses of Trigonella foenum graecum Linn. to induced cadmium and lead stress. J Plant Biol 48(1):64–84

Alvarez D, Laguna G, Rosas I (1998) Macroscopic and microscopic symptoms in Abies religiosa exposed to ozone in a forest near Mexico City. Environ Pollut 103:251–259

Bacic T, Popovic Z (1998) Preliminary report on epicuticular wax surface condition on stomata of Abies alba Mill. needles from Risnjak National Park in Croatia. Acta Biol Cracov Ser Bot 40:25–31

Bacic T, Krstin L, Rosa J, Popovic Z (2005) Epicuticular wax on stomata of damaged silver fir trees (*Abies alba* Mill.). Acta Soc Botanicorum Pol 74:159–166

Berg SV (1989) Leaf cuticles as potential markers of air pollutant exposure in trees. In: Biologic markers of air pollution stress and damage in forests. Committee on biological markers of air pollution damage in trees. National Research Council, National Academy Press, Washington DC

Bermadinger E, Grill D, Golob P (1988) Influence of different air pollutants on the structure of needle wax of spruce (Picea abies (L.) Karsten). Geojournal 17:289–293

Bermadinger-Stabentheier E (1995) Physical injury, re-crystallization of wax tubes and artefacts: identifying some causes of structural alteration to spruce needle wax. New Phytol 130(1):67–74

Choi Y-E, Harada E, Wada M, Tsuboi H, Morita Y, Kusano T, Sano H (2001) Detoxification of cadmium in tobacco plants: formation and active excretion of crystals containing cadmium and calcium through trichomes. Planta 213:45–50

Christodoutakis NS, Fasseas C (1990) Air pollution effects on the leaf structure of Laurus nobilis, an injury resistant species. Bull Environ Contam Toxicol 44:276–281

Da Silva CL, Oliva M, Azevedo AA, Araújo JM, Aguiar RM (2005) Micromorphological and anatomical alterations caused by simulated acid rain in Restinga plants: Eugenia uniflora and Clusia hilariana Water. Air Soil Pollut 168:129–143

Dineva SB (2004) Comparative studies of the leaf morphology and structure of white ash Fraxinus americana L. and London plane tree Platanus acerifolia Willd growing in polluted area. Dendrobiology 52:3–8

Dixon RA, Paiva NL (1995) Stress-induced phenylpropanoid metabolism. Plant Cell 7:1085–1097

Evelyn J (1661) Fumifugium or the Inconvenience of the Aer and Smoake of London Dissipated: together with some remedies humbly proposed. W. Godbid, London

Fink S (1991) Unusual patterns in the distribution of calcium oxalate in spruce needles and their possible relationships to the impact of pollutants. New Phytol 119:41–51

Freeman DC, Brown ML, Dobson M, Jordan Y, Kizy A, Micallef C, Hancock LC, Graham JH, Emlen JM (2003) Developmental instability: measures of resistance and resilience using pumpkin (Cucurbita pepo L.). Biol J Linn Soc 78:27–41

Gostin I (2008) Relationship between pollution and fluctuant asymmetry in some Fabaceae species. In: Stadler J, Schöppe F, Frenzel M (eds) Proceedings Verhandlungen der Gesellschaft für Ökologie, vol 38, p 609

Gostin I (2009) Air pollution effects on the leaf structure of some Fabaceae species. Notulae Bot Hortic Agrobot 37(2):49–56

Gostin I (2010) Structural changes in silver fir needles in response to air pollution. Analele Univ Oradea Fascicula Biol 17(2):300–305

Gostin I (2015) Relationship between pollution and fluctuant asymmetry in some Fabaceae species. In: Proceedings of the 2nd CommScie International Conference "Challenges for Sciences and Society in Digital Era", December 2015, Iasi, Romania, pp 111–114

Gostin I, Ivanescu L (2007) Structural and micromorphological changes in leaves of Salix alba under air pollution effect. Int J Energy Environ 1:219–226

Graham JH, Freeman DC, Emlen JM (1993) Antisymmetry, directional asymmetry, and dynamic morphogenesis. Genetica 89:121–137

Graham JH, Raz S, Hel-Or H, Nevo E (2010) Fluctuating asymmetry: methods, theory, and applications. Symmetry 2:466–540

Grill D, Pfeifhofer H, Halbwachs G, Waltinger H (1987) Investigations on epicuticular waxes of differently damaged spruce needles. Eur J For Pathol 17:246–255

Günthardt-Goerg MS, Vollenweider P (2007) Linking stress with macroscopic and microscopic leaf response in trees: new diagnostic perspectives. Environ Pollut 147:467–488

Gupta MC, Iqbal M (2005) Ontogenetic histological changes in the wood of mango (Mangifera indica L. cv Deshi) exposed to coal-smoke pollution. Environ Exp Bot 54(3):248–255

Gupta GP, Kumar B, Singh S, Kulshrestha UC (2015a) Urban climate and its effect on biochemical and morphological characteristics of Arjun (Terminalia arjuna) plant in National Capital Region Delhi. Chem Ecol 31(6):524–538

Gupta GP, Singh S, Kumar B, Kulshrestha UC (2015b) Industrial dust sulphate and its effects on biochemical and morphological characteristics of Morus (Morus alba) plant in NCR Delhi. Environ Monit Assess 187:67

Gupta GP, Kumar B, Singh S, Kulshrestha UC (2015c) Deposition and impact of urban atmospheric dust on two medicinal plants during different seasons in NCR Delhi. Aerosol Air Qual Res. doi:10.4209/aaqr.2015.04.0272

Heumann HG (2002) Ultrastructural localization of zinc in zinc-tolerant Armeria maritima ssp. halleri by autometallography. J Plant Physiol 159:191–203

Hódar JA (2002) Leaf fluctuating asymmetry of Holm oak in response to drought under contrasting climatic conditions. J Arid Environ 52(2):233–243

Iqbal MZ (1985) Cuticular and anatomical studies of white clover leaves from clean and air-polluted areas. Pollut Res 4:59–61

Iqbal M, Abdin MZ, Mahmooduzzafar M, Yunus M, Agrawal M (1996) Resistance mechanisms in plants against air pollution. In: Yunus M, Iqbal M (eds) Plant response to air pollution. Wiley, Chichester, pp 195–240

Ivanescu L, Gostin I (2007) Cito-histological changes due to the action of atmosphere pollutants on three species of gymnosperms. Int J Energy Environ 1(2):95–100

Ivanescu L, Gostin I (2008) Phenological data concerning the influence of atmosphere pollutants on some species of woody plants. 6th IASME/WSEAS international conference on heat transfer, thermal engineering and environment (HTE'08). Rhodes, Greece, August 20–22, pp 328–332

Kerstiens G (1994) Air pollutants and plant cuticles: mechanisms of gas and water transport, and effects on water permeability. In: Percy K, Cape JN, Jagels R, Simpson CJ (eds) Air pollutants and the leaf cuticle. Springer, Berlin

Kozioł MJ, Whatley FR (1984) Gaseous air pollutants and plant metabolism. Butterworths, London

Krause GHM, Arndt U, Brandt CJ, Bucher J, Kenk G, Matzner E (1986) Forest decline in Europe; development and possible causes. Water Air Soil Pollut 31:647–668

Krausman P (1999) Some basic principles of habitat use. In: Launchbaugh KL, Sanders KD, Mosley JC (eds) Grazing behavior of livestock and wildlife. Idaho forest wildlife and range exp. sta. bull. 70. Univ Idaho, Moscow

Leary RF, Allendorf FW (1989) Fluctuating asymmetry as an indicator of stress: implications for conservation biology. Trends Ecol Evol 4:214–217

Lendzian KJ, Unsworth MH (1983) Ecophysiological effects of atmospheric pollutants. In: Lange OL, Nobel PS, Osmond CB, Zigler H (eds) Hysiological plant ecology. Springer, Berlin, pp. 466–491

Maranho LT, Galvão F, Preussler KH, Muniz GIB, Kuniyoshi YS (2006) Efeitos da poluição por petróleo na estrutura da folha de Podocarpus lambertii Klotzsc ex Endl., Podocarpaceae. Acta Bot Bras 20:615–624

Maruthi Sridhar BB, Diehl SV, Han FX, Monts DL, Su Y (2005) Changes in plant anatomy due to uptake and accumulation of Zn and Cd in Indian mustard (Brassica juncea), environmental. Exp Bot 54:131–141

Maruthi Sridhar BB, Han FX, Diehl SV, Monts DL, Su Y (2007) Effects of Zn and Cd accumulation on structural and physiological characteristics of barley plants. Braz J Plant Physiol 19(1):15–22

McKenzie JA, Clarke GM (1988) Diazinon resistance, fluctuating asymmetry and fitness in the Australian sheep blowfly, Lucilia cuprina. Genetics 120:213–220

Møller AP, Swaddle JP (1997) Asymmetry, developmental stability and evolution. Oxford University Press, New York

Mudd JB, Kozlowski T (eds) (1975) Responses of plants to air pollution. Academic, New York

Palmer AR, Strobeck C (1986) Fluctuating asymmetry: measurement, analysis and patterns. Annu Rev Ecol Syst 17:391–421

Parsons P (1992) Fluctuating asymmetry: a biological monitor of environmental and genetic stress. Heredity 68:361–364

Patel JD, Devi GS (1984) Ultrastructural variations in leaves of Streblus asper growing near a fertilizer complex. Phytomorphology 34:140–146

Psaras GK, Christodoulakis NS (1987) Air pollution effects on the ultrastructure of Phlomis fruticosa mesophyll cells. Bull Environ Contam Toxicol 38:610–617

Rai R, Rajput M, Agrawal M, Agrawal SB (2011) Gaseous air pollutants: a review on current and future trends of emissions and impact on agriculture. J Sci Res 55:77–102

Rettig JE, Fuller RC, Corbett AL, Getty T (1997) Fluctuating asymmetry indicates levels of competition in an even-aged poplar clone. Oikos 80:123–127

Soikkeli S, Karenlampi L (1984) Cellular and ultrastructural effects. In: Treshow M (ed) Air pollution and plant life. Wiley, Chichester

Stockhardt JH (1871) Untersuchungen fiber die schgdlichen Einwirkungen des Hiitten – und Steinkohlenrauches auf das Wachstum der Pflanzen, insbesondere der Fichten und Tannen. Tharandt forstl Jb 21:218–254

Sukumaran D (2014) Effect of air pollution on the anatomy some tropical plants. Appl Ecol Environ Sci 2(1):32–36

Tracy M, Freeman DC, Emlen J, Graham JH, Hough RA (1995) Developmental instability as a biomonitor of environmental stress. In: Butterworth FM, Corkum LD, Guzmán-Rincón J (eds) Biomonitors and biomarkers as indicators of environmental change. Plenum Press, New York, pp 313–416

Velickovic M, Perisic S (2006) Leaf fluctuating asymmetry of common plantain as an indicator of habitat quality. Plant Biosyst 140(2):138–145

Velikova V, Yardanov I, Edreva A (2000) Oxidative stress and some antioxidant systems in acid rain-treated bean plants. Plant Sci 151:59–66

Viskari EL (2000) Epicuticular wax of Norway spruce needles as indicator of traffic pollutant deposition. Water Air Soil Pollut 121:327–337

Vollenweider P, Ottiger M, Günthardt-Goerg MS (2003) Validation of leaf ozone symptoms in natural vegetation using microscopical methods. Environ Pollut 124:101–118

Vollenweider P, Cosio C, Günthardt-Goerg MS, Keller C (2006) Localization and effects of cadmium in leaves of a cadmium-tolerant willow (*Salix viminalis* L.): part II microlocalization and cellular effects of cadmium. Environ Exp Bot 58(1–3):25–40

Vollenweider P, Bytnerowicz A, Fenn M, Menard T, Günthardt-Goerg MS (2008) Structural changes in Ponderosa pine needles exposed to high ozone concentrations in the San Bernardino Mountains near Los Angeles, CA. [Abstract]. In: Schaub M, Kaennel Dobbertin M, Steiner D (eds) Air pollution and climate change at contrasting altitude and latitude. 23rd IUFRO conference for specialists in air pollution and climate change effects on forest ecosystems, Murten, Switzerland

Wild A, Schmitt V (1995) Diagnosis of damage to Norway spruce (Picea abies) through biochemical criteria. Physiol Plant 93:375–382

Wolf H (2003) EUFORGEN technical guidelines for genetic conservation and use for silver fir (Abies alba). International Plant Genetic Resources Institute, Rome. http://www.euforgen.org/fileadmin/bioversity/publications/pdfs/925.pdf

Biomonitoring and Remediation by Plants

11

Gyan Prakash Gupta and Umesh Kulshrestha

Abstract

Dealing with environmental pollution promises to be one of man's most urgent problems in the years to come. This chapter deals with different components of air pollution biomonitoring and their remediation by using different plant species of herbs, shrubs, and trees as green technology. Various methods of biomonitoring apply the whole or part of an organism to measure the exposure of a plant as well as accumulation of a pollutant. They have the great advantage to show clearly the effects of air pollutants as bioindicator plants. Bioindicators can reveal the impact and the cumulative effects of different pollutants. Phytoremediation is a set of processes such as rhizodegradation, phytostabilization, phytofiltration, phytoextraction, phytodegradation, and phytovolatilization. Through these processes plants remediate the pollutants, partially and sustainably from the atmosphere. Atmospheric gases (NO_2, SO_2, O_3, etc.), heavy metals, and VOC pollutants are reduced by absorbing and metabolizing them into less toxic compounds by site-specific plants or through the changes in the plant genome by overexpression of pollution-fighting genes through genetic engineering.

Keywords

Biomonitoring • Phytoremediation • Air pollution • Bioindicator

11.1 Introduction

Urban air pollution is one of the important environmental concerns worldwide. Atmospheric aerosols as well as gaseous pollutants such as SO_2, NO_2, CO, O_3, etc. pose severe health effects both for humans and plants (Maatoug 2010).

G.P. Gupta • U. Kulshrestha (✉)
School of Environmental Sciences, Jawaharlal Nehru University, New Delhi 110067, India
e-mail: umeshkulshrestha@gmail.com

U. Kulshrestha, P. Saxena (eds.), *Plant Responses to Air Pollution*,
DOI 10.1007/978-981-10-1201-3_11

Certain nonradioactive metals such as mercury (Hg), zinc (Zn), and arsenic (As) and radioactive metals such as strontium (Sr) and cesium (Cs) also have toxic effects for health and the environment. Plants are considered a very good responder for selected air pollutants. The science of air pollutant monitoring through plants is known as biomonitoring. In its historical outlook, biomonitoring is a very old technique to study air pollution. Long ago in the seventeenth century, John Evelyn, in his book *Fumifugium* which was published in 1661, first time mentioned about air pollution damage to vegetation. In the nineteenth century, Nylander (1866) used lichen population as air pollution indicator. During the past century in 1958, biological indicators were used for the same purpose in the Los Angeles basin in the USA. Later, Heck (1966) and Heggestad and Darley (1969) reported air pollution effects on the tobacco plants in California. In the Netherlands, Van (1969) used plant indicators to understand the effect of SO_2 and HF. However, biomonitoring received attention after the work of Schonbeck et al. (1970) who reported certain unique characteristics of biological indicators. These physiological and biological features of plants gather information related to impact of air pollution which cannot be accessed through chemical methods of air pollution monitoring.

In biomonitoring through plants, the changes occurring in plant morphology and physiology or biochemical changes are used as the indicators of pollution impact. Generally, we use passive and active methods to observe the changes and response of plants. In the passive method, plant growth is seen as a natural process of development. In the active method, the presence of different air pollutants is determined by plantation of an experimental plant for its response and genotype at a location in which plant response is assessed indirectly by recording biochemical and physiological changes in the plant. For example, parameters such as foliar injuries, stomatal pore size, chlorophyll content, etc. are used to observe the response of a bioindicator against air pollutants. Bioindicators can reveal the impact and the cumulative effects of different pollutants. The statement of Tingey (1989) justifies the importance of a bioindicator in an appropriate manner. He says "There is no better indicator of the status of a species or a system than a species or system itself." The biomonitor can be used to monitor the effects of air pollutants with the changes in the temporal and spatial variations. However, while using biomonitoring, their standardization method is very sensitive especially for the development of air quality standards to prevent harmful effects on the plant health and ecosystem.

Since plants interact with air and air pollutants, their role becomes crucial because such interactions can alter the atmospheric environment including local meteorology and pollutant levels. Hence, the plants can be used as a phytoremediator of air pollutants which is a very useful tool for purifying air in urban areas. Both gaseous and particulate pollutants can be removed by the plants. It has been found that some plant species are useful to monitor a single pollutant as they are sensitive to a specific pollutant, while certain species are useful to monitor mixtures of pollutants.

11.1.1 Significance of Biomonitoring

The biomonitoring of air pollution by plants offers some important results from different abilities:

1. Biomonitoring provides very important information about the effects of pollution as these effects are exhibited as visible injuries in sensitive plants while as an accumulated pollutant in less sensitive species. Even tolerant plants also accumulate pollutants which can be used as indicators.
2. Sometimes, very low concentrations of air pollution are difficult to measure with chemical and physical methods, but plants can accumulate such pollutants to a level which can easily be analyzed.
3. Some plants show an integrated response against the pollution and climate; the risk potential of particular pollutants or the mixture of pollutants can be estimated more realistically.

4. Biomonitoring becomes highly useful in monitoring different levels of the organization of plants ranging from the individual plant (or even single cell to leaf) to the plant cluster and up to the ecosystem level e.g. during the shift in species composition at the community level is a result of an integration of different factors over a longer period experienced by plant species; response is estimated more realistically by a biomonitor rather than any physicochemical method.
5. Unlike physicochemical monitoring, biomonitoring offers monitoring of a large-scale pattern of pollutant distribution and temporal changes without involving maintenance cost. Many of these attributes render biomonitoring as being suitable for both developing and developed countries.

11.1.2 Definition of Biomonitoring and Terminology

Biomonitoring involves several different terminologies, such as bioindicators, biosensors, bioaccumulators, and biointegrators, which have been described below:

Bioindicators: The individual plants which have visible symptoms, e.g., chlorosis, necrosis, and disturbed physiology, are known as bioindicators. These provide information about the quality of the environmental conditions.

Biosensors: These are those plants which respond to the presence of air pollutants; these are also known as biomarkers. These plants have nonvisible effects. Detection of effects at molecular, cellular, biochemical, or physiological levels needs microscopic and physiological techniques, as well as biochemical analysis.

Bioaccumulators: These plants which can accumulate air pollutants such as aerosols, dust, and gaseous molecules into their tissues are called bioaccumulators. These are also known as accumulative indicators.

Ecological indicators: It is a slightly different category than the above described. The concept of an ecological indicator is related to the population and community loss of plants. Plants related to these categories are also known as biointegrators. Ecological indicators highlight the changes in composition of the species in an ecosystem along with their appearance and disappearance and the variation in their density in a given area.

11.2 Biomonitoring by the Deposition/Accumulation of Air Pollutants in Plant Tissue

Various methods of biomonitoring apply the whole or part of an organism to measure exposure of a plant as well as accumulation of a pollutant. Based on the specific response of plants against pollutants, it can be classified as a good bioindicator, while tolerant species are generally used as a bioaccumulator. If the rate of clearance of a pollutant from the organism is known, we can do direct quantitative assessment of exposure by analyzing plant tissues. Various known methods have been used to assess the effects of pollutants such as:

11.2.1 Biomonitoring of Nitrogen Deposition

Biomonitoring has been found a very useful technique for the measurement of foliar nitrogen, atmospheric NH_3, and total nitrogen deposition. Bobbink et al. (1993) measured total tissue nitrogen (N) for many years in all types of plant tissues to find out atmospheric nitrogen deposition. Pitcairn et al. (1998) found this method of assessment very useful for tissue nitrogen content. They observed that the foliar nitrogen concentrations decreased with increasing distance from the livestock buildings. These workers noticed that there was a close relationship between foliar N, atmospheric NH_3 concentrations, and total N deposition for each selected moss, herb, and tree species.

11.2.2 Biomonitoring for Sulfur Deposition

The plants absorb sulfur dioxide (SO_2) from the atmosphere primarily through their leaves (Laisk et al. 1988). This can be an alternative source for normal growth in the situations where soil S is low. However, excess sulfur from the atmosphere might have an adverse impact on plant growth when S in soil is present in an adequate quantity. Manninen and Huttunen (1995) found that young Scots pine needles accumulated S proportionally to the ambient SO_2 load. Haapala et al. (1996) found that S content in pine needles was observed decreasing with the increasing distance from the known pollutant source. Sulfur content of needle tissues of Sitka and Norway spruce was observed to be higher corresponding to higher dry deposition of sulfur (Innes 1995).

11.2.3 Biomonitoring for Dust Deposition

Biomonitoring can be used for estimating dust deposition. Gupta et al. (2015b, c) have reported the average dustfall deposition fluxes on arjun leaves at two different sites. At the urban site, they recorded dust fluxes at 57±3, 120±5, and 117±7 mg/m²/day during monsoon, winter, and summer seasons, respectively, while at the industrial site, these workers found dustfall fluxes at 151±4, 286±6, and 259±8 mg/m²/day during monsoon, winter, and summer seasons, respectively.

11.2.4 Biomonitoring of the Chemical Component of Dust

Once collected on the leaves, the dustfall can be analyzed for the ionic content process in a water-soluble fraction. Fluxes of ions such as SO_4^{2-} and NO_3^- can be estimated as discussed below:

11.2.4.1 Dustfall Fluxes of SO_4^{2-} and NO_3^- on Foliar Surfaces

As mentioned earlier, biomonitoring of ionic species in the aqueous extract of dustfall recorded higher fluxes of (SO_4^{2-}+NO_3^-) as reported by Gupta et al. 2015a at an industrial site in Delhi which have been attributed to the local emissions. Chemical analysis of dustfall can provide information about the deposition rates; covariation of ionic species is important to know their sources and transformations. The fluxes of both SO_4^{2-} and NO_3^- were much lower at the urban background site as compared to the industrial site (Table 11.1). Table 11.1 gives the dustfall fluxes SO_4^{2-} and NO_3^-. The deposition of both the ionic species is in accordance with the ambient concentration of SO_4^{2-} and SO_2 for SO_4^{2-} fluxes while with NO_3^- and NO_2 for NO_3^- fluxes, respectively.

Their seasonal levels also varied according to the abundance of respective aerosols and gaseous species in ambient air. As given in Table 11.1, SO_4^{2-} fluxes were much higher than the NO_3^- fluxes. The reasons for lower NO_3^- fluxes may be due to formation of nitrate (NO_3^-) or nitrite (NO_2^-) after NO_2 entry through the epidermal layer and the substomatal chamber reaching the

Table 11.1 Dustfall fluxes of SO_4^{2-} and NO_3^- on arjun foliar surface and SO_4^{2-} and NO_3^- aerosol and ambient SO_2 and NO_2 concentrations at industrial and urban sites

		Dustfall flux of SO_4^{2-} (mg/m²/d) on foliar surfaces	Dustfall flux of NO_3^- (mg/m²/d) on foliar surfaces	Ambient SO_4^{2-} aerosols (μg/m³)	Ambient NO_3^- aerosols (μg/m³)	Ambient SO_2 (μg/m³)	Ambient NO_2 (μg/m³)
Industrial	Monsoon	1.74	0.91	1.63	3.03	24	26
	Winter	3.29	1.73	4.08	6.34	31	38
	Summer	2.63	0.95	3.11	5.42	42	77
Urban	Monsoon	5.58	0.56	3.61	2.69	36	34
	Winter	15.99	0.30	14.43	8.92	44	53
	Summer	7.52	1.85	9.78	6.91	64	115

mesophyll cell and reacting with the mesophyllic cell wall of the ascorbic acid (Ramge et al. 1993). In order to fight against stress and to show increased tolerance, ascorbic acid production was higher at the industrial site than that of the urban background site which may be due to enhanced oxidative stress for activity in situ. In this process, reaction to ascorbic acid plays an important role. Due to this reason, the plants having higher levels of ascorbic acid in their leaves show a higher rate of NO_2 uptake (Teklemariam and Sparks 2006) which may result in higher NO_3^- and NO_2^- accumulation in cells producing greater acidity in the leaves. As reported by Gupta et al. (2015a, b, c), higher fluxes of SO_4^{2-} may be due to the dry deposited SO_2 and its oxidation of SO_4^{2-} in a polluted and dusty atmosphere (He et al. 2014; Kulshrestha et al. 2003; Seinfeld and Pandis 1998). In addition, SO_2 adsorption onto the dust particles (already settled on the surface of leaves) can also react in the formation of $CaSO_4$ (Gupta et al. 2015a). $CaCO_3$ rich soil dust in India significantly scavenges atmospheric SO_2 in the form of $CaSO_4$ (Kulshrestha et al. 2003, Kulshrestha 2013). Furthermore, the oxidation of SO_2 is catalyzed by NO_2 forming SO_4^{2-} formation (He et al. 2014). High SO_4^{2-} formation due to NO_3^- present on foliar surface which enhances the hygroscopicity of mineral particles on the leaves increases the possibility of SO_2 oxidation. As reported by Singh (2014), gas/aerosol partitioning coefficients were favoring NO_2 existence in the gas phase as compared to the aerosol phase (NO_3^-) with respect to SO_2/SO_4^{2-} existence.

11.2.5 Biomonitoring of Deposition of Heavy Metals

Mosses and lichens have been commonly used as biomonitoring species for metal pollution. Some workers have found that *Lolium multiflorum* ssp. *italicum* cv. "Lemma" is a very good biomonitor for trace elements, fluoride, sulfur, and organics at urban sites in Europe as this species grows faster and provides reliable and quick information (Rodriguez et al. 2010; Klumpp et al. 2009). However, higher plant species are also good biomonitors for selected trace elements.

Heavy metals, such as Pb, Zn, and Cu, were found to be accumulated in plant leaves of two urban plant species, viz., plane and cypress, which were used for air quality assessment in Tiaret city in Algeria (Maatoug 2010). Results showed that the ratio of fresh weight to dry weight (FM/DM) of leaves is affected by trace metal concentrations. The study showed that FM/DM ratio had an inverse relationship with metal concentrations. Experimentally observed lower ratios were attributed to poor air quality near the site of investigation.

11.2.6 Biomonitoring of Gaseous Pollutants Due To Changes Through Physiological and Biochemical Changes in Plants

11.2.6.1 Monitoring of Crops, Vegetables, and Trees Damaged by Air Pollution

Air pollution may damage the crops, vegetables, and trees in various manners. Agricultural productivity in urban areas has been found low due to high emissions and concentrations of SO_2, NO_2, and O_3 (Rai et al. 2011; Nandy et al. 2014). Air pollution effects can be seen as visible symptoms where the plant has acute injury. Even the growth response effect in some vegetables by air pollution has been reported by Ghouse and Khan 1984. Ashmore and Marshall (1999) found that the agriculture crop was severely injured by ozone. Dhir et al. (1999) found that the crop of *Achyranthes aspera* Linn. exhibited air pollution effects on morphology and some physiology of the plant. It was also reported that air pollution affects the growth and reproductive behavior of mustard plants (Saquib and Khan 1999).

Coal-smoke pollution was found to affect the stomatal conductance, photosynthetic rate, and pigment content in *Ruellia tuberosa* (Nighat and Mahmooduzzafar 2000). Acute injury was noticed due to SO_2 showing bifacial chlorosis and marginal and interveinal necrosis on leaves (Legge and Krupa 2002). Evidence of oxidative stress was found to affect the plant physiology which exhibited spots on barley leaves during spring and winter seasons (Wu and Tiedmann 2002). Elevated ozone exposure was found to affect the photosynthesis, growth, and productivity of soybean (Morgan et al. 2003). Severe effects of O_3 may result in death of plant organs. It also affects plant growth leading to a low yield (Ashmore 2005).

Air pollution effect on yield and quality of mung beans has been reported by Agrawal et al. (2006) for peri-urban areas of Varanasi in India. Ambient air pollution was found to reduce the yield by 43 %, 39 %, and 18 % in three wheat cultivars, respectively, at mean concentrations of 70, 28, and 15 ppb of O_3, NO_2, and SO_2, respectively, during different seasons at Lahore in Pakistan (Wahid 2006). At ambient levels of SO_2 (7.8 ppb), NO_2 (40.6 ppb), and O_3 (42.1 ppb), Rai et al. (2007) noted 20.7 % reduction in wheat yield in Varanasi, India. According to Heath (2008), O_3 generates ROS which causes damage to plants. The initial site of injury is the plasma membrane due to which permeability and fluidity of cells are changed. During changing of climate, elevated levels of CO_2 and O_3 affect the rice production (Ainsworth 2008). Higher air pollution load (SO_2 6.5 ppb and NO_2 9 ppb) caused reduction in wheat and mustard crop yield in Haridwar, India, as reported by Chauhan and Joshi (2010).

Gupta et al. (2015a, b, c) have reported a significant impact on biochemical constituents such as chlorophyll, carotenoids and soluble sugar, ascorbic acid, and proline on Arjun (*Terminalia arjuna*) and Morus (*Morus alba*) plants due to the urban dust deposition. They reported these changes in biochemical constituents in accordance with the stress level at two sites. The biochemical changes were noticed more prominently at the industrial site as compared to the urban background site due the greater oxidative stress caused due to higher accumulation of dustfall, particulate SO_4^{2-}, etc. at the industrial site.

11.2.6.2 Biomonitoring of Dust Particles Through Foliar Surface Morphology Observation

As mentioned earlier, both gaseous and particulate pollutants affect physiological and morphological characteristics of plants (Rai et al. 2010). According to Kim et al. (2000) and Chaturvedi et al. (2013), sedimentation of coarse particles has more impact on the upper surfaces of leaves. Deposition of dust particles has been noticed more at the industrial site than the residential site (Gupta et al. 2015a, b, c), whereas the finer particles have more impact on lower surfaces of leaves (Fowler et al. 1989; Beckett et al. 2000). Larger-sized particles pile up on the foliar surfaces, while the smaller particles enter through the stomata affecting the photosynthesis, gaseous exchange, and water retention further affecting the growth and yield of plants (Tomasevic and Anicic 2010; Rai et al. 2010). Dust particle deposition can cause reduced stomatal diffusive resistance and an increased leaf temperature (Fluckinger et al. 1979). More damage in the upper and lower epidermis and reduction in palisade parenchyma cells have been reported by Stevovic et al. (2010) at a polluted site as compared to a nonpolluted site. They found that erosion of the epicuticular wax and cuticle rupture were more frequent on the adaxial side, whereas loss of sinuosity on the anticlinal wall of the epidermal cells and stomatal deformity and obstruction were seen on the abaxial side of the leaves (Rocha et al. 2014).

The foliar surface morphological analysis verifies the deposition of industrial and road dust which effects damage in the cuticle and epidermal layer at adaxial site and at abaxial site damages in guard cells and stomatal pores. The abrasive effect of dust deposition restructures the cuticle and affects the epicuticular wax. Dust depositions also rupture epidermal hairs along with swollen guard cells (Gupta et al. 2015a, b, c). Also, dust deposition causes changes in the size of the stomata. Gupta et al. 2015a, b, c reported that the size of the stomatal pore was smaller at the industrial site (9.52 × 2.43 μm), whereas it was larger (13.5 × 4.74 μm) at the urban site. Even SO_2 and NO_2 cause wax degradation and surface needle erosion (Grodzinska-Jurczak and SzarekLukaszewska 1999). The length (L), breadth (B), and L/B ratio of leaves are significantly changing according to the automobile air pollution scenario at the site (Nandy et al. 2014). They observed pigmentation, chlorosis, and necrosis symptoms in roadside plant species such as *Alstonia scholaris*, *Neolamarckia cadamba*, *Ficus benghalensis*, and *Ficus religiosa*.

11.3 Phytoremediation of Atmospheric Pollutants

The process of cleaning up the environment using plants is called phytoremediation. Phytoremediation has its origin from the Greek word **phyto**, which means "plant," and the Latin word **remediare**, which means "to remedy." This term is usually used to describe the removal of contaminants by using plants from any system, i.e., air, water, soil, etc. (Chhotu and Fulekar 2009). When it comes to remediation of pollutants, plants have always been considered as effective removers of pollutants for soil, water, and air. Plant-based remediation is highly low cost and is efficient in the removal of pollutants/contaminants. Interestingly, as transgenic are being tested in the field and the associated risks assessed, their use appears to be more accepted and less regulated than has been the case for transgenic crops. Approaches such as molecular, biochemical, and physiological are being applied to identify the plant mechanisms by which they accumulate gaseous, particulate matter and metal aerosols. They have also exerted effort to use transgenic plants for phytoremediation in order to find out genes responsible for this unique biological property. During the two decades, phytoremediation work has received great attention from researchers worldwide.

Among various plants, trees are having superiority for air filtration as compared to shrubs and grasses. Due to this probability, trees have been termed as the "lungs of cities." Removal of air pollution increases with leaf surface area, and hence evergreen trees are considered as highly effective pollution filters as the evergreen trees have a large surface area and year-round coverage. Therefore, trees tend to be better filters than shrubs and grasses. Several species of ornamental shrubs as well as herbaceous plants have been identified as phytoremediators to purify the indoor and outdoor air quality.

11.3.1 Principles of Phytoremediation

Phytoremediation is based on rhizodegradation, phytostabilization, phytofiltration, phytoextraction, phytodegradation, and phytovolatilization (Morikawa and Erkin 2003) (Fig. 11.1).

Phytostabilization: it is the process in which contaminants are sequestrated in the root zone.

Rhizodegradation: it is the process of degradation of pollutants by root secretions and rhizospheric microorganisms. The rhizosphere has more densities of microorganisms in a narrow region of 1–3 mm from root surface than in bulk soil.

Phytoextraction: it involves extraction of pollutants from soil and their translocation from the roots to shoots.

Phytodegradation: it is the process of degrading the pollutants which entered from the soil, water, or air to the plant.

Phytovolatilization: in this process, pollutants are volatiled out of the stoma where they are degraded by hydroxyl radicals.

Phytofiltration: it is the process of removal of particulate matters through the surface of plants.

11.3.2 Two-Compartment Model for Phytoremediation

In order to explain atmosphere-foliage bioaccumulation, a two-compartment model has been used (Keymeulen et al. 1995; Mackay et al. 2006). This model considers a relatively fast initial uptake followed by a period of a slower uptake. In this model, the leaf is divided into two compartments. The first compartment has the nonliving plant cuticle. In the first compartment, occurrence of physicochemical sorption of airborne lipophilic compounds takes place, while the second compartment has the leaf interior. In the second compartment, sorption and metabolism of organic compounds take place.

The final fate and disposition of organic pollutants (including xenobiotic compounds) within plants have been explained by using the "green liver" model. The process of detoxification

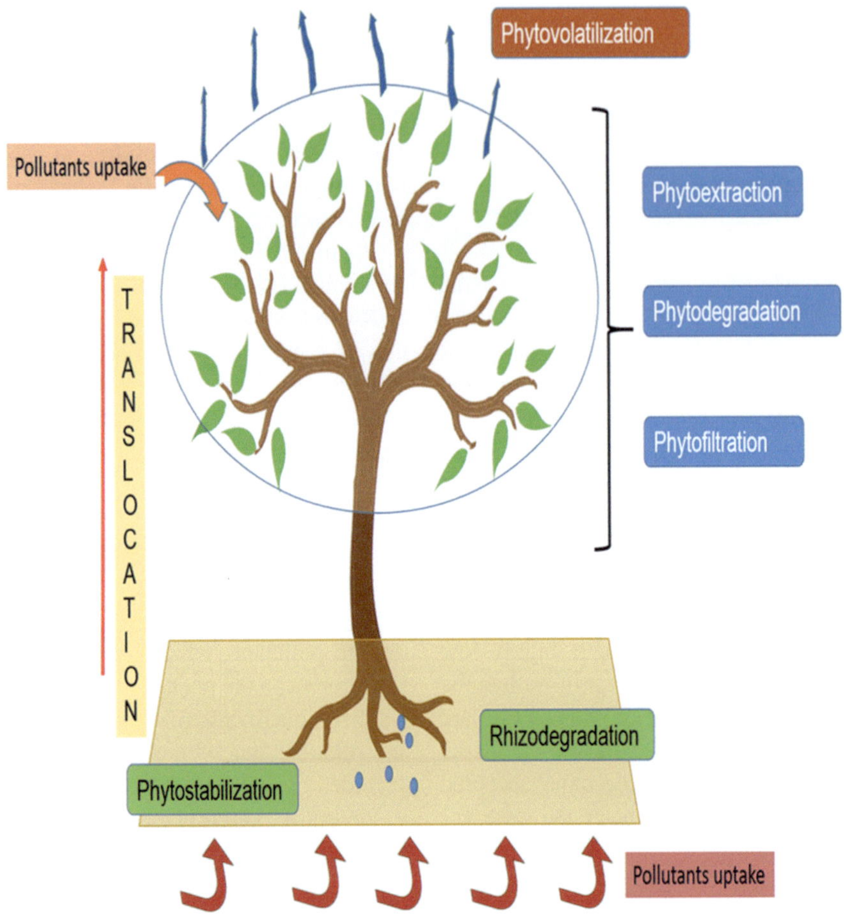

Fig. 11.1 Different parts of plants are responsible for different processes of phytoremediation

metabolism of foreign compounds in plants is very similar to the metabolism of xenobiotic compounds in the human liver system, and, hence, the term "green liver" is used in this model (Burken et al. 2000).

11.3.3 Properties of Air Phytoremediators: Ornamental Plants and Woody Trees

The relevant criteria for the ornamental and woody trees to be phytoremediators for air pollution are they should be evergreen trees having broad leaves, very rough barks, low water requirement, ecological compatibility, high capacity for absorption of pollutants, minimum care needs, agro-climatic suitability, and good height and spread. The canopy of these plants is suitable for an aesthetic effect. Besides their attractive flower color and conspicuous foliage, tolerance and dust removal capacity also make these trees effective phytoremediators of air pollution (Brethour et al. 2007; Kumar et al. 2013).

Generally, ornamental and woody trees are effectively used as phytoremediator tools for reducing air pollution (Zhai 2011). These types of plants are part of the greenbelt in urban areas and can be used for phytoremediation of air pollutants. Such plants remediate air pollution via absorption and degradation of urban air pollutants resulting in reduced levels of air pollutants (Brown 1997; Yoneyama et al. 2002; Burchett

Table 11.2 Average air pollution removal and value for all urban trees in the USA

Pollutant	Removal (metric tons)	Value (million US$)
Ozone (O_3)	305,100	2060
Particulate matter (PM_{10})	2014,900	969
Nitrogen dioxide (NO_2)	97,800	660
Sulfur dioxide (SO_2)	7090	117
Carbon monoxide (CO)	22,600	22
Total	**711,300**	**3828**

Nowak et al. (2006)

et al. 2008). In general, trees are ideal in the remediation of heavy metals because they can withstand even after an accumulation of a higher amount of pollutants (Shah and Nongkynrih 2007). Some fast-growing trees, e.g., eucalyptus, pine, and poplar, are found to be very effective in air pollution removal. Also hardwood trees, e.g., rosewood and mahogany, are used for remediation of polluted air. According to a study, around 711,000 metric tons of air pollution is removed by urban trees every year in the USA (Nowak et al. 2006). Table 11.2 gives the estimated removal of specific pollutants by trees (Nowak et al. 2006).

11.3.4 Examples of Phytoremediation of Various Air Pollutants

11.3.4.1 Particulate Matter

Vegetation canopies effectively protect and remediate the dust of the atmosphere this process is called phytofiltration. According to the studies, an 8 meter wide greenbelt may reduce two to three times of dustfall (Novoderzhikina et al. 1966). Phytofiltration becomes more efficient if the vegetation is selected based on suitable morphological characteristics such as leaf orientation on the main axis, leaf size and shape, leaf surface nature, the presence or absence of trichomes on the leaf, and wax deposition on the leaf which help in capturing atmospheric dust.

11.3.4.2 Heavy Metals

As trees are good bioaccumulators, they can remove and store a huge amount of heavy metals. In a study, concentrations of five metals (Cu, Cd, Cr, Pb, and Ni) were determined in tree leaves. These were collected from 13 different areas in Greece. Geographical distribution patterns were investigated, and factors affecting toxic element accumulation in trees were discussed (Sawidis et al. 2012). The researcher found that the order of average heavy metal content in the tree leaves was Cu>Pb>Ni>Cr>Cd. High Cr, Cu, and Ni are found in *Citrus aurantium* leaves which are probably because of stomatal uptake. The conifer tree *Pinus brutia* which has a rough leaf surface showed high accumulation of Cd and Pb. Broad-leaved *Olea europaea* which has a thick waxy cuticle forms a smooth sheet due to which entry of metals is restricted through the epidermis. The trichomes also act as protective screen factors to keep away air pollutants. Sometimes, even the presence of a certain metal within the foliar cells can reduce the uptake or toxicity of other metals (Sawidis et al. 2012).

11.3.4.3 Inorganic Pollutants

NO_X and particulate N

Air pollution of NO_X is caused by vehicular exhaust, industries, and biomass burning. NO_X is one of the precursors of photochemical reactions. NO_X can deposit on plant leaves through wet or dry deposition. During the dry deposition of both gases and partciles on the leaves surafecs NH_4^+ and NO_3^- particulates are also deposited (Davidson and Wu 1990; Bobbink et al. 1992). Most of NO_2 enters into plants through stomata where it is metabolized to organic compounds such as amino acid. In this process, plant enzymes like nitrate and nitrite reductase or glutamine synthetase play an important role (Davies, 1986; Allen, 1988).

SO_2

Most of SO_2 in the atmosphere is emitted by fossil fuel combustion. Atmospheric SO_2 enters into plants through the stoma where it is utilized as nutrient. SO_2 is changed into SO_4^{2-}/SO_3^{2-} in cell walls which finally produce cysteine or other organic compounds; if the SO_2 concentration is beyond limits, it causes severe acidity for plants and generates stress conditions.

11.3.4.4 Organic Pollutants

Plants have been reported to reduce the ambient organic pollutants such as formaldehyde, benzene, and toluene. Formaldehyde is a ubiquitous air pollutant removed by plant leaves as reported by NASA's research in the 1980s. It is found that spider plants (*Chlorophytum comosum* L.) metabolize formaldehyde in shoots in the form of organic acids, free sugars, amino acids, and lipids (Giese et al. 1994). Godish and Guindon (1989) reported that up to 50% of formaldehyde is removed by spider plants.

Benzene and toluene can also be removed from the ambient air by plants (Ugrekhelidze 1997). In a test chamber study, *Opuntia microdasys* was found to remove 2 ppm of toluene from air completely in 55 h. *D. deremensis* took 120 h to remove toluene from the air in the test chambers. In a study reported by Mosaddegh et al. (2014), toluene in test chambers was removed at 1.47 and 0.67 mg/m^3 day^1 rate for *Opuntia microdasys* and *D. deremensis*, respectively.

Yang et al. (2009) found that *Asparagus densiflorus*, *Hedera helix*, *Hemigraphis alternata*, and *Hoya carnosa* had the greatest removal efficiencies among 28 species tested. He found that *Tradescantia pallida* was superior in removing benzene, toluene, TCE, and α-pinene VOCs. *Ficus benjamina* was found as an effective remover of octane and α-pinene from air.

Among the tested plants, *Chlorophytum comosum* was superior in removing HCHO and SOx from the air at 1830 μg day^{-1} and 2120 μg day^{-1} rate, whereas *Spathiphyllum wallisii* for NOx was effective at 3200 μg day^{-1} rate (El-Sadek et al. 2012). Based on the air pollution tolerance index (APTI), Nugrahani et al. (2012) have identified the species of landscape ornamental and herbaceous shrubs which can be used as bioindicators of urban air pollution. These are *Mussaenda philippica*, *Heliconia psittacorum*, *Ipomoea batatas*, *Jatropha pandurifolia*, *Bougainvillea* sp., *Hymenocallis speciosa*, *Codiaeum variegatum*, *Cordyline terminalis*, *Canna indica*, and *Sansevieria trifasciata*.

Prescod (1990) suggested that orchids can be used during the daylight hours for efficient removal of multiple pollutants. According to his report, carbon dioxide and xylene (Anonymous 2007) are also effectively removed by plants during the night due to a unique metabolic process of orchids (and bromeliads) because their stomata (Ibrahim et al. 2008) open during nighttime.

11.4 Applications of Genetic Engineering for Phytoremediation

11.4.1 Transgenic Plants for NO_2 and VOC Pollution Control

Genetically engineered plants can provide adequate sinks of air pollutants is an important to develop. These plants are termed as "wonder plants" which can clean up the specific pollutants from the atmosphere. Primary metabolism of nitrate involves enzymes such as nitrate reductase (NR), nitrite reductase (NiR), and glutamine synthetase (GS) which play a key role in NO_2-nitrogen metabolism in the plants. All genes for NR, NiR, and GS are nuclear encoded. Takahashi et al. (2001) found positive correlations for NiR gene overexpression and NO_2 assimilation in transgenic *Arabidopsis* plants. Doty et al. (2007) developed transgenic poplar (*Populus tremula* × *Populus alba*) plants with greatly increased rates of metabolism for the removal of VOC pollutants from the atmosphere through the overexpression of the cytochrome P450 2E1 enzyme.

11.4.2 Gas-Gas-Converting Plants That Convert Nitrogen Dioxide to Gaseous Nitrogen

Genetically engineered plants can convert NO_2 to N_2O or to N_2 (gas-gas-converting plants). Generally, NO_3^- or (NO_2^-) is converted to N_2 or N_2O in denitrification process through denitrifier bacteria and fungi, respectively (Shoun et al. 1992; Zumft 1997). Goshima et al. (1999) have demonstrated transgenic tobacco showed reduced NiR activity by the expression of NiR cDNA in an antisense orientation and emitted N_2O when fed with 15N-labeled nitrate and nitrite (Vaucheret et al. 1992; Takahashi et al. 2001). This invention of a gas-gas-converting plant that converts NO_2, via N_2O, to N_2 will be a highly useful approach for air pollutant control.

11.5 Conclusion

Biomonitoring has become a complement of traditional techniques of air quality measurements. The identification of pollution within sensitive organisms also allows detection and degradation of air quality before the biota is severely affected. In this context, sensitive plants can be considered as real bioindicators of pollution stress generated by gases as well as particulate matter. Phytoremediation of air pollution is proved to be useful in reducing air, water, and soil pollutants. This technique is cost effective as compared to physicochemical methods. Since it is a natural phenomenon, it does not require energy supply. Plantation increases the aesthetic value of the environment too. This approach needs to be included in urban town planning as a mandatory component which will help in improving air quality of megacities in the future.

References

Agrawal M, Singh B, Agrawal SB, Bell JNB, Marshall F (2006) The effect of air pollution on yield and quality of mung bean grown in peri-urban areas of Varanasi. Water Air Soil Pollut 169:239–254

Ainsworth EA (2008) Rice production in a changing climate: a meta-analysis of responses to elevated carbon dioxide and elevated ozone concentration. Glob Chang Biol 14:1642–1650

Allen S, Raven JA, Sprent JI (1988) The role of long-distance transport in intracellular pH regulation in Phaseolus vulgaris growth with ammonium or nitrate as nitrogen source, or nodulated. J Exp Bot 39:513–528

Anonymous (2007) Air quality strategy for England, Scotland, Wales and Northern Ireland. http://www.official-documents.gov.uk/document/cm71/7169/7169

Ashmore MR (2005) Assessing the future global impacts of ozone on vegetation. Plant Cell Environ 28:949–964

Ashmore MR, Marshall FM (1999) Ozone impacts on agriculture: an issue of global concern. Adv Bot Res 29:32–49

Beckett KP, Freer-Smith PH, Taylor G (2000) Particulate pollution capture by urban trees: effect of species and windspeed. Glob Chang Biol 6:995–1003

Bobbink R, Heil GW, Raessen MBAG (1992) Atmospheric deposition and canopy exchange processes in heathland ecosystems. Environ Pollut 75:29–41

Bobbink R, Boxman D, Fremstad E, Heil G, Houdijk A, Roelofs J (1993) Nitrogen eutrophication and critical load for nitrogen based upon changes in flora and fauna in semi-natural terrestrial ecosystems. In: Critical loads for nitrogen, Proceedings of a UN-ECE workshop at Lo¨keberg, Sweden. 6–10 April 1992, Nordic Council of Ministers, Copenhagen, Denmark, pp 111–159

Brethour C, Watson G, Sparling B, Bucknell D, Moore T (2007) Literature review of documented health and environmental benefits derived from ornamental horticulture products. Final report. George Morris Centre. http://www.Deenenlandscaping.Com/UserFile/file/Morris_Report.pdf

Brown SK (1997) Volatile organic compounds in indoor air: sources and control. Chem Aust 64:10–13

Burchett M, Torpy F, Tarran J (2008) Interior plants for sustainable facility ecology and workplace productivity. In: Proceeding of Ideaction'08–Enabling Sustainable Communities; Gold Coast, Qld, 2008

Burken JG, Shanks JV, Thomposn PL (2000) Phytoremediation and plant metabolism of explosives and nitroaromatic compounds. In: Spain JC, Hughes JB, Knackmuss HJ (eds) Biodegradation of nitroaromatic compounds and explosives. Lewis, Washington, DC, pp 239–275

Chaturvedi RK, Prasad S, Rana S, Obaidullah SM, Pandey V, Singh H (2013) Effect of dust load on the leaf attributes of the tree species growing along the roadside. Environ Monit Assess 185:383–391. doi:10.1007/s10661-012-2560-x

Chauhan A, Joshi PC (2010) Effect of ambient air pollutants on wheat and mustard crops growing in the vicinity of urban and industrial areas. N Y Sci J 3:52–60

Chhotu DJ, Fulekar MH (2009) Phytoremediation of heavy metals: recent techniques. Afr J Biotechnol 8:921–928

Davidson CI, Wu C-L (1990) Dry deposition of particles and vapors. In: Lindberg SE, Page AL, Norton SA (eds) Acidic precipitation. Springer, New York, pp 103–215

Dhir B, Sharma MP, Mahmooduzzafar, Iqbal M (1999) Form and function of Achyranthes aspera Linn. under air pollution stress. J Environ Biol 20:19–23

Davies DD (1986) The fine control of cytosolic pH. Physiol Plant 67:702–706
Doty SL, James CA, Moore AL, Vajzovic A, Singleton GL, Ma C, … Meilan R (2007) Enhanced phytoremediation of volatile environmental pollutants with transgenic trees. Proc. Natl Acad Sci 104:16816–16821
El-Sadek M, Koriesh E, Fujii E, Moghazy E, Abd Elfatah Y (2012) Correlation between some components of interior plants and their efficiency to reduce formaldehyde, nitrogen and sulfur oxides from indoor air. Int Res J Plant Sci 3:222–229
Fluckinger W, Oertli JJ, Fluckiger W (1979) Relationship between stomatal diffusive resistance and various applied particle sizes on leaf surface. Z Pflanzenphysiol 91:773–775
Fowler D, Cape JN, Unsworth MH, Crowther HMJM et al (1989) Deposition of atmospheric pollutants on forests. Philos Trans R Soc London B Biol Sci 324:247–265
Ghouse AKM, Khan FA (1984) Effect of air pollutants on the growth responses of Solanum nigrum L. Acta Bot Ind 12:93–94
Giese M, Bauer-Doranth U, Langebartels C, Sandermann H (1994) Detoxification of formaldehyde by the spider plant (Chlorophytum comosum L.) and by soybean (Glycine max L.) cell-suspension cultures. Plant Physiol 104:1301–1309
Godish Tand Guindon C (1989) An assessment of botanical air purification as a formaldehyde mitigation measure under dynamic laboratory chamber conditions. Environ Pollut 89(61):13–20
Goshima N, Mukai T, Suemori M, Takahashi M, Caboche M, Morikawa H (1999) Emission of nitrous oxide (N_2O) from transgenic tobacco expressing antisense NiR mRNA. Plant J 19:75–80
GrodzinskaJurczak M, SzarekLukaszewska G (1999) Evaluation of SO_2 and NO_2 related degradation of coniferous forest stands in Poland. Sci Total Environ 241:115
Gupta GP, Singh S, Kumar B, Kulshrestha UC (2015a) Industrial dust sulphate and its effects on biochemical and morphological characteristics of Morus (*Morus alba*) plant in NCR Delhi. Environ Monit Assess 187:67
Gupta GP, Kumar B, Singh S, Kulshrestha UC (2015b) Urban climate and its effect on biochemical and morphological characteristics of Arjun (*Terminalia arjuna*) plant in National Capital Region Delhi. Chem Ecol.http://dx.doi.org/10.1080/02757540.2015.1043286
Gupta GP, Kumar B, Singh S, Kulshrestha UC (2015c). Chemistry and impact of urban atmospheric dust on two medicinal plants during different seasons in NCR Delhi. Aero Air Qual Res-Index. 10.4209/aaqr.2015.04.0272
Haapala H, Goltsova N, Seppala R, Huttunen S, Kouki J, Lampp J, Popovichev B (1996) Ecological condition of forests around the eastern part of the Gulf of Finland. Environ Pollut 91:253–265
He H, Wang Y, Ma Q, Ma J, Chu B, Ji D, Tang G, Liu C, Zhang H, Hao J (2014) Mineral dust and NOx promote the conversion of SO_2 to sulfate in heavy pollution days. Sci Rep 4:4172. doi:10.1038/srep04172
Heath RL (2008) Modification of the biochemical pathways of plants induced by ozone: what are the varied route to changes? Environ Pollut 155:453–463
Heck WW (1966) The use of plants as indicators of air pollution. Air Water Pollut Intern 10:99–111
Heggestad HE, Darley EF (1969) Plants as indicators of the air pollutants ozone and PAN. In: Proceedings of the First European Congress on the influence of air pollution on plants and animals. Wageningen, 1968. Pudoc, Wageningen, pp 329–335
Ibrahim A, Abd Elaziz F, Toma Z, Zhang J (2008) Indoor air pollution and child health in a rural area in Egypt. Epidemiology 19:377
Innes JL (1995) Influence of air pollution on the foliar nutrition of conifers in Great Britain. Environ Pollut 88:183–192
Keymeulen R, Schamp N, Langenhove HV (1995) Uptake of gaseous toluene in plant leaves: a two compartment model. Chemotherapy 31:3961–3975
Kim E, Kalman D, Larson T (2000) Dry deposition of large, airborne particles onto a surrogate surface. Atmos Environ 34:2387–2397
Klumpp A, Ansel W, Klumpp G, Breuer J, Vergne P, Sanz MJ, … Calatayud V (2009) Airborne trace element pollution in 11 European cities assessed by exposure of standardised ryegrass cultures. Atmos Environ 43:329–339
Kulshrestha U (2013) Acid rain. In: Jorgensen SE (ed) Encyclopedia of environmental management. Taylor & Francis, New York, pp 8–22
Kulshrestha MJ, Kulshrestha UC, Parashar DC, Vairamani M (2003) Estimation of SO_4 contribution by dry deposition of SO_2 onto the dust particles in India. Atmos Environ 37:30573063
Kumar RS, Arumugam T, Anandakumar CR, Balakrishnan S, Rajavel DS (2013) Use of plant species in controlling environmental pollution-a review. Bull Env Pharmacol Life Sci 2:52–63
Laisk A, Pfanz H, Heber U (1988) Sulfur dioxide fluxes into different cellular compartments of leaves photosynthesizing in a polluted atmosphere. II Consequences of SO_2 uptake as revealed by computer analysis. Planta 173:241–252
Legge AH, Krupa SV (2002) Effects of sulphur dioxide. In: Bell JNB, Treshow M (eds) Air pollution and plant life. Wiley, West Sussex, pp 130–162
Maatoug M (2010) Cartographie de la pollution atmosphérique par le plomb d'origine routière à l'aide de transplantation d'un lichen bioaccumulateur *Xanthoria parietina* dans la ville de Tiaret (Algérie). Rev Pollut Atmos 205:93–101
Mackay D, Foster KL, Patwa Z, Webster E (2006) Chemical partitioning to foliage: the contribution and legacy of Davide Calamari. Environ Sci Pollut Res 30:786–791

Manninen S, Huttunen S (1995) Scots pine needles as bioindicators of sulphur deposition. Can J For Res 25:1559–1569

Morgan PB, Ainsworth EA, Long SP (2003) How does elevated ozone impact soybean? A meta-analysis of photosynthesis, growth and yield. Plant Cell Environ 26:1317–1328

Morikawa H, Erkin ÖC (2003) Basic processes in phytoremediation and some applications to air pollution control. Chemosphere 52:1553–1558

Mosaddegh MH, Jafarian A, Ghasemi A, Mosaddegh A (2014) Phytoremediation of benzene, toluene, ethylbenzene and xylene contaminated air by Dracaena deremensis and Opuntia microdasys plants. J Environ Health Sci Eng. doi:10.1186/2052-336X-12-39

Nandy A, Talapatra SN, Bhattacharjee P, Chaudhuri P, Mukhopadhyay A (2014) Assessment of morphological damages of leaves of selected plant species due to vehicular air pollution, Kolkata, India. Int Lett Nat Sci 9:76–91

Nighat F, Mahmooduzzafar MI (2000) Stomatal conductance, photosynthetic rate and pigment content in Ruellia tuberosa leaves as affected by coal-smoke pollution. Biol Plant 43:263–267

Novoderzhikina YG, Andrianova LA, Zheldakkova GG (1966) Effect of plantings on the sanitary and hygienic conditions of densely polluted settlement. In: Nuttonson M (ed) AICE survey of USSR, vol 2. American Institute of Cropecology, Silver Spring, pp 25–31

Nowak DJ, Craneand DE, Stevens JC (2006) Air pollution removal by urban trees and shrubs in the United States. Urban For Urban Green 4:115–123

Nugrahani P, Prasetyawati ET, Sugijanto PH (2012) Ornamental shrubs as plant palettes elements and bioindicators based on air pollution tolerance index in Surabaya city, Indonesia. Asian J Exp Biol Sci 3:298–302

Nylander W (1866) Les lichens du Jardin de Luxembourg. Bull Soc Bot France 13:364–371

Pitcairn CER, Leith ID, Sheppard LJ, Sutton MA, Fowler D, Munro RC, Tang S, Wilson D (1998) The relationship between nitrogen deposition, species composition and foliar nitrogen concentrations in woodland flora in the vicinity of livestock farms. Environ Pollut 102:41–48

Prescod AW (1990) Growing indoor plants as air purifiers. Pappus 9:13–20

Rai R, Agrawal M, Agrawal SB (2007) Assessment of yield losses in tropical wheat using open top chambers. Atmos Environ 41:9543–9554

Rai A, Kulshrestha K, Srivastava PK, Mohanty CS (2010) Leaf surface structure alterations due to particulate pollution in some common plants. Environment 30:18–23

Rai R, Rajput M, Agrawal M, Agrawal SB (2011) Gaseous air pollutants: a review on current and future trends of emissions and impact on agriculture. J Sci Res 55:77–102

Ramge P, Badeck FW, Plochl M, Kohlmaier GH (1993) Apoplastic antioxidants as decisive elimination factors within the uptake process of nitrogen dioxide into leaf tissues. New Phytol 125:771–785

Rocha DI, Luzimar CS, Eduardo GP, Bruno FS, Elisa RG, Marco AO (2014) Early detection of injuries in leaves of *clusia hilariana* schltdl. (clusiaceae) caused by particulate deposition of iron. Rev Árvore Viçosa MG 38:423–432

Rodriguez JH, Pignata ML, Fangmeier A, Klumpp A (2010) Accumulation of polycyclic aromatic hydrocarbons and trace elements in the bioindicator plants Tillandsia capillaris and Lolium multiflorum exposed at PM10 monitoring stations in Stuttgart (Germany). Chemosphere 80:208–215

Saquib M, Khan FA (1999) Air pollution impacts on the growth and reproductive behaviour of mustard. J Environ Biol 20:107–110

Sawidis T, Krystallidis P, Veros D, Chettri M (2012) A study of air pollution with heavy metals in Athens city and Attica basin using evergreen trees as biological indicators. Biol Trace Elem Res 148:396–408

Schönbeck H, Buck M, Van Haut H, Scholl G (1970) Biologische Meßverfahren für Luftverunreinigungen. VDI Ber 149:225–234

Seinfeld JH, Pandis SN (1998) Atmospheric chemistry and physics. Wiley-IntersciencePress, New York

Shah K, Nongkynrih JM (2007) Metal hyperaccumulation and bioremediation: review. Biol Plant 51:618–634

Shoun H, Kim DH, Uchiyama H, Sugiyama J (1992) Denitrification by fungi. FEMS Microbiol Lett 73:277–281

Singh S (2014) Chemistry and source identification of fine aerosols and role of their precursors in outdoor and indoor rural environment in North India, PhD thesis. Jawaharlal Nehru University, New Delhi

Stevovic S, Mikovilovic VS, Calic DD (2010) Environmental impact on morphological and anatomical structure of Tansy. Afr J Biotechnol 9:2413–2421

Takahashi M, Sasaki Y, Ida S, Morikawa H (2001) Nitrite reductase gene enrichment improves assimilation of nitrogen dioxide in Arabidopsis. Plant Physiol 126:731–741

Teklemariam TA, Sparks JP (2006) Leaf fluxes of NO and NO_2 in four herbaceous plant species: the role of ascorbic acid. Atmos Environ 40:2235–2244

Tingey DT (1989) Bioindicators in air pollution research – applications and constraints. Biologic markers of air-pollution stress and damage in forests. National Academies Press, Washington, DC, pp 73–80. ISBN 978-0-309-07833-7

Ugrekhelidze D, Korte F, Kvesitadze G (1997) Uptake and transformation of benzene and toluene in plant tissues. Ecotoxicol Environ Saf 37:24–29

Tomasevic M, Anicic M (2010) Trace element content in urban tree leaves and SEM-EDX characterization of deposited particles. Phys Chem Technol 8:1–13

Van RA (1969) The use of indicator plants to estimate air pollution by SO_2 and HF. In: Proceedings of the First

European Congress on the influence of air pollution on plants and animals. Wageningen, 1968. Pudoc Wageningen, pp 319–328
Vaucheret H, Kronenberger J, Lepingle A, Vilaine F, Boutin JP, Caboche M (1992) Inhibition of tobacco nitrite reductase activity by expression of antisense RNA. Plant J 2:559–569
Wahid A (2006) Influence of atmospheric pollutants on agriculture in developing countries: a case study with three new varieties in Pakistan. Sci Total Environ 371:304–313
Wu Y, Tiedmann V (2002) Evidence for oxidative stress involved in physiological leaf spot formation in winter and spring barley. Phytopathology 92:145–155
Yang DS, Pennisi SV, Son K, Kays SJ (2009) Screening indoor plants for volatile organic pollutant removal efficiency. Hortic Sci 44:1377–1381
Yoneyama T, Kim HY, Morikawa H, Srivastava HS (2002) Metabolism and detoxification of nitrogen dioxide and ammonia in plants. In: Omasa K et al (eds) Air pollution and plant biotechnology–prospects for phytomonitoring and phytoremediation. Springer, Tokyo, pp 221–234. 31
Zhai G (2011) Phytoremediation: right plants for right pollutants. J Bioremediation Biodegrad 2:3. http://dx.doi.org/10.4172/2155-6199.1000102e
Zumft G (1997) Cell biology and molecular basis of denitrification. Microbiol Mol Biol Rev 61:533–616

Air Pollution Control Policies and Regulations

12

Ranjit Kumar and Pratima Gupta

Abstract

Air pollution and climate change have worsened the health status of human beings and affected plants and vegetation adversely. This has posed serious threats to the entire ecosystem. An increasing demand of energy and dependencies on fossil fuels and nonrenewable energy resources may result into a further increase in air pollution. Air pollutants, gases (SO_2, NOx, O_3, hydrocarbons, etc.) and particles (mixture of various solid elements and dust particles), damage plants and reduce crop production. These pollutants cause an acute problem and affect vegetation directly and indirectly. The effect of pollutants depends upon the concentration of polluting species as well as on the age of plants. The day-by-day deteriorating quality of air and environment has necessitated a well-planned strategy to mitigate the menace of air pollution. It required proper understanding of causes, impacts, and control of air pollution. This chapter highlights types of pollutants affecting plants/vegetation and their sources and control technologies adopted to reduce pollution level. Euro/Bharat stage norms for reduction in emission from automobiles are illustrated. Regulations and legislations adopted are enumerated. The Clean Air Act envisaged guidelines for industries regarding emission. It has been seen and observed that in spite several rules and regulations, not much has been achieved in controlling air pollution particularly in countries like India. India needs to make tough legislations and ensure its implications. Thus, the need of envisaging new rules/regulations along with pollution control standards must be enacted and implemented honestly to protect plants/vegetation from air pollution and climate change.

R. Kumar (✉) • P. Gupta
Department of Chemistry, Faculty of Science, Dayalbagh Educational Institute (Deemed University), Dayalbagh, Agra-5 282005, Uttar Pradesh, India
e-mail: rkschem@rediffmail.com

U. Kulshrestha, P. Saxena (eds.), *Plant Responses to Air Pollution*,
DOI 10.1007/978-981-10-1201-3_12

Keywords
Air pollutants • Impacts • Control • Euro/Bharat norms • Clean Air Act • Vegetation

12.1 Introduction

Air pollution globally has come into existence due to accumulation of various types of pollutants located on the entire surface of the planet. Air pollution may be defined as any unfavorable change in the atmosphere which can be present in such concentrations that affect the man and ecosystem. These substances include gases of SO_x, NO_x, CO, hydrocarbons, particulate matters, radioactive materials, and many others. It affects the surrounding air quality in a great way, making it difficult for humans as well as other living beings to survive. Increase in the concentration of these species in the atmosphere disturbs the atmospheric equilibrium set by nature itself, and consequent results are seen in the form of ozone layer depletion, greenhouse effect, acid rain, damage to vegetation and building materials, global warming, and climate change. Climate change is basically a change in environmental/climatic conditions for a longer period of time. It is a result of global warming which is mainly due to increase in the concentration of air pollutants, viz., carbon dioxide (greenhouse gas), black carbon, organic carbon, and other light-absorbing particles and gases. Its harmful effects can be noticed in the form of long and extremely hot summers, short and extremely cold winters, very scarce precipitation, change in the seasonal rainfall pattern, frequent droughts, and floods. It has also led the onset of various deadly epidemic and endemic diseases. Change in crop pattern and production depreciation is also its aftereffects. Hence, it's leading the overall deterioration of the earth's ecosystem and human life.

An air pollutant is the impurities added into the air through different sources which results in change in the composition of air. Change in the composition also results in change in the environment, making it difficult for the inhabitants to survive, and adversely affects the entire ecosystem. Natural sources include soil, dust, sea spray, forest fire, etc., while anthropogenic sources include area sources (residential/offices), point sources (industrial establishments), and mobile or line sources (automobile/transportation services). Although natural source contribution is larger than anthropogenic sources, the effects of anthropogenic sources are much as they are localized, while natural sources are wide so pollutants get diluted. Pollutants are classified as primary pollutants and secondary pollutants. Primary pollutants are those pollutants which are directly emitted into the atmosphere, while secondary pollutants are those pollutants which get formed in the atmosphere from the chemical reaction of primary pollutants. Pollutants are also classified as criteria pollutants (those for which a sufficient dataset is available and standards have been set) and non-criteria pollutants (those for which standards have not been set). The main source of local pollution is the emission from automobiles and biomass burning which leads to climate change. Developed and developing countries have witnessed excessive degradation of air quality, which is deteriorating further each day throughout the world. They have gone through rapid industrialization, high urban population growth, and an indiscriminate increase in the population of automobiles, industries, and power generation units, leading to an excessive emission (Word Bank 1996). In the last century, the high urbanization in the developing countries has been observed. Urbanization produces a challenge of pollution and congestion that is prevailing in many of the developing countries (Asian, African, and Latin American) (Ashmore 2005). The urban population in 1960 was not more than 22 % which rose to 34 % in 1990, and it is estimated to be 50 % of the global population by 2020 in the developing world (World Bank 2009).

The growth of human urbanization and industrialization has been very closely connected with the growth of power production. Power consumption has become the criterion of becoming

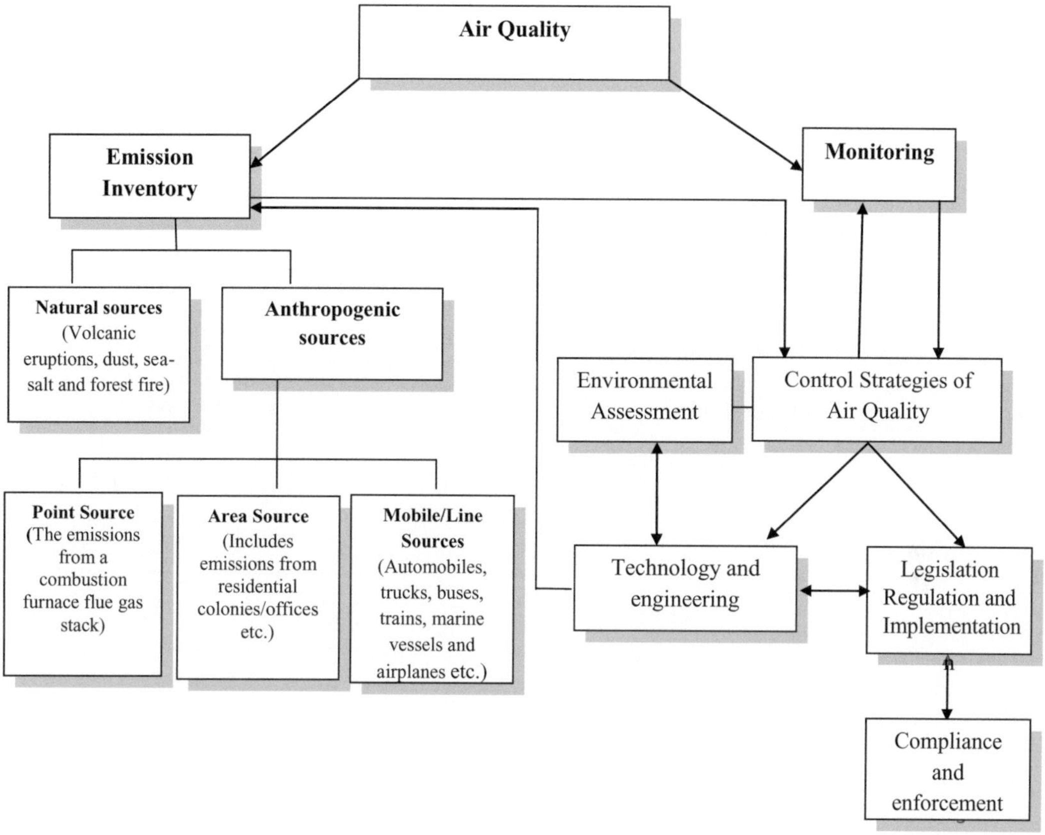

Fig. 12.1 Schematic representation of air pollution and its sources and control

developed. This leads to an increase in per capita power consumption. The per capita energy consumption (kWh) in Sweden is 15,656, 13,228 in the USA, 6,158 in the UK, 1,208 in China, and 789 in India (IEA 2013). The dominant sources of power are coal, diesel, gas, hydro, nuclear, and renewable energy resources which contribute 58 %, 1 %, 9 %, 18 %, 2 %, and 12 % (IEA 2013), respectively, to the total energy production. The International Energy Agency has estimated and reported that by 2035 energy demands will double up and in spite of efforts in reducing dependencies on coal and efforts of a clean energy source, coal will remain a significant raw material of energy over the world. Hence, there will be an increase in the concentration of air pollution in the atmosphere. Figure 12.1 summarizes the emission inventory and required control strategy to maintain the healthy atmosphere.

The environment is severely damaged by air pollution. Agricultural crops are affected due to air pollution. The major air pollutants affecting and damaging plants are SO_2, NOx, O_3, fluorine, suspended particulate matter, etc. Emissions reduce the ability of plants to uptake carbon dioxide which leads to the heat imbalance of the entire ecosystem. Plants are excessively affected by air pollution directly or indirectly. Direct exposure of a plant to several gaseous pollutants in the atmosphere in addition to trace elements, dust particles through deposition in the soil which are inherited by plants, is a direct effect of pollution. The indirect effect of air pollution is uneven rainfall/droughts, extreme temperatures, loss of soil fertility, etc. The effects of particulate matter on the ecosystem are also direct and indirect. Particulate matter when chemically reacting with soil changes its chemical properties affecting a plant's production capacity and growth (Wall and Moore 1999). Indirect effects of pollutants are very difficult to monitor and recognize as they show their true nature over a longer period and are also chronic in nature (Garner 1994). The environmental impacts of air pollutants are given in Table 12.1.

A high concentration of carbon dioxide is consumed in the process of photosynthesis of plants and trees reducing a large part of greenhouse gases. Greenhouse gas reduction by any form is necessary for minimizing its effect on global warming and various other harmful effects. It also

Table 12.1 Air pollutants and their environmental/climatic impacts

Environmental impacts			
Pollutant	Direct impact	Pollutant	Indirect impact
SO_2	SO_2 increases the acidity of soil and causes injury to plant species losses in aquatic and terrestrial systems	Arsenic	Decomposition in soil reduces the fertility
O_3	Damages vegetation	Boron	Affects older leaves than younger leaves
	Reduces photosynthesis		Produces injury symptoms
	Decreases crop yields		
	Reduces biodiversity		
	Decreases plant uptake of CO_2		
	Contributes to the warming of the atmosphere		
NO_2	Acidification and nutrient enrichment	Cadmium	Reduces seed germination
	Leads to biodiversity losses		Decrease in plant nutrient
			Reduces shoot and root length
Carbon monoxide (CO)	Contributes to the formation of CO_2 and ozone	Chromium	Effect on plant growth, growth of roots, stems, and leaves, which may affect total dry matter production and yield
Ammonia (NH_3)	Contributes to eutrophication of surface water	Cobalt	Reduces shoot and root length
	Contributes to the formation of nitrate and sulfate particles		Decrease in chlorophyll content
			Reduction in plant nutrient content
			Decrease in plant sugar and protein content
Volatile organic compounds (VOCs)	Contributes to ozone formation	Lead (Pb)	Deposit in soils and adversely affect terrestrial and aquatic systems
	Contributes to the formation of CO_2 and ozone		
CO_2	Most potential global warming pollutant	Manganese	Chlorosis
			Decrease in chlorophyll
	Causes the greenhouse effect		Slower plant growth

(continued)

Table 12.1 (continued)

Environmental impacts			
Pollutant	Direct impact	Pollutant	Indirect impact
Ethylene	Leaf abnormalities	Nickel	Affects nutrient absorption by roots
	Flower dropping		Inhibits photosynthesis
	Abscission		Causes ultrastructural modifications
	Water stress in plant species		
Fluoride	Tip and margin burns	Mercury (Hg)	Accumulates into water bodies, resulting in exposure to humans and wildlife
	Dwarfing		
	Leaf abscission		
	Red spots on peach fruits		
Particulate matter	Impairs visibility	Other toxic	Some toxic air pollutants accumulate in the food chain
	Damages structures and/or soil properties	Air pollutants	Some toxic air pollutants contribute to loss of vegetation
	Other impacts include changing the rainfall pattern		

Table 12.2 Symptoms of plant diseases due to air pollution

Pollutant	Symptoms	Effect on plants
SO_2	Bleached spots, bleached areas between veins, middle-aged leaves' mesophyll cell chlorosis, insect injury	Middle-aged leaves most sensitive
NO_2	Irregular shape of leaves and white or brown collapsed lesions on inter-tissue and near-leaf margin	Middle-aged leaves most sensitive
Ozone (O_3)	Spots in leaves, pigmentation, leaves' tips become brown and necrotic	Oldest leaves most sensitive, youngest least sensitive
Fluoride	Tip and margin burns, dwarfing, leaf abscission, fungal disease, drought, and wind may produce similar markings to suture red spot in fruits	Youngest leaves most sensitive
Ethylene	Sepal withering, leaf abnormalities, flower dropping, abscission, and water stress may produce similar markings	Young leaves recover
Chlorine	Bleaching between veins, tip and margin burn, leaf abscission	Mature leaves most sensitive
Ammonia	Green appearance of leaves becoming brown or green black	Mature leaves most sensitive
Mercury	Chlorosis and abscission, brown spotting, yellowing of veins	Oldest leaves most sensitive

plays a wide role in climate change. The carbon stored in plants is a striking balance between carbon fixing of photosynthesis and respiration. The carbon stock in plants' biomass is higher and accumulated for a longer period due to the complex structure of vegetation. Peatlands formed by dead organic matter also store a large amount of carbon. Plants that are repeatedly exposed to pollutants get absorbed, accumulated, and integrated into their systems. The changes in the biochemical processes and accumulation of metabolites have been observed. It depends upon plants' sensitivity level (Agbaire and Esiefarienrhe 2009). The condition of plant species is also an indicator of the overall impact of pollution (Rai et al. 2007). Acute effects result from short-term exposure while chronic effects result from exposure to lower concentration for months to several years. The most exposed parts of a plant's body are the leaf surfaces. Pollutants can cause leaf injury, decrease in photosynthesis activity, premature senescence, stomatal damage, permeability of the membrane, and reduced growth and yield in sensitive plant species (Tiwari et al. 2006). The effects of different polluting species on plants and their symptoms are summarized in Table 12.2.

Air pollution and climate change have badly affected agriculture and plants. Many studies have been carried out on air pollution, plant disease, and agriculture production (Agarwal et al. 2006). However, the reduced intensity of plants depends upon the concentration of pollutants, exposure time, climatic conditions, species of plants, etc. The long-term effect of air pollution can be seen when the ecosystem is not able to sustain more pollution. Deterioration of air quality is due to a lack of civic consciousness. Now, air pollution control has become a necessity. Several methods/techniques have been utilized for the control of air pollution, and a number of rules and regulations have been adopted from time to time to combat the menace of air pollution.

12.2 Control Method of Air Pollution

Air pollution and climate change have posed extinction threats to the survival of human beings and plants and damage to materials. About seven million people get affected every year in the world from air pollution while climate change is estimated to cause US $ 5 billion global crop losses per year. It claims 3 % of the GDP of India. Air pollution is an emergent issue. Hence, air pollution control is required. Air pollution can be controlled by reducing the emission of air pollutants from industrial and domestic sources into the atmosphere through source emission control and pollution control equipment and by making rules and regulations to adhere with. The government can regulate air pollution through air quality and emission control standards by instructing the parties to adhere with the norms and to adopt emission control technologies (i.e., technology that reduces or eliminates emissions). These emission control technologies may use control devices such as electrostatic precipitators, baghouses, wet scrubbers, cyclones, soft X-ray, venturi scrubbers, carbon absorbers, etc. Figure 12.2 shows a schematic presentation of air pollution control methods; it can be controlled through various modern technologies by collection, measurement, and treatment of pollutants. However, control of contaminants at their source level is a desirable and effective method.

12.2.1 Source Emission Control

Mobile sources and industries have been recognized as one of the most important sources of air pollution since the earliest days of the Clean Air Act. Hence, it has been a prime target for emission control. Unauthorized release of smoke in the air without any refinement process is one of the main sources of air pollutants in the air. Formation of these pollutants can only be prevented and their emission can be minimized if they are stopped at the source itself. Most importantly, implementation of the new preventive methods to air pollution is to be looked at and proper monitoring is necessary. The following two methods are used for controlling the emission of air pollution.

12.2.1.1 Replacement of Raw Material

Replacement of highly polluting raw material by cleaner and high quality raw material harmful pollutant will get reduced. Reduction in use of high sulfur containing raw material in electric utilities. The use of natural gas (CNG), propane, and ethanol reduces the air pollution from automobiles.

12.2.1.2 Modification of Equipment

Air pollution can be considerably reduced by modifications in the existing equipment. In petroleum refineries, loss of hydrocarbon vapors from storage tanks due to evaporation, temperature changes, or displacement during filling can be controlled by making a modification in design, etc.

12.2.2 Pollution Control Equipment

Airborne particles are highly concentrated at the source but when their distance from the source increases, it gets diluted by diffusion into the atmosphere. Sometimes pollution control is not

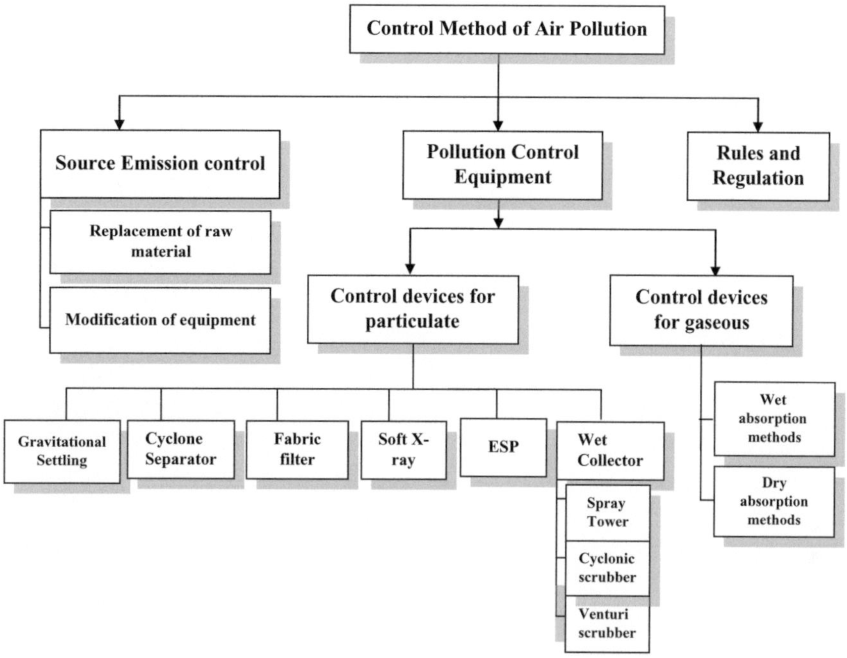

Fig. 12.2 Flow diagrams of the control methods of air pollution

able to reduce the emission of air pollutants at the source. It becomes necessary to install pollution control equipment to remove the pollutants from the main gas stream. The need for technological advancement is the need of the hour. Steps are to be taken in all fronts in order to stop this menace of air pollution. Common types of equipment can be used for collecting fine particulate matter including settling chambers, cyclones, scrubbers, electrostatic precipitators, baghouse filters, and adsorption, condensation, and combustion methods. The air pollution control system can be grouped generally into two classes.

12.2.2.1 Control Devices for Particulate Contaminants

Several equipment have been developed and utilized for particulate matter control. Table 12.3 shows different important methods for the control particulate species.

12.2.2.2 Control Devices for Gaseous Contaminants

Major gaseous air pollutants are carbon monoxides (CO), nitrogen oxides (NOx), sulfur oxides (SOx), and volatile organic compounds (VOCs). In general, the concentration of pollutants in a waste air stream is relatively low, but the emissions can still exceed the regulatory limits. Removal of gaseous pollutants can be achieved by the following methods.

Absorption

This is one of the most common mass transfer phenomena, especially for controlling emissions from small sources. The liquid must be either to serve as a solvent for the pollutant or to capture it by means of a chemical reaction. Absorption techniques are also dependent on corrosive nature, reactivity, shape, density, and size. Mass transfer is a diffusion process wherein the pollutant gas moves from points of higher concentration to points of lower concentration. Hence, pollutants get transferred to a particular location where equipment such as spray towers, packed columns, cyclone scrubbers, and venture scrubbers are employed to absorb pollutant gases. Once particulate matter gets collected, particulates adhere to each other, forming agglomerates that can readily be removed from

Table 12.3 Control methods/devices for air pollution prevention

Control method/device	Procedure
Gravitational settling chamber	Gravitational settling chambers on the force of gravity to remove large abrasive particles
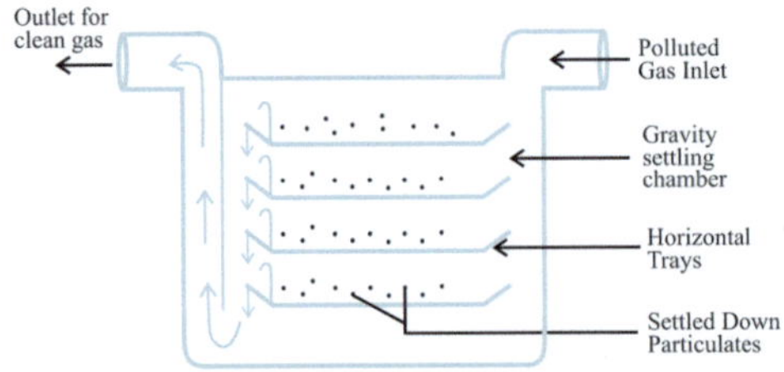	The velocity of the gas is reduced in the gas stream chamber as gas expands in the chamber
	Hoppers collect the large particles which settle down from the gas. But only larger particle removal is possible
	The chamber must be properly sealed to prevent the entrance of air and allow proper exit of dust particles
Cyclone separator	It works between gas inlets and dust particle outlets in a conical chamber using the principle of centrifugal force to separate particulate matters
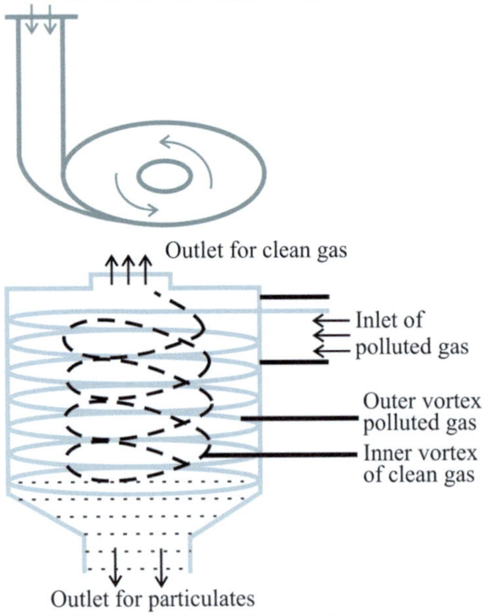	The cyclone generated within rotates the gases resulting in collision of heavier dust particles on the walls which lose speed and fall down. The smaller particles lose their speed on reaching the narrower section of the cone
	The settled dust particles are collected at the bottom of the container
	They are effective in dealing with large as well as small dust particles
Fabric filter	They are also known as baghouses as gases containing dust particles are made to pass through a layer of cloth
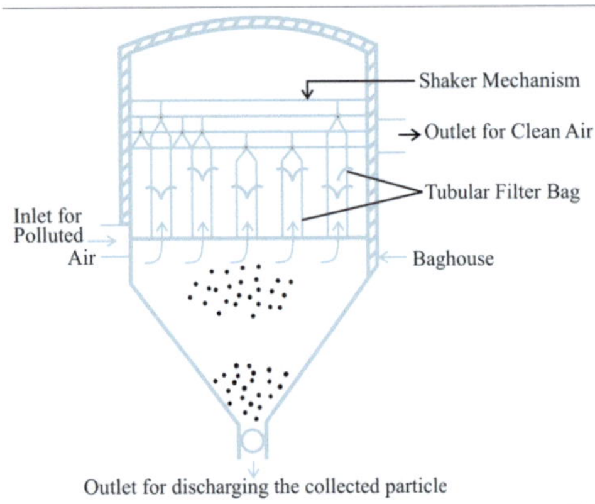	The solid dust particles get collected on the cloth while clean air passes through
	The dust carpet forms another filtering surface for the next installment of gases
ESP (Electrostatic Precipitator)	The gas-containing particulate matter is passed between two electrodes

(continued)

Table 12.3 (continued)

Control method/device	Procedure
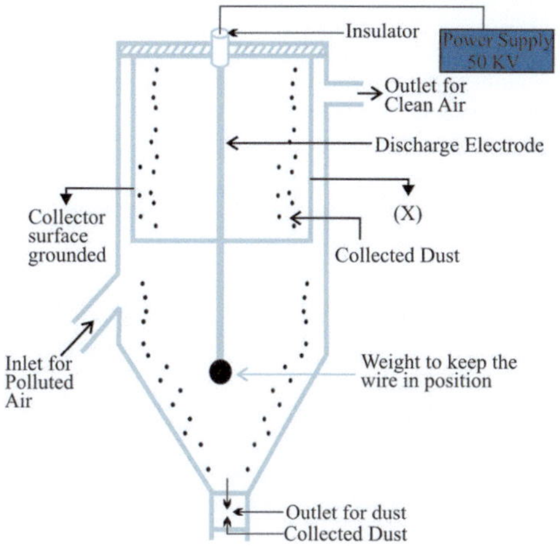	Here gas is charged by one electrode and another electrode attracts the oppositely charged dust particle
	The chemical deposited on an electrode depends upon the type of dust particle to be separated. They have high efficiency and can work on very high temperatures and pressures
Wet collector	
(a) Spray tower	A stream of water is passed through a nozzle into a spray tower from above while gases are made to enter from below
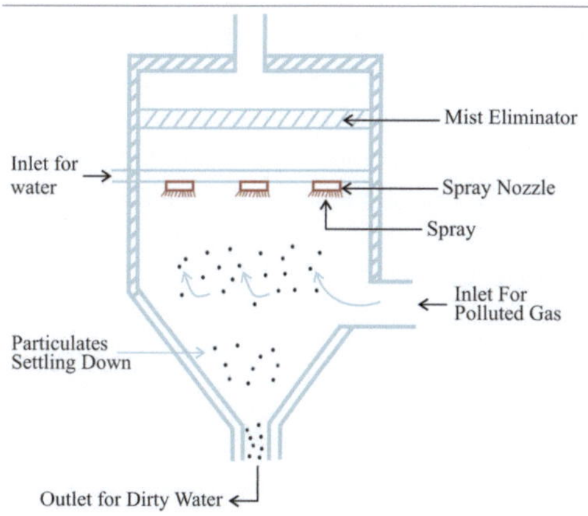	Hence the liquid pushes dust particles at the bottom of the tower
	Dust particles are drained out through an outlet
(b) Venturi scrubber	A venturi scrubber uses an absorbing liquid stream and turbulence in the settling chamber to separate solid particulates
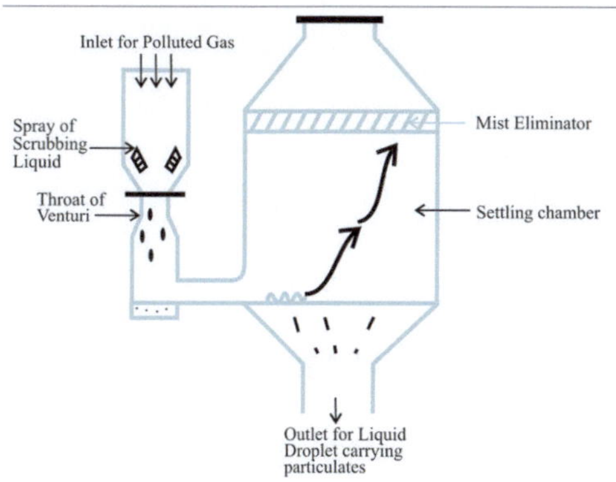	The particulates are forced to settle at the bottom by the liquid
	They efficiently remove small dust particles

(continued)

Table 12.3 (continued)

Control method/device	Procedure
(c) Cyclone scrubber	Many high-pressure spray nozzles are used to convert the dry cyclone chamber to a wet cyclone chamber which forms the contact space for dust particles and contacting liquid
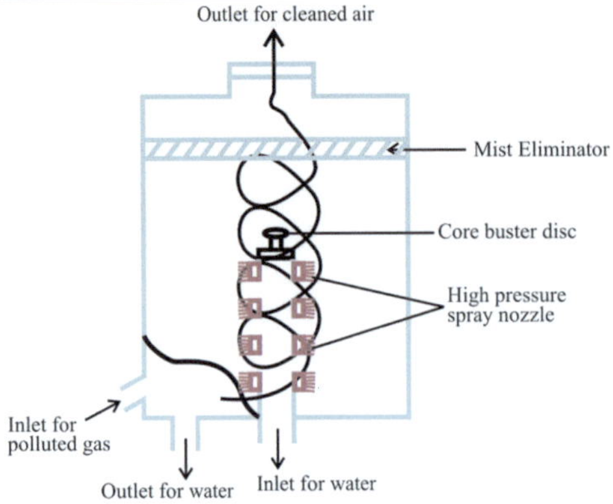	High-speed water streams generated through high-pressure spray nozzles force dust particles toward the walls under the influence of a cyclone

the equipment and disposed of usually in a landfill. Absorption is used extensively in the separation of corrosive and hazardous pollutants from waste gases.

Adsorption

It involves transferring pollutants from a gas phase to a contacting solvent. Gas adsorption, contrasted to absorption, is a surface phenomenon. Here gas molecules are attracted to the surface of a solid. Odor control at various types of chemical-manufacturing and food-processing facilities uses gas adsorption as method in the recovery of a number of volatile solvents like benzene and in the control of VOCs at industrial facilities. Activated carbon is one of the most common adsorbent materials as it is very porous, hence having an extremely high ratio of surface area to volume. Adsorption systems are configured either as stationary bed units or as moving bed units.

Condensation

In this process, condensers are used as pretreatment devices which convert gas/vapor to liquid. The use of condensers with absorbers, adsorbers, and incinerators reduces the volume of the total gas. Condensers can be used as contact condensers and surface condensers for controlling pollution. In contact condensers when gas comes in contact with cold liquid, it gets compressed. In surface condensers when the polluted gas comes in contact with a cooled surface, cooled liquid or gas is circulated at the surface of the tube.

Combustion

The combustion process is also called an incineration. It can be used to convert VOCs and other gaseous hydrocarbon pollutants to carbon dioxide and water. Sufficient turbulence, or mixing, is a key factor in combustion because it reduces the required burning time and temperature. A process called direct flame incineration can be used when the waste gas is itself a combustible mixture and does not need the addition of air or fuel. Afterburners are used to control odor, destroy toxic compounds, or reduce the amount of photochemical reactive substance released into the air. These are usually used in petroleum refineries, paint-drying facilities, and paper mills.

12.2.3 Rules and Regulations

Governments worldwide and international organizations such as the World Health Organization and European Union are facing a growing problem of air quality effects induced by gaseous and

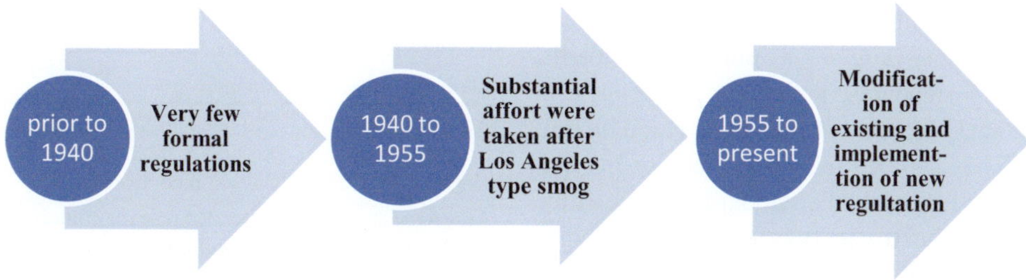

Fig. 12.3 History of air pollution regulation

particulate pollutants arising from industries, motor vehicle emissions, etc. The most ambient air pollutants in the urban region from high levels of vehicle emission are particulate matter ($PM_{2.5}$ and PM_{10}), NO_2, and O_3. Air quality regulations are often designed specifically to protect human health as well as plant health by limiting or eliminating airborne pollutant concentration. Hence, the need for environmental regulations and for policy of these regulations arises to control air pollution (Fig. 12.3).

The USA started to introduce several legislative amendments which soon became a movement by a series of new laws and regulations of air quality. The Air Pollution Control Act, July 14, 1955, was the first clean air act which addresses a national environmental problem. The objective of this act was based on the impact of air pollution. The act also conserves the primary responsibilities and rights of the states and local government in controlling air pollution. Several amendments have been made in the Air Pollution Act of 1955. The first and second amendments of the Air Pollution Act came in 1960 and 1962, which extended research funding for 4 years and enforced the provision of the original act. Additionally, the Clean Air Act was enacted in 1963 which defined the criteria of air quality. This act focused on vehicle exhaust and emission standards from local sources. The 1965 amendments of the Motor Vehicle Air Pollution Control Act came into existence which established automobile emission standards. The amendments of the 1967 act established the timeline for State Implementation Plans (SIPs) and emission inventories, ambient monitoring techniques, and control techniques. The amendments in 1970 were newly created. The US-EPA set the National Ambient Air Quality Standards and rewritten version of the original Clean Air Act. It was the monumental year of the Clean Air Act regulation of various states for achieving and maintaining the National Ambient Air Quality Standards. Air pollution has been regulated a series of laws and regulations throughout the world after achieving the Clean Air Act of 1970. Therefore, a large number of laws and environmental protection regulation acts came into existence to control air pollution (Tables 12.4).

Identification of substances and energies which cause pollution should be controlled by enacting air quality regulations. As before, there were no norms to tell about the constituents of air pollution, so it was necessary for the nations to identify pollutants and enact air pollution regulations accordingly. Air quality standards are legal norms which determine the quality of indoor and outdoor air pollutants. They form a base for criteria pollutants that are acceptable in ambient air and minimize their effect on human health. The secondary effects such as vegetation and building damages are also considered. Periodic and up-to-date scientific data and monitoring of air quality with their exposure time are necessary to constitute air quality standards. The US-EPA has developed the National Ambient Air Quality Standards (NAAQS) for pollutants which are considered harmful to human health and the environment. NAAQS set a threshold for sulfur dioxide, particulate matter (PM_{10} and $PM_{2.5}$), carbon monoxide, ozone, nitrogen oxide (NO_x), and lead (Pb) in ambient air. NAAQS forms the basis for other international air quality agencies to formulate their standards accordingly.

The air pollution control strategies for enacting a regulatory program are emission standards, air quality standards, taxing on pollution, cost benefit analysis, and risk standards based on pollution. They can be used individually or in combination to formulate the standards of air quality. Emission standards are the legal norms used to set the permissible limit of different types of air pollutants to be emitted into the atmosphere from a specific source in a particular time. The purpose of formulating emission standards is to protect ambient air quality. Several regulatory approaches and methods exist depending on the source industry and type of pollutant to determine appropriate standard emission. Specific control technologies based on feasibility, availability, and cost are also mandated by a regulatory body. Air quality standards prescribe the maximum limit of pollutants in a particular geographical area in given time periods. National Ambient Air Quality Standards (NAAQS) are designed in two ways:

- Primary standards: Protecting public health and safety is the primary objective of this standard. They are implemented for a stipulated time period regardless of the cost.
- Secondary standards: Protecting the environment from air pollution is the main objective of this standard. The time period for achieving the goal of secondary standards will be determined by state and local governments.

Several air quality standards around the world have come into effect in compliance with applicable NAAQS using these two standards. Likewise US-EPA's NAAQS, countries like China and India have also set their own air quality standards. Table 12.5 presents air quality standards for different polluting species set by WHO (World Health Organization), US-EPA (United States Environmental Protection Agency), and CPCB (Central Pollution Control Board), India.

In India, air pollution control strategies have been made since long back even before independence. In India, the Air (Prevention and Control of Pollution) Act was enacted in 1981. Under this

Table 12.4 Summary of legislation involved in air pollution control

Air pollution acts and amendments	
Air pollution Control Act of 1995	Provided research technical assistance relating to air pollution control
Amendments of 1960	Extended research funding for four more years
Amendments of 1962	Enforced the principal provisions of the original act for the air
Clean Air Act of 1963	Provided, strengthened, and accelerated programs for the prevention and amendments of air pollution
Amendments of 1965	Controlled motor vehicle air pollution
Amendments of 1966	Implied the local air pollution control program
Amendments of 1967	Involved state implementation plans and recommended control technologies
Amendments of 1969	Extended authorization for research on low emission fuel and automobiles
Clean Air Act of 1970	Provided an effective program which improved the quality of air throughout the country
Amendments of 1977	Involved motor vehicle emission standards
Clean Air Act of 1990	Solved the problem of the past as well as dealt with new issues

Table 12.5 Air quality standards for different countries

Pollutant	US-EPA	WHO	EU	India
CO (8 h)	9	–	10	2.0
NO_2 (annual)	53	40	40	40
NO_2 (daily)	–	–	–	80
SO_2 (annual)	–	–	–	50
SO_2 (daily)	–	20	125	80
PM_{10} (annual)	–	20	40	60
PM_{10} (daily)	150	50	50	100
$PM_{2.5}$ (annual)	11–13	10	25	40
$PM_{2.5}$ (daily)	30–35	25	–	60
Ozone (8 h)	120–140 (60–70 ppb)	160	120	100
Lead (annual)	–	0.5–1.0	0.5	0.5

act, rules were notified by the government of India on November 18, 1982. The Environment (Protection) Act of 1986 was legislated by the Indian parliament in response to the effect of the Bhopal gas tragedy. This act covers the whole of India and empowers the government of India to necessitate any action for improving and protecting the environment. The Motor Vehicle Act of 1988 marks another milestone that limits the smoke-containing harmful gases such as CO, Pb, and particulate matter emitted from automobiles.

12.2.4 Bharat/Euro Emission Standards

European Union emission standards introduce the agreeable limits of automobile emission for every light-duty vehicle with the number of amendments which were adopted in 2004. In European emission standards, some important regulations were implemented, viz., Euro 1 standards (passenger cars only), Euro 2 standards (for passenger cars and motorcycles), Euro 3–4 standards (for any vehicle and motorcycle), and Euro 5–6 standards (for light-passenger and commercial vehicles). An advancement technology and ever-increasing fuel cost will result in a rising manufacturing cost of automobiles which indirectly reduces the pollution level by the help of norms. Bharat standards are based on European regulations of 2000. The government of India instituted the Bharat stage emission standards for regulating the air pollutants from internal combustion of engine equipment. The Central Pollution Control Board under the Ministry of Environment, Forest and Climate Change is responsible to set the standards and time frame for its implementation. Various strict norms have been adopted since then including manufacturing of new vehicles, adhering with the standards. Since 2010, the country has been using Euro 4 based norms and the proposed date of adopting Euro 6 norms by skipping Euro 5 is 2020. In 1989, idle emission limit where first adopted by Indian emission regulation which were replaced by mass emission limit for petrol 1991 and diesel 1992. They were made stricter in the 1990s. India has adopted European emission and fuel regulation standards since 2000 for four-wheel light-duty and heavy-duty vehicles. But India has its own standards for two- and three-wheel vehicles. It is necessary for all transport vehicles to carry a fitness certificate which needs to be renewed every year after first the 2 years of a vehicle's registration (Tables 12.6, 12.7, and 12.8).

12.2.5 Steps of Enforcement of Environmental Legislation

The environmental legislation is the collection of laws regulating environmental factors. These laws and regulations regulate the interaction between human and nature and its effect on human and plant health. The National Environmental Policy Act (NEPA) was one such law enacted by the USA which has been reciprocated throughout the world for the protection of the environment. This act came in response to the Air Pollution Act of

Table 12.6 Bharat emission standards in India

Year	Emission norms in India
1989	First emission norms for diesel vehicles
1991	Mass emission norms and idle CO limits for gasoline vehicles
1992	Mass emission norms for diesel vehicles
1996	Revised mass emission norms for gasoline and diesel vehicles (catalytic converter for cars)
1998	Introduced cold start norms
2000	Bharat stage II norms for Delhi like Euro I norms, modified IDC (Indian driving cycle)
2001	Bharat stage II like Euro II norms for all metros, emission norms for CNG and LPG vehicles
2003	Bharat stage II (like Euro II norms) for 13 major cities
2005	From April 1 Bharat stage III equivalent to Euro III norms for 13 major cities
2010	Bharat stage III emission norms for entire country two wheelers, three wheelers, and four wheelers Bharat stage IV (like Euro IV) for 13 major cities only for four wheelers
2020	Proposed date for the country to adopt Euro VI norms for cars, skipping Euro V

Table 12.7 Bharat emission standards

Emission standards for diesel trucks, buses, and light-duty diesel vehicles(g/kWh)									
Year	Reference	CO		HC		NOx		PM	
		Diesel truck and bus	Light diesel vehicle	Diesel truck and bus	Light diesel vehicle	Diesel truck and bus	Light diesel vehicle	Diesel truck and bus	Light diesel vehicle
1992	–	17.3–32.6	17.3–32.6	2.7–3.7	2.7–3.7	–	–	–	–
1996	–	11.20	5.0–9.0	2.4	–	14.4	2.0–4.0	–	–
2000	Euro 1	4.5	2.72–6.90	1.1	–	8.0	0.97–1.70	0.36[b]	–
2005[a]	Euro 2	4.0	1.0–1.5	1.1	–	7.0	0.7–1.2	0.15	–
2010[a]	Euro 3	2.1	0.64	0.66	–	5.0	0.56	0.10	0.05
		5.45	0.80	0.76	–	5.0	0.72	0.16	0.07
			0.95		–		0.86		0.10
2010††	Euro 4	1.5	0.50	0.46	–	3.5	0.30	0.02	0.025
		4.0	0.63	0.55	–	3.5	0.39	0.03	0.04
			0.74		–		0.46		0.06

[a]For selected regions (Mumbai, Kolkata, Chennai, Bengaluru, Hyderabad, Ahmedabad, Pune, Surat, Kanpur, Lucknow, Sholapur, Jamshedpur, and Agra)
[b]0.612 for engines below 85 kW
††for selected regions

Table 12.8 Emission standards for three-wheel and two-wheel gasoline vehicles, g/km

Year	CO	HC	HC + NOx			
	Three-wheel	Two-wheel	Three-wheel	Two-wheel	Three-wheel	Two-wheel
1991	12–30	12–30	8–12	8–12	–	
1996	6.75	5.50	–	–	5.40	3.60
2000	4.0	2.0	–	–	2.00	2.00
2005 (BS II)	2.25	1.5	–	–	2.00	1.5
2010.04 (BS III)	1.25	1.0	–	–	1.25	1.0

1955 where it was decided by the government to deal with the problem on a wider scale. This act identifies air pollution as a national problem and announced the steps needed to be taken to avoid its harmful effects. The Clean Air Act was enacted to promote public health and welfare. This act also includes harmful effects of automobile exhaust and emission from stationary establishments. It also encourages the development of various emission standards to deal with it. Organizations need to be made aware about environmental legislations and regulations. The government is the chief enforcement body of these environmental legislations which pave the way for less polluting small-scale industries. In India, the Air (Prevention and Control of Pollution) Act of 1981 was also in commitment with the responsibility to enforce rules and regulations. This set of organizations was also renamed as the Central and State Pollution Control Boards. The central government should take the entire responsibility of protecting and improving the environmental condition as per the Environmental Protection Act of 1986.

The three main issues involving environmental legislation are:

1. *Preventive steps*

It deals with risks and uncertainties involved in the implementation of environmental regulations. The principle implies "prevention is better than cure." Taking preventive measures beforehand will certainly reduce risk factors and frame back plans.

2. *Taxing the polluter*

In this step, a polluter who is responsible for the deterioration of air quality has to pay for its monitoring and policy-making cast, but this will mark the exit of small-scale industries only as big industries can bear their cost easily. This will not serve the purpose of the law but controlling automobile emission by taxing heavy polluters can be a formidable step to control air pollution on roads.

3. *Freedom of information*

Freedom of information refers to providing public access to environmental information held by public authorities. This may be in two ways:

- Public authorities should proactively make environmental information available. Members of the public are entitled to request public authorities for environmental information.
- Public authorities should include government departments, local authorities, police forces, and universities. It should also cover some other bodies that do public work that affects the environment. For simplicity, all organizations subject to this are referred to as "public authorities" in this guide. There should be a legislation regarding enforcement.

1. Anybody should have access to data regarding the environment unless it may question the sovereignty or integrity of the nation.
2. The legislation should also focus on providing security to the environmental activists so that they may not be harmed by the parties against whom they seek information. Lately, there have been cases of loss of lives of environmental activists.
3. A panel also needs to be formed which invites the suggestion of different activists, reporters, experts, and concerned public regarding pollution-controlling methods and other preventive ways to stop this menace.

12.2.6 Case Study

The National Green Tribunal was instituted in response to the aftermath of several tragedies and mainly due to the Bhopal gas tragedy and the following discussion. It was set up by an act of parliament known as the National Green Tribunal Act of 2010. It aims to protect natural resource, safeguard environment, and protect existing forests including providing rehabilitation and compensation for loss of life and property. The enforcement of legal rights of the natural environment is also its responsibility. Since its realization, it has taken several steps to weed out pollution and the causes adversely affecting the air quality. In one of its recent landmark decisions, it has announced a blanket ban on the movement of diesel vehicles older than 10 years in the national capital. It has also ordered a heavier penalty on people burning solid waste. These steps were in response to the recent list of most polluted cities in the world released by WHO in which Delhi tops the list. Opposing the central government and canceling the coal block allocation for Hasdeo-Arand forest and Meghalaya rat-hole mining are few other strong steps taken by it. But these have proved minuscule as the graph of pollution is continuously rising. Lack of support from center and state governments is the biggest hindrance in their strong steps. The state pollution board's inefficiency to implement orders passed by NGT is another roadblock in the realization of its goals in the stipulated time period. Limited staff and lack of support from the government limit its vigilance throughout the country. NGT is overloaded with a large number of pending cases though it tries to quickly solve cases. The biggest hurdle in the path of NGT is that it doesn't have *suo moto* power to take and force decisions. The importance of NGT increases with the rise of pollution-related health hazards. It needs to be more vigilant in taking a strong decision and enforcing its execution. It should work in tandem with the central and state governments to secure the environment more efficiently.

12.3 IPCC and IPPC Recommendation for Plant Health

The Intergovernmental Panel on Climate Change (IPCC) is intended to assess scientific, technical, and socioeconomic information concerning climate change and its potential effects and options for adaptation and mitigation all over the world. Climate Change 2007 is the Fourth Assessment Report of the United Nations' Intergovernmental Panel on Climate Change (IPCC).

The important findings of the report were "warming of the climate system is unequivocal," and "most of the observed increase in global average temperatures since the mid-twentieth century may be due to the observed increase in anthropogenic greenhouse gas concentrations." The IPCC has estimated that stabilizing atmospheric greenhouse gases at between 445 and 535 ppm CO_2 equivalent would result in a reduction of average annual GDP growth rates of less than 0.12 %. Forty to fifty percent of Earth's land surface is occupied for agricultural production consisting of croplands. Agriculture releases significant amounts of CO_2, CH_4, and N_2O into the atmosphere (Cole et al. 1997; IPCC 2001). The major sources of CO_2 emission are from microbial decay or burning of plant litter and soil organic matter (Smith and Conen 2004; Janzen 2004). The decomposition of organic materials in oxygen-deprived conditions forms CH_4. It also gets formed from stored manures and from rice grown under flooded conditions (Mosier et al. 1998). Microbial alteration of nitrogen in soils and manure is mainly responsible for N_2O production (Oenema et al. 2005). The substantial fraction of these mitigation costs may be offset by benefits to health as a result of reduced air pollution. There will be further cost savings from other benefits from increase in energy security, increase in agricultural production, and reduction in pressure on natural ecosystems.

The International Plant Protection Convention (IPPC) is an international treaty relating to plant health. The purpose of the IPPC is to secure common and effective actions to prevent the spread and introduction of pests of plants and plant products and to promote appropriate measures for their control (Article I of the IPPC). One important objective of the IPPC is to protect cultivated/unmanaged plants, wild flora, habitats, and ecosystems with respect to invasive alien species (IAS) that are plant pests. The many national obligations related to IAS under the IPPC are:

- Establishment of NPPO (National Plant Protection Organization)
- Risk analysis of pests
- Surveillance of plants and plant products
- Pest eradication or control

The relationships between countries are facilitated by the IPPC framework. It encourages them to cooperate at a regional level and gives them guidance in developing their own national plant protection systems. The IPPC was further amended in 1997 and on October 2, 2005. All contracting parties to the IPPC follow these amendments. The 1997 amendments update the convention and reflect the role of the IPPC in relation to the Agreement on the Application of Sanitary and Phytosanitary Measures of the World Trade Organization (the WTO-SPS Agreement). One hundred fifty-eight countries are contracting parties to the IPPC as of December 2006. Many countries agree to cooperate in the development of ISPMs (International Standards for Phytosanitary Measures) under the IPPC (i.e., Article X). These ISPMs provide guidance to countries to meet their IPPC obligations (as of April 2006, 27 ISPMs have been adopted). Each country is required to establish a national plant protection organization (NPPO) under the IPPC (i.e., Article IV). The key responsibilities of an NPPO are:

- Growing plants (cultivated and noncultivated) and plant and plant product surveillance in storage/transportation
- Conducting pest risk analysis
- Protecting endangered areas
- Meeting phytosanitary measures and internationally moving consignment (of plants and plant products) disinfection

- Issuing disinfection certificates relating to the phytosanitary regulations of the importing country
- Inspecting consignments

References

Agbaire PO, Esiefarienrhe E (2009) Air pollution tolerance indices (apti) of some plants around otorogun gas plant in Delta State, Nigeria. J Appl Sci Environ Manage 13(1):11–14

Agarwal PK, Banerjee B, Daryaci MG, Bhatia A, Bala A, Rani S (2006) InfoCrop: a dynamic simulation model for the assessment of crop yields, losses due to pests and environmental impact of agro-ecosystem in tropical environments. II. Performance of the model. Agr Syst 89:47–67

Ashmore MR (2005) Assessing the future global impacts of ozone on vegetation. Plant Cell Environ 28:949–964

Cole CKV, Duxbury J, Freney J, Heinemeyer O, Minami K, Mossier A, Paustian K, Rosenberg N, Sampson N, Saucerbeck D, Zhao Q (1997) Global estimates of potential mitigation of greenhouse gas emissions by agriculture. Nutr Cycl Agroecosyst 49:221–228

CPCB (2009) Ambient air quality data. Central Pollution Control Board, New Delhi. http://www.cpcb.nic.in/bulletin/del/2009html

Garner JHB (1994) Nitrogen oxides, plant metabolism, and forest ecosystem response. In: Alscher RG, Wellburn AR (eds) Plant responses to the gaseous environment: molecular, metabolic and physiological aspects. Chapman and Hall, London, pp 301–314

International Energy Agency (IEA) (2013) World Energy Outlook in International Press. London

IPCC (2001) Climate change 2001: the scientific basis. In: Houghton JT, Ding Y, Griggs DJ, Noguer M, van der Linden PJ, Dai X, Maskell K, Johnson CA (eds) Contribution of working group 1 to the third assessment report of the intergovernmental panel on climate change. Cambridge University Press, USA, 881 pp

Janzen HH (2004) Carbon cycling in earth systems; a soil science perspective. Agric Ecosyst Environ 104:399–417

Mosier AR, Duxbury JM, Freney JR, Heinemeyer O, Minami K, Johnson DE (1998) Mitigating agricultural emissions of methane. Clim Change 40:39–80

NRC (2001) National Research Council 2001: global air quality. An imperative for long-term observational changes. National Academy Press, Washington, DC

Oenema O, Wrage N, Velthof GL, van Groenigen JW, Dolfing J, Kuikman PJ (2005) Trends in global nitrous oxide emissions from animal production systems. Nutr Cycl Agroecosyst 72:51–65

Rai R, Agrawal M, Agrawal SB (2007) Assessment of yield losses in tropical wheat using Rao, C. S., Environmental pollution control engineering. New Age International Publishers. Revised 2nd edn, 2006

Smith KA, Conen F (2004) Impacts of land management on fluxes of trace greenhouse gases. Soil Use Manage 20:255–263

Tiwari S, Agrawal M, Marshall FM (2006) Evaluation of ambient air pollution impact on carrot plants at a sub urban site using open top chambers. Environ Monit Assess 119:15–30

U.S. Environmental Protection Agency (USEPA) (1996) Office of air quality planning and standards, review of national ambient air quality standards for particulate matter: policy assessment of scientific and technical information, Report No. EPA-452/R-96-013 (USEPA, Washington, D.C, V-2-V-24, V-27-V-28, V-71

US-EPA (2007) Latest finding on national air quality: status and trends through 2006. www.gov.epa/air/airtrends/2008

US-EPA (2012) Our nation's air, status and trends through 2010, EPA-454/R-12-001. http://www.epa.gov/airtrends/2010

Wall DH, Moore JC (1999) Interactions underground: soil biodiversity, mutualism, and ecosystem processes. Bioscience 49:109–117

WHO/UNEP (1992) Urban air pollution in megacities of the world. World Health Organization and the United Nations Environment Programme. Blackwell Scientific, Oxford

Word Bank (1996) Livable cities for the 21st century. Directions in development. World Bank, Washington, DC

World Bank (2009) The world bank annual report 2009. Year in review

World Health Organization (WHO) (2003) Health aspects of air pollution with particulate matter, ozone and nitrogen dioxide. Report EUR/03/5042688 of working group, Bonn, Germany, 13–15 January 2003. WHO Regional Office for Europe, Copenhagen

Pollution and Plants: Changing Policy Paradigms

13

Vandana Maurya

Abstract

Today, urban areas are afflicted with numerous problems, viz. increasing population, unplanned urbanisation, limited resources, industrialisation, deforestation and over-increasing pollution from different sectors which are disrupting the climate cycle. The world is witnessing huge upsurge in GHG emission level and now working more meticulously then never before to mitigate and adapt to this changing scenario. The transportation, industry and energy sectors are coming up as killer sector impacting economies negatively. This chapter gives the current Indian pollution scenario and discusses various interventions at policy levels, i.e. National Forest Policy (Available at www.moef.gov.in. Accessed on 3 Jan 2016, 1988), GIM, Compensatory Afforestation and REDD+. It recognises the need for policy development at sectoral level and its proper evaluation and implementation.

Keywords

Industrialization • Deforestation • GHGs emissions • Policy • Evaluation and implementation

13.1 Introduction

Just few days before the Paris summit, China's capital Beijing issued its first ever "red alert" for pollution and issued restriction on certain vehicles, construction work and other pollution-generating activities which worked as a warning system to all participating countries of the Paris summit to come to an agreement to deal with climate change and related issues. One hundred and ninety six countries came forward at Paris Climate Conference COP-21 which aimed at achieving a legally binding international agreement to keep average global temperature no more than 2 °C above pre-industrial temperatures. This agreement resulted in generation of US$100 billion per year as base by 2020, to help developing

V. Maurya (✉)
Centre for Science Policy, School of Social Sciences, Jawaharlal Nehru University, New Delhi 110067, India
e-mail: maurya.vandana09@gmail.com

U. Kulshrestha, P. Saxena (eds.), *Plant Responses to Air Pollution*,
DOI 10.1007/978-981-10-1201-3_13

nations in both mitigation and adaptation activities of climate change.

The air pollution is emerging out as global concern because of its short-term and long-term effect on economies. Rapid urbanisation, increasing population, access to fossil fuels and technology and non-availability and non-accessibility of cleaner fuels play an important role in piling up of toxic gases in the atmosphere. As per estimates of IPCC (2014), recent anthropogenic emissions of GHGs are influencing climate system and have widespread impact on human and natural systems. Of the 49 (±4.5) GtCO2eq/year in total anthropogenic GHG emissions in 2010, CO2 remains the major anthropogenic GHG accounting for 76 % (38 ± 3.8 GtCO2eq/year), while 16 % (7.8 ± 1.6 GtCO2eq/year) come from methane (CH4), 6.2 % (3.1 ± 1.9 GtCO2eq/year) from nitrous oxide (N2O) and 2.0 % (1.0 ± 0.2 GtCO2eq/year) from fluorinated gases (IPCC 2014).

Since 1850, temperature is progressively increasing, and the last three decades (1983–2012) were the warmest of the last 1,400 years in Northern Hemisphere (IPCC 2014). Cumulative anthropogenic CO_2 emissions in the atmosphere were 2040±310 GtCO2 in between 1750 and 2011, and about half of the anthropogenic CO_2 emissions have occurred in the last 40 years. Increased uptake of CO_2 by ocean has resulted in 26 % increase in acidity of ocean.

Due to the continuous increase in temperature, glaciers have continued to shrink worldwide. The annual mean Arctic sea ice extent has decreased at the range 3.5–4.1 % per decade over the period of 1979–2012. The ocean surface (upper 75 m) is warmed by 0.11 (0.09–0.13) degree Celsius per decade over the period of 1971–2010, and global mean sea level rose by 0.19(0.17–0.21) m during the period of 1901–2010 (IPCC 2014).

The energy sector contributed approximately 35 % of total anthropogenic GHG emission in 2010, and annual GHG emissions from the energy sector has witnessed an increase from 1.7 % per year from 1990 to 2000 to 3.1 % per year from 2000 to 2010 due to increased energy demand (Bruckner et al., 2014). In 2010, the direct GHG emission from the transportation sector produced was 7.0 GtCO2eq and was responsible for approximately 23 % of total energy-related CO_2 emissions (6.7 GtCO2) due to ever-increasing demands of mobility and movement of goods. Transport emission is expected to increase in the coming future and can reach around 12 GtCO2eq/year by 2050. Buildings accounted for 19 % of energy-related GHG emissions (included electricity related) (Lucon et al., 2014), whereas industry contributed over 30 % of global GHG emission in 2010 dominated by Asian region (Fischedick et al., 2014). Annual GHG emissions from agricultural production in 2000–2010 were estimated at 5.0–5.8 GtCO2eq/year, while annual GHG flux from land use and land-use change activities accounted for approximately 4.3–5.5 GtCO2eq/year (Smith et al. 2014).

13.2 Air Pollution: A Stinking Killer

Estimates by UNEP find that more than 1 billion people are exposed to outdoor air pollution annually, and 3.5 million deaths take place each year from outdoor air pollution. This death rate has witnessed a growth by 4 % worldwide, by 5 % in China and by 12 % in India between 2005 and 2010. Indoor air pollution is also linked to premature deaths and poor health conditions especially in developing nations. As per CSE's Body Burden 2015 report, air pollution is the fifth leading cause of death and results in 620,000 premature deaths caused by stroke, chronic obstructive pulmonary disease, ischemic heart disease, lower respiratory infection and trachea, bronchus and lung cancer, etc. in India. A study was conducted by Central Pollution Control Board (CPCB) based on source apportionment studies conducted in six cities, i.e. Delhi, Mumbai, Bangalore, Chennai, Kanpur and Pune. It found that air pollution is an emerging issue and requires immediate attention to control. PM10 and PM 2.5 in ambient air are significantly high in all locations due to natural and anthropogenic reasons especially in winter and post-monsoon season. Many cities have shown high EC to OC ratio (EC/OC) at various locations (CPCB 2011). Increasing population, urbanisation, migration to cities, changing behaviour and lifestyles, increased consumption rates, accessibility to technology, increasing income, usage of quiescent and inefficient technology and non-availability and non-affordability of clean and green technology and fuel are major reasons for increasing air pollution levels in India. It poses a huge economic burden on affected countries; an estimate by OECD shows that cost of air pollution was US$1.4 trillion in China and US$0.5 trillion in India in 2010 (UNEP 2014).

13.2.1 Transportation Sector

The transportation sector is one of the major sources of air pollution in India. It has witnessed a growth of total registered vehicles from 5.3 million in 1981 to 159 million in 2012. It has contributed up to 50 % of fine particulate matter concentrations (PM 2.5) (TERI 2014). Automotive vehicles emit several pollutants which can have negative impact on health of human and plants. Long-term emission can lead to climate change. IPCC (2014) estimated that GHG emissions from the transportation sector have doubled since 1970 and in 2010 produced 7.0 GtCO2eq of direct GHG emissions (including non-CO2 gases) which can reach up to 12 GtCO2eq/year by 2050 (Sims et al. 2014). Petrol-/gasoline-driven vehicles mainly emit hydrocarbons and carbon monoxide, whereas diesel-based vehicles emit oxides of nitrogen and particulate matter (CPCB 2010). Increasing number of vehicles, presence of old technology, poor vehicle maintenance, inadequate infrastructure and low quality fuel are major barriers towards achieving the goal of clean air.

13.2.2 Domestic Sector

In India, the domestic sector is emerging as the major consumer of electricity after the industrial sector. Its electricity consumption has witnessed a increase of 882,592GWh during 2013–2014 from 411,887 GWh during 2005–2006 (CAGR is 8.84 %). The number of households (rural and urban) increased at a rate of 2.3 % per annum which led to increase in the energy consumption. In the residential sector, lightning takes around 40 % of electricity in households, followed by fans (31 %) and other appliances (28 %) (Phadke et al. 2014). Studies suggest that energy consumption will increase in bottom third of the population rapidly. It will increase from 13 to 19 % in urban areas and 11–23 % in rural areas (ESMAP, 2010).

A study indicates that total residential energy use is expected to increase 65–75 % in 2050 due to increase in population and economic growth compared to 2005, which can lead to increase in emissions to by nine to ten times (Ruijven et al. 2011). The demand for residential energy services is due to many factors, i.e. population, household size, income, geography and demography, and is also affected by energy consumption behaviour of consumer, and this behaviour is shaped by consumer's socio-demographics and social contextual variables like cost, legal regulations, policies, availability of technology and information, role models, pricing and social conditions.

13.2.3 Industrial Sector

In 2013–2014, the industrial sector consumes 52.72 % of the total final energy consumption, and within the industrial sector, iron and steel industry accounts for 18.22 % and is the most polluted sector. As stated earlier, industries emit the highest amount of GHGs and contribute towards climate change as these feed on dirty fuels, i.e. fossil fuels. In industry-wise consumption, the electricity generation sector consumes the highest quantity of raw coal and lignite and is witnessing an increasing trend in the last decade. While cement and paper industry is observing a negative growth with CAGR of −2.46 % and −5.49 %, respectively. In case of lignite, paper and textile industries have noticed a steady growth with CAGR of 12.74 % and 10.99 %, respectively (MoSPI 2015) (Table 13.1).

Table 13.1 Consumption trends of raw coal and lignite (in MTs) from different industries

Industry	Raw coal (2005–2006)	Raw coal (2013–2014)	CAGR (2005–2006 to 2013) in %	Lignite (2005–2006)	Lignite (2013–2014)	CAGR (2005–2006 to 2013) in %
Electricity	306.04	427.23	3.78	23.36	36.48	5.08
Steel and washery	19.66	23.13	1.82	–	0.03	–
Cement	14.97	11.96	−2.46	0.79	1.40	6.54
Paper	2.77	1.67	−5.49	0.23	0.66	12.74
Textiles	0.29	0.36	2.51	1.11	2.83	10.99
Others[a]	51.85	107.54	8.44	4.86	2.51	-7.10

Source: energy statistics 2015

[a]Other industry includes sponge iron, colliery consmn., fertilisers, jute, bricks, coal for soft coke, colliery, fertilisers and other industry consumption

Table 13.2 Change in consumption of conventional source of energy in India (Source: energy statistics, 2015 (MoSPI))

S. no	Energy source	2005–2006	2013–2014	CAGR in % (2005–2006 to 2013–2014)
1	Coal (in MT)	407.04	571.89	3.85
2	Lignite (in MT)	30.23	43.90	4.23
3	Crude oil (in MMT)	130.11	222.50	6.14
4	Natural gas (in BCM)	31.33	34.64	1.12
5	Electricity (in GWh)	411,887	967,950	9.95

13.2.4 Energy Sector

During the Twelfth Plan period (2012–2017), the total capacity addition is targeted to achieve 88,537 MW comprising 26,182 MW in the central sector, 15,530 MW in the state sector and 46,825 MW in the private sector. Till December 2014, 56.5 % (50,058.22 MW) of cumulative capacity addition is achieved (GoI 2015). With increasing demands from different sectors, consumption of conventional sources of energy had witnessed a steady growth in the last decade (Table 13.2). Electricity has witnessed the highest CAGR of 9.95 % during 2005–2006 to 2014–2015. The CAGR of coal, lignite, crude oil and natural gas was 3.85, 4.23, 6.14 and 1.12 % during the same period.

This increase is due to increased demand from the industrial, transportation and electricity generation sector. In 2013–2014, the highest consumption of electricity is from the industrial sector (43.83 %), followed by domestic (22.46 %), agriculture (18.03 %), commercial (8.72 %), traction and railways (2 %) and other sectors (5 %), while, per capita energy consumption (PEC) has increased with CAGR of 4.3 % from 13694.83 M joules in 2005–2006 to 19522.15 M joules in 2013–2014 (MoSPI 2015).

13.3 Need for Policies for Plants

The forest plays a very important role in maintaining and disrupting the carbon cycle. Cumulative CO_2 emissions from Forestry and Other Land Use (FOLU) since 1750 increased from 490 ± 180 GtCO2 in 1970 to 680 ± 300 GtCO2 in 2010. The agriculture, forestry and other land use (AFOLU) sector accounts for ~10–12 GtCO2eq/year of net anthropogenic GHG emissions mainly from deforestation, agricultural emissions from soil and nutrient management and livestock. Therefore, policies governing both mitigation and adaptation are important for reducing emissions. REDD+ is one of the effective measures to reduce emissions (Nabuurs et al. 2007). In India, various plans and policies are implemented to maintain qualitative forest cover for inclusive and sustained development. These plans and policies are discussed in given section.

13.3.1 National Forest Policy, 1988

National Forest Policy (1988) aimed at maintaining environmental stability through preservation and restoration of ecological balance, conserving limited natural forest, checking deterioration of resources, limiting desertification, increasing quality and quantity of forest cover, utilising forest produce efficiently, meeting forest requirement of rural or tribal population and disseminating knowledge and information.

It envisages green cover to have a minimum of one-third of local land area of the country. While, in hilly and mountainous regions, it should be two-third of the area. Further, it calls for need-based and time-bound programme for afforestation and for the development of fuel wood and fodder on all degraded and denuded lands of the country. It encourages plantation of trees alongside roads, railway lines, rivers, streams and canals and other unutilised lands under public or private ownership. It mandates development of green belts in urban/industrial areas as well as in arid tracts. It asks to modify land laws to facilitate and motivate individuals to improve the quality of forests (NFP 1988). National Forest Policy (1988) mandates plantation of trees alongside roads, railway lines, rivers, streams and canals and unutilised lands under state/corporate and institutional or private ownership and emphasises green belt development in urban/industrial and arid tracts.

Green belt is referred to as the buffer zone created beyond which industrial activity may not be carried or can be said as planned open spaces safeguarded from developmental activities. It aims at protecting sensitive areas to maintain ecological balance and can act as sink for harmful gases released from urban areas and maintains ecological, economical and productive services. The size of green belt varies from case to case on the type of industrial activity carried out.

There is no such regulation or policy for formation of green belts in India, but increasing the green cover is comprehensively undertaken by National Forest Policy 1988, GIM, REDD and later REDD+. Environment Management Plan (EMP) of the Ministry of Environment and Forest (MoEF) mandates building of 1–1.5 km of green belt around community buildings and townships and around power plants; it should cover about 33 % of plant area. Under GIM, forest/tree cover is to be increased to the extent of 5 million hectare (mha) and improvement of quality of forest/tree cover on another 5 mha of forest/non-forest lands. REDD aims at incentivising protection, management and conservation of forest resources for reducing carbon stock in atmosphere.

13.3.1.1 Green Belts as Sinks of Gases

Green plants form a surface capable of absorbing air pollutants and act as sink for pollution but only within tolerance limits of constituent plants. For designing green belts, source-oriented approach and receptor-oriented approach are used. Source-oriented approach is suitable if pollutants of single industry need to be contained, whereas in urban-industrial complexes with multiple sources of pollution, receptor-oriented approach is feasible (CPCB, 2000). As plants also get affected by increased concentration of pollution, negative effects of pollutants on plant health are enumerated in Table 13.3.

13.3.2 Green India Mission

India introduced National Action Plan for Climate Change (NAPCC) in 2009 with eight missions with special focus on renewable energy, increasing forest cover, sustainable habitat and agriculture, energy efficiency, water conservation, knowledge dissemination about climate change and maintenance of Himalayan ecosystem. Guiding principles for NAPCC are sustainable and inclusive development, efficient use of natural resources, deployment of appropriate technology for mitigation and adaptation of GHGs, involvement of major stakeholders with better outreach and implementation (GoI 2009).

Table 13.3 Effect of pollutants on plants' health (Source: CPCB guidelines for green belts)

S. no	Gases	Sources	Effects
1.	SO2	Emissions from industries, households, vehicles and natural decomposition	Formation of marginal and interveinal chlorotic, bronzed or necrotic areas. Necrotic areas extend and are visible on both epidermal surfaces
2.	NOx	Emission from industries, fertiliser industries, burning of fuels, natural decomposition	Discoloured spots of grey-green or light brown colour and later turns into bleached or necrotic spots in interveinal areas of leaves
3.	Fluorine	Combustion of fossils, smelting of ores and rock cycles	Chlorosis of leaf tip and with increased concentration of the injury may extend along margins and inwards along veins. Injured brown or dead areas of leaves become necrotic, leading to premature leaf fall
4.	CO and CO_2	Emission from vehicles, industries, burning of fuels, oil refineries	No phytotoxic effect
5.	NH_3	Decomposition of organic matter of different origins including excreta fertiliser breakdown, coal combustion and releases from industries	Blackening and bleaching of leaves, spotting, brown lesions between veins and colour change of fruits
6.	HC	Emissions from automobiles, oil refining processes and chemical factories	Distortion of foliage, excessive curvature in growth, chlorosis, senescence and flower abscission
7.	SPM	Emission from industries, households, factories, burning of fossil fuels, mining, thermal power plants and dust	Finer particles clog stomatal apertures and prevent gaseous exchange by leaves, increase weight and temperature of leaves. Dust deposition on stigmatic surface of flowers reduces effective pollination and hence fruit yields

Increasing forest cover is one of the major steps towards mitigation of climate change as it neutralises ~11% of India's GHG emissions. Therefore, under the aegis of NAPCC, national mission for a green India or Green India Mission (GIM) was launched with a budget of Rs 46,000 crores (approx. US$ 10 billion) over a period of 10 years. It proposes to take holistic view of afforestation, enhancing quality and quantity of biodiversity and restoring ecosystem services with more decentralisation. The local community, especially the young (community foresters), will be major stakeholder, and Gram Sabha will assess mission implementation at a village level. The main goals are as follows:

- To increase forest/tree cover to the extent of 5 million hectare (mha) and improve quality of forest/tree cover on another 5 mha of forest/non-forest lands
- To improve or enhance of ecosystem services, provisioning services and non-timber forest products
- To increase forest-based livelihood income of about three million households

It has adopted an integrated cross-sectoral approach covering public and private lands. Local communities had a key role in planning, decision-making, implementing and monitoring of mission goals. GIM is converged with related mission of NAPCC and other programmes like Compensatory Afforestation Management and Planning Authority (CAMPA) through converging compensation afforestation fund and Mahatma Gandhi National Rural Employment Guarantee Scheme (MGNREGS), a scheme administered by Ministry of Rural Development for wider and better coordination in developing forest and their fringe areas in holistic and sustainable manner to address climate change (MoEF 2015).

13.3.3 Reducing Emissions from Deforestation and Forest Degradation (REDD)

REDD is a global endeavour and aims at incentivising protection, management and conservation of forest resources. REDD+ goes further and incentivises positive elements of conservation, sustainable management of forests and enhancement of forest carbon stocks. Under this, financial value is created for carbon stored and enhancement of biomass and soil of standing forests. It incorporates benefits of livelihood improvement, biodiversity conservation and food security services. Global REDD+ mechanism is win-win situation for India as it encourages conservation, inclusive development, reduction of emissions and increase of forest cover. Incentives from REDD+ would be passed to local communities for sustained efforts to conserve forest. It is estimated that a REDD+ programme for India could provide capture of more than 1 billion tonnes of additional CO_2 over the next 3 decades and provide more than US$ 3 billion as carbon service incentives under REDD+ (MoEF 2010).

13.3.4 Compensatory Afforestation

To compensate the diversion of forest land, central government had mandated the transfer and mutation of nonequivalent non-forest land in favour of State Forest Department (SFD) for creation of compensatory afforestation from the funds provided by user agency and its effective implementation, monitoring and evaluation which is done by the Compensatory Afforestation Management and Planning Authority (CAMPA). CAMPA works at central as well as at state level and aims to promote afforestation and regeneration activities as a way to compensate for forest land diverted to non-forest land.

13.4 Changing Policy Paradigm

There are various instruments in policymaking, i.e. economic, regulatory, voluntary and informational. Economic instrument for climate change mitigation includes taxes and subsides, for example, in India there is tax on using coal and subsidy for encouraging renewable energy usage. Regulatory approaches include rules and penalty for non-compliance, e.g. NAAQS for air pollutants, while information measures aim at dissemination of complete, reliable, cost-effective and quick information among the public, e.g. BEE Star Labels. These measures go hand in hand and increase the effectiveness (in terms of cost and output) of sector-specific policies. Table 13.4 have listed various policy instruments in different sectors.

Looking at present situation, it becomes inevitable to frame rightly timed policies which can result in most effective outcomes. Policies do not follow one-size-fit-all scenario; therefore, policies may and should differ at international, national and regional levels, and policymakers should make sure that flexible and relevant policy is made which can be implemented to yield best results. Studies have found

Table 13.4 Sector policy instruments

Policy instruments	Energy	Transport	Buildings	Industry	AFOLU	Human settlements and infrastructure
Economic instruments – taxes (carbon taxes may be economy wide)	Carbon taxes	Fuel taxes	Carbon and/or energy taxes (either sectoral or economy wide)	Carbon tax or energy tax	Fertiliser or nitrogen taxes to reduce nitrous oxide	Sprawl taxes, impact fees, exactions, split-rate property taxes, tax increment finance, betterment taxes, congestion charges
		Congestion charges, vehicle registration fees, road tolls		Waste disposal taxes or charges		
		Vehicle taxes				
Economic instruments – tradable allowances (may be economy wide	Emission trading (e.g. EU ETS)	Fuel and vehicle standards	Tradable certificates for energy efficiency improvements (white certificates)	Emissions trading	Emission credits under CDM	Urban-scale cap and trade
	Emission credits under the Kyoto protocol's clean development mechanism (CDM)			Emission credits under CDM	Compliance schemes outside Kyoto protocol (national schemes)	
	Tradable green certificates			Tradable green certificates	Voluntary carbon markets	
Economic instruments – subsidies	Fossil fuel subsidy removal	Biofuel subsidies	Subsidies or tax exemptions for investment in efficient buildings, retrofits and products	Subsidies (e.g. for energy audits)	Credit lines for low carbon agriculture, sustainable forestry	Special improvement or redevelopment districts
		Vehicle purchase subsidies				
	Feed-in tariffs for renewable energy	Feebates	Subsidised loans	Fiscal incentives (e.g. for fuel switching)		
	Capital subsidies and insurance for first-generation carbon dioxide capture and storage (CCS)					

Regulatory approaches	Efficiency or environmental performance standards	Fuel economy performance standards	Building codes and standards	Energy efficiency standards for equipment	National policies to support REDD+ including monitoring, reporting and verifying	Mixed use zoning
		Fuel quality standards				
	Renewable portfolio standards for renewable energy	GHG emission performance standards	Equipment and appliance standards	Energy management systems (also voluntary)	Forest law to reduce deforestation	Development restrictions
	Equitable access to electricity grid	Regulatory restrictions to encourage modal shifts (road to rail)	Mandates for energy retailers to assist customers invest in energy efficiency	Voluntary agreements (where bound by regulation)	Air and water pollution control GHG precursors	Affordable housing mandates
	Legal status of long-term CO_2 storage	Restriction on use of vehicles in certain areas		Labelling and public procurement regulation	Land-use planning and governance	Site access controls
		Environmental capacity constraints on airports				Transfer development rights
		Urban planning and zoning restrictions				Design codes
						Building codes
						Street codes
						Design standards
Information programmes		Fuel labelling	Energy audits	Energy audits	Certification schemes for sustainable forest practices	
		Vehicle efficiency labelling	Labelling programmes	Benchmarking	Information policies to support REDD+ including monitoring, reporting and verifying	
			Energy advice programmes	Brokerage for industrial cooperation		

(continued)

Table 13.4 (continued)

Policy instruments	Energy	Transport	Buildings	Industry	AFOLU	Human settlements and infrastructure
Government provision of public goods or services	Research and development	Investment in transit and human-powered transport	Public procurement of efficient buildings and appliances	Training and education	Protection of national, state, and local forests	Provision of utility infrastructure such as electricity distribution, district heating/cooling and wastewater connections, etc.
		Investment in alternative fuel infrastructure				
	Infrastructure expansion (district heating/cooling or common carrier)			Brokerage for industrial cooperation	Investment in improvement and diffusion of innovative technologies in agriculture and forestry	Park improvements
		Low-emission vehicle procurement				Trail improvements
						Urban rail
Voluntary actions			Labelling programmes for efficient buildings	Voluntary agreements on energy targets or adoption of energy management systems or resource efficiency	Promotion of sustainability by developing standards and educational campaigns	
			Product eco-labelling			

Source: IPCC report (2014)

Table 13.5 Evaluation of major forest plans and policies based on various IPCC policy instruments

Policy instruments energy	National forest policy (1988)	GIM	Compensatory afforestation	REDD+
Economic instruments – taxes (carbon taxes may be economy wide)	0	0	Y	0
Economic instruments – tradable allowances may be economy wide	0	Y	Y	Y
Economic instruments – subsidies	Y	0	0	Y
Regulatory approaches	Y	Y	Y	Y
Information programmes	Y	Y	Y	Y
Government provision of public goods or services	Y	Y	Y	Y
Voluntary actions	Y	Y	Y	Y

Source: author's work
Y stands for yes and 0 stands no

that even soft approaches can also help in mitigating climate change (Verma and Kulshresta 2015) (Table 13.5).

Sector-specific policies are more preferable over economy-wide and market-based policies and help to overcome market failure. Therefore, the need of the hour is to formulate sector-specific policies with national and regional interest with more flexibility and decentralisation rather than formulation of policies formed under international pressures.

13.5 Conclusion

This chapter examines the current situation of emissions and finds that GHG emission is increasing at steady pace, and relevant policy intervention can help in mitigation. The chapter begins by drawing attention towards the recently concluded UNFCCC Paris summit which aimed at achieving a legally binding international agreement to keep average global temperature no more than 2 °C above pre-industrial temperatures. General consensus on legally binding international agreements shows the concern of climate change all over the globe.

The next section has discussed the present scenario of air pollution in India from different sectors. The industrial and energy sectors are highly fuel guzzling and led to huge emissions. In urban areas, emissions from transportation and industries pollute the air up to dangerous levels. Huge upsurge in number of vehicles, faulty technologies, non-availability of clean fuels and absence of stringent standards and norms are major reasons of increasing pollution in the transportation sector. This can be improved by introduction of recent technology, better infrastructure, strict implementation of standards and norms, penalty for non-compliance and better governance. While in the domestic sector, access to clean and affordable fuel, access to modern and efficient technology, availability of reliable and affordable information, changing behaviour towards consumption and development of standards can reduce the emissions. At supply side, the energy sector can be revamped in different ways by improving technology and quality of fuel, increasing efficiency of plants, switching to renewable energy sources, decentralising more, reducing T&D loss and improving governance.

Forests are seen as best carbon sequesters as they can remove 5–11 tons of CO_2/ha/year depending on location and productivity. They not only sequester CO_2 but also act as natural cleanser and provide timber and various other goods and services. The last section has discussed the need for policymaking and emphasises that sector-specific policies should be formed and implemented as it is more effective and efficient rather than economy-wide or market-wide policies. Then various plans and policies present for reducing emission by plantation or afforestation are discussed and analysed. These include NAPCC, National Forest Policy 1988, REDD and REDD+, CAMPA and Green India Mission.

References

Body burden report. Available at http://cseindia.org/content/cse%E2%80%99s-inaugural-state-india%E2%80%99s-health-report-connects-most--environmental-factors-some-gravest. Accessed on 30 Dec 2015

Bruckner T, Bashmakov IA, Mulugetta Y, Chum H, de la Vega Navarro A, Edmonds J, Faaij A, Fungtammasan B, Garg A, Hertwich E, Honnery D, Infield D, Kainuma M, Khennas S, Kim S, Nimir HB, Riahi K, Strachan N, Wiser R, Zhang X (2014) Energy systems. In: Edenhofer O, Pichs-Madruga R, Sokona Y, Farahani E, Kadner S, Seyboth K, Adler A, Baum I, Brunner S, Eickemeier P, Kriemann B, Savolainen J, Schlömer S, von Stechow C, Zwickel T, Minx JC (eds) Climate change 2014: mitigation of climate change. Contribution of working group III to the fifth assessment report of the a Intergovernmental Panel on Climate Change. Cambridge University Press, Cambridge/New York

CAMPA. Available at www.moef.org. Accessed on 2 Jan 2016

CPCB (2000) Guidelines for development of green belts, programme objective series: PROBES/75/1999–2000. CPCB, New Delhi

CPCB (2010) Status of vehicular pollution control programme in India, CPCB, New Delhi

CPCB (2011) Air quality monitoring, emission, inventory and source apportionment study for Indian cities, CPCB, New Delhi

ESMAP (2010): Briefing note 006/10, Energy intensive sector of Indian economy: path to low carbon development. Accessed on 25 May 2014

Fischedick M, Roy J, Abdel-Aziz A, Acquaye A, Allwood JM, Ceron J-P, Geng Y, Kheshgi H, Lanza A, Perczyk D, Price L, Santalla E, Sheinbaum C, Tanaka K (2014) Industry. In: Edenhofer O, Pichs-Madruga R, Sokona Y, Farahani E, Kadner S, Seyboth K, Adler A, Baum I, Brunner S, Eickemeier P, Kriemann B, Savolainen J, Schlömer S, von Stechow C, Zwickel T, Minx JC (eds) Climate change (2014) mitigation of climate change. Contribution of working group III to the fifth assessment report of the Intergovernmental Panel on Climate Change. Cambridge University Press, Cambridge/New York

GoI (2015) Economic survey of India 2014–15, GoI, New Delhi

GoI (2009) National Action Plan for climate change, GoI, New Delhi accessed from http://envfor.nic.in/ccd-napcc on 3 Jan 2016

Green belts. Available at http://greencleanguide.com/regulatory-provisions-for-green-belt-development-in-india/. Accessed on 2 Jan 2016

Green belts, http://www.legalserviceindia.com/articles/greeneco.htm. Accessed on 2nd Jan 2016

IPCC (2014) Summary for policymakers. In: Edenhofer O, Pichs-Madruga R, Sokona Y, Farahani E, Kadner S, Seyboth K, Adler A, Baum I, Brunner S, Eickemeier P, Kriemann B, Savolainen J, Schlömer S, von Stechow C, Zwickel T, Minx JC (eds) Climate change 2014: mitigation of climate change. Contribution of working group III to the fifth assessment report of the Intergovernmental Panel on Climate Change. Cambridge University Press, Cambridge/New York

Lucon O, Urge-Vorsatz D, Zain Ahmed A, Akbari H, Bertoldi P, Cabeza LF, Eyre N, Gadgil A, Harvey LDD, Jiang Y, Liphoto E, Mirasgedis S, Murakami S, Parikh J, Pyke C, Vilariño MV (2014) Buildings. In: Edenhofer O, Pichs-Madruga R, Sokona Y, Farahani E, Kadner S, Seyboth K, Adler A, Baum I, Brunner S, Eickemeier P, Kriemann B, Savolainen J, Schlömer S, von Stechow C, Zwickel T, Minx JC (eds) Climate change (2014) mitigation of climate change. Contribution of working group III to the fifth assessment report of the Intergovernmental Panel on Climate Change. Cambridge University Press, Cambridge/New York

MOEF (2010) India's forest and REDD+ Available at http://www.moef.gov.in/. Accessed on 3 Jan 2016

MOEF (Ministry of Environment and Forests) Report. (2015). pp 1–119

MoSPI (2015) Energy Statistics, MoSPI, GOI, New Delhi

Nabuurs GJ, Masera O, Andrasko K, Benitez-Ponce P, Boer R, Dutschke M, Elsiddig E, Ford-Robertson J, Frumhoff P, Karjalainen T, Krankina O, Kurz WA, Matsumoto M, Oyhantcabal W, Ravindranath NH, Sanz Sanchez MJ, Zhang X (2007) Forestry. In: Metz B, Davidson OR, Bosch PR, Dave R, Meyer LA (eds) Climate change 2007: mitigation. Contribution of working group III to the fourth assessment report of the intergovernmental panel on climate change. Cambridge University Press, Cambridge/New York

National forest policy (1988) Available at www.moef.gov.in. Accessed on 3rd Jan 2016

Phadke A, Abhyankar N, Shah D (2014) Avoiding 100 new power plants by increasing efficiency of room air conditioners in India: opportunities and challenges, pp 1–14. Available at superefficient.org/Resources/~/media/files/EEDAL%20papers%20-% 202013/031_shah_finalpaperEEDAL13.pdf. Accessed on 20th May 2014

Ruijven et al (2011) Model projections for household energy use in India. Energy Policy 39:7747–7761

Sims R, Schaeffer R, Creutzig F, Cruz-Núñez X, D'Agosto M, Dimitriu D, Figueroa Meza MJ, Fulton L, Kobayashi S, Lah O, McKinnon A, Newman P, Ouyang M, Schauer JJ, Sperling D, Tiwari G (2014)

Transport. In: Edenhofer O, Pichs-Madruga R, Sokona Y, Farahani E, Kadner S, Seyboth K, Adler A, Baum I, Brunner S, Eickemeier P, Kriemann B, Savolainen J, Schlömer S, von Stechow C, Zwickel T, Minx JC (eds) Climate change (2014) mitigation of climate change. Contribution of working group III to the fifth assessment report of the Intergovernmental Panel on Climate Change. Cambridge University Press, Cambridge/New York

Smith P, Bustamante M, Ahammad H, Clark H, Dong H, Elsiddig EA, Haberl H, Harper R, House J, Jafari M, Masera O, Mbow C, Ravindranath NH, Rice CW, Robledo Abad C, Romanovskaya A, Sperling F, Tubiello F (2014) Agriculture, Forestry and Other Land Use (AFOLU). In: Edenhofer O, Pichs-Madruga R, Sokona Y, Farahani E, Kadner S, Seyboth K, Adler A, Baum I, Brunner S, Eickemeier P, Kriemann B, Savolainen J, Schlömer S, von Stechow C, Zwickel T, Minx JC (eds) Climate change 2014: mitigation of climate change. Contribution of working group III to the fifth assessment report of the Intergovernmental Panel on Climate Change. Cambridge University Press, Cambridge/New York

TERI (2014) Advancement of fuel quality and emissions norms to improve urban air quality in India, Policy Brief, Issue 14, TERI

UNEP (2014) Air pollution: world's worst environmental health risk. Available at http://www.unep.org/urban_environment/Issues/urban_air.asp. Accessed on 30 Dec 2015

Verma K, Kulshresta U (2015) CO_2 emissions and soft approaches of mitigation for NCR-Delhi, Climate Change, World Focus. Accessed on 5 Jan 2016

Tropospheric O_3: A Cause of Concern for Terrestrial Plants

14

Richa Rai, Aditya Abha Singh, S.B. Agrawal, and Madhoolika Agrawal

Abstract

Tropospheric ozone (O_3) is a phytotoxic pollutant causing risk to food production, pasture, and forest communities. In the present scenario, unsustainable resource utilization has turned this secondary pollutant into a major component of global climate change. The background levels of O_3 are very high, and IPCC projections have shown that it will increase by 20–25 % in 2050 and 40–60 % in 2100, causing severe consequence on global food security. Ozone enters plants through stomata, where it can be dissolved in the apoplastic fluid. Ozone has several potential effects on plants: direct reaction with cell membranes, generation of ROS and H_2O_2 (which alter cellular function by causing cell death), induction of premature senescence, negative impact on photosynthetic machinery and up- or downregulation of antioxidants, defense reactions, and variations in metabolic pathways. Tropospheric O_3 causes changes in tree diameter, wood quality, herbivory pattern, forage quality, and crop yield and quality. In this chapter, we make an attempt to present an overview of O_3 concentrations throughout the globe and its impact on agricultural crops, forest, and grassland ecosystems. We summarized the information available on plant responses to O_3 at physiological, cellular, and biochemical levels; crop yield; and forest and grassland communities at present concentrations and also under projected future concentrations.

Keywords

Tropospheric ozone • IPCC • Photosynthetic machinery • ROS • Crop yield

14.1 Introduction

During the last few decades, tropospheric ozone (O_3) has become one of the most wide spread toxic pollutants around the globe (IPCC 2013; Booker et al. 2009), negatively affecting the crop

R. Rai • A.A. Singh • S.B. Agrawal
M. Agrawal (✉)
Laboratory of Air Pollution and Global Climate Change, Department of Botany, Banaras Hindu University, Varanasi 221005, India
e-mail: madhoo.agrawal@gmail.com

U. Kulshrestha, P. Saxena (eds.), *Plant Responses to Air Pollution*,
DOI 10.1007/978-981-10-1201-3_14

productivity and hence a major threat to global food security (Rai and Agrawal 2012). Since the industrial revolution, the global average O_3 concentrations have increased from around 20–30 ppb to present-day values of 30–50 ppb with significant variability in its spatial distribution (IPCC 2013). Due to substantial escalation in industrialization and urbanization, ground-level O_3 is considerably higher in the Northern Hemisphere than the Southern Hemisphere. Its abundance in mid-latitudes of Northern Hemisphere has increased twice or more since the preindustrial era with present background O_3 concentrations ranging from 35 to 50 ppb (Cooper et al. 2010; IPCC 2013). At the global scale, O_3 concentrations are higher in Central Europe, Eastern China, and the Eastern USA (Royal Society 2008). In Europe, the highest O_3 levels occur in Central and Southern Europe (Royal Society 2008). High levels of O_3 occur in the tropics, and worryingly, there are projections of further increase in O_3 concentrations (Harmens 2014).

In plants, O_3 predominantly penetrates in internal environment of leaf tissues through stomatal openings, where it generates a cascade of reactive oxygen species (ROS) in surrounding aqueous medium that cause membrane damage, alteration of gene expression, impairment of photosynthetic proteins, degradation of chlorophyll, and alterations in metabolic activities (Booker et al. 2009; Fuhrer 2009; Singh et al. 2014a). Oxidative stress due to enhanced ROS production under O_3 stress triggers an array of complex antioxidant defense strategies which may be enzymatic or nonenzymatic (Blokhina et al. 2003). Ozone is also known to cause reduction in photosynthesis (Rai and Agrawal 2012; Ainsworth et al. 2012). Thus, reduced photosynthesis, induced defense system, and secondary metabolic pathways conclusively lead to decreased carbon assimilation and altered carbon partitioning, resulting in reduced total biomass accumulation and yield (Singh et al. 2015).

The negative impacts of O_3 on growth and yield of agricultural crops affecting food production across the globe have been widely reported (Ashmore 2005; Fuhrer 2009; Emberson et al. 2009; Feng et al. 2008, 2009; Sarkar and Agrawal 2010a, b; Rai and Agrawal 2014; Singh et al. 2014b). Using various approaches, Booker et al. (2009) estimated that the current yield losses due to O_3 are in the range of 5–15 %. Global crop yield losses for four major crops (wheat, rice, soybean, and maize) due to ambient O_3 in the year 2000 were estimated to be worth $14–26 billion (Van Dingenen et al. 2009). For the same year in the European Union, an estimated crop yield of $ 6.7 billion was calculated for the arable crops. The increasing O_3 concentrations have also been implicated as one of the factors contributing in forest decline (Royal Society 2008), affecting the forest productivity linked to direct economic losses (Percy et al. 2007). Concurrently, O_3 poses a threat to seminatural ecosystems including grasslands, reducing the primary productivity of wild plants as well as species biodiversity (Agathokleous et al. 2015; Ainsworth et al. 2012).

Apart from the crop yield and primary productivity losses in natural ecosystems, second-tier O_3 effects include changes in herbivory pattern and alterations in plant interactions with diseases, pests, and insects (Ashmore 2005). Infection of a foliar pathogen on trembling aspen increased under elevated O_3 in the Aspen Free Air CO_2 Enrichment (FACE) experiment due to the changes in the leaf surface properties (Karnosky et al. 2005). In the same experiment, elevated O_3 also affected the performance of forest pests, which were related to changes in plant chemistry or the increased abundance of natural enemies. Chronic exposure to O_3 stimulates the carbon fluxes from the primary to secondary metabolic pathways, leading to the synthesis of secondary products (Iriti and Farror 2009), which may cause alteration in forage nutritive value, phytopathology, and natural enemy interaction and may perhaps promote establishment of invasive species (Booker et al. 2009). Ozone also affects competitive ability in different species which in the long term results in changes in the species and genetic composition as well as functioning of seminatural plant communities or ecosystems with impacts on nutrient cycling and carbon sequestration (Fuhrer et al. 2003; US EPA 2006; Harmens 2014; IPCC 2014). A number of studies on real and artificial plant communities have revealed that O_3 can modify the events of plant competition and hence species composition (Bender et al. 2006), and the performance of more sensitive species tends to be reduced further by O_3 in the communities due to competition compared to monoculture (Fuhrer et al. 2003). Rising O_3 concentration is also a threat to forest growth and species composition (Ashmore 2005; Wittig et al. 2009; Paoletti et al. 2010). Though the vulnerability to natural ecosystems such as forests and grasslands to rising O_3 is established, studies on these aspects are scarce.

The present chapter highlights the physiological and defense-related responses affecting biomass and yield in crops, forest tree species, and grasses under O_3 stress. Ozone-induced secondary responses in terms of changes in herbivory pattern, phytopathology, pest interactions, and competitive hierarchies were also discussed. Literature collated for the chapter are chiefly based on field studies, Open Top Chambers (OTCs), and FACE conducted in near natural environmental conditions.

14.2 Ozone Formation

Being a secondary pollutant, O_3 is formed by various precursor gases such as NOx, CO, CH_4, and non-methane VOCs, emitted from a wide range of sources that are either natural in origin or driven by human activities. Energy generation, industrial processes, transport, agriculture, biomass burning, and land use changes are significant sources of O_3 precursor gases (Royal Society 2008).

Ozone is produced in planetary boundary layer (PBL), free troposphere, and in the stratosphere. In the stratosphere, O_3 is produced due to photolysis of O_2 by ultraviolet radiation into atomic oxygen to form O_3. However, in the troposphere, O_3 formation occurs due to photolysis of NO_2. In the free troposphere, O_3 formation depends on reaction of CH_4, CO, and non-methane organic compounds with NOx. These reactions are principally controlled by sunlight and temperature. Nitrogen dioxide diminishes when O_3 reaches its peak. O_3 concentration peaks during the late morning and early afternoon hours (Krupa and Manning 1988).

14.3 Variations in Ozone Concentrations: Present Scenario and Future Projections

14.3.1 Worldwide O_3 Distribution

Tropospheric O_3 concentrations are persistently increasing worldwide (Mittal et al. 2007; Rai and Agrawal 2012). Between late nineteenth century and 1980, concentrations of background O_3 in the mid-latitudes of Northern Hemisphere doubled to about 30–35 ppb and have since increased by another 5 ppb reaching up to 35–40 ppb (Royal Society 2008). The peak values of O_3 continued to exceed the WHO guideline value of 50 ppb in many countries, including Latin America, North America, Europe, and Africa (WHO 2006). In Southern Hemisphere, an increase of 30 ppb in O_3 concentrations was reported in South America and Africa (Zeng et al. 2008). Springtime O_3 increased by 0.46 ppb $year^{-1}$ during 1985–2007 over Western coastal USA(Cooper et al. 2010). In rural agricultural areas of the USA, the mean O_3 concentration was reported between 50 and 60 ppb (US EPA 2006). Jaffe and Ray (2007) reported a significant increase in O_3 with a mean trend of 0.26 ppb $year^{-1}$ in seven remote rural sites in Western USA between the years 1987 and 2004.

Despite reductions in emissions of anthropogenic O_3 precursors, an increase in background O_3 in lower European troposphere is reported (Chevalier et al. 2007). The report published by Royal Society (2008) documented that background O_3 concentrations in Europe are still rising and predicted to rise at least till 2030 partly due to hemispheric transport of O_3 precursors from developing areas of the world. All countries of Europe are experiencing periodic O_3 episode each year with several days of peak O_3 concentrations exceeding 50 ppb and sometimes exceeding 90 ppb (Hayes et al. 2007a, b). At Mediterranean region from 2000 to 2010, out of 214 monitored stations, 58 % of rural stations displayed an average reduction of 0.43 % $year^{-1}$ in O_3 concentration, while an increase of 0.64 % $year^{-1}$ was recorded in urban and 0.46 % $year^{-1}$ in suburban stations (Sicard et al. 2013).

Monks (2005) pooled together the monitoring data and modeling studies along whole of the Europe and showed that European emissions of O_3 and its precursors had decreased over the past three decades with larger reductions in Russia. Very recently, Saitanis et al. (2015) found that hourly O_3 concentrations often exceeded 70 ppb at Tripolis plateau located in Greece. The highest 1 h O_3 concentration of 240 ppb was reported in France (Pellegrini et al. 2011). From 1987 to 2003, the increase in mean O_3 concentration measured at an Atlantic coastal station (Mace Head) in Ireland was 0.49 ppb $year^{-1}$ (Simmonds et al. 2004) and 0.31 ppb $year^{-1}$ from 1987 to 2007 (Derwent et al. 2007). An average increase of 0.14 ppb $year^{-1}$ in O_3 concentration was reported from 13 rural sites in United Kingdom over the period 1990–2006 (Jenkin 2008).

A modeling study carried by van Tienhoven et al. (2006) documented that maximum hourly O_3 concentration over 50 ppb is common over central Zimbabwe. Emberson et al. (2009) reported that large parts of South Asia experience up to 50–90 ppb mean 7 h (M 7) O_3 concentration. Monitoring studies suggest that mean monthly O_3 concentration of 50 ppb occurred commonly in several parts of Asian continent, especially during growing season of important agricultural crops (EANET 2006; Xu et al. 2008). Ozone concentrations were 41.7 ppb in Xiaoji, China (Pang et al. 2009), 71 ppb in Lahore, Pakistan (Wahid 2006a, b), and 48.1 and 47.1 ppb in Osaka and Tokyo, respectively, in Japan (Sadanaga et al. 2008). Wang et al. (2009) monitored the variations in O_3 concentrations from 1994 to 2007 at a costal site in Hong Kong and reported a 0.87 ppb $year^{-1}$ increase by comparing means during 1994–2000 and 2001–2007.

Yamaji et al. (2006) using Community Multiscale Air Quality model has calculated highest O_3 concentrations ranging from 55 to 70 ppb during May and June in the boundary layer over East China and Japan. With the fast industrialization and urbanization in the last two decades, O_3 concentration is rising at a higher rate in China than other countries and the mean of the daily 24 h average O_3 concentration reached more than 50 ppb during the crop growing season in some regions (Zhao et al. 2009; Tang et al. 2013). During May–September, 2010 in Beijing, China, the daily mean and hourly peak O_3 concentrations at urban and exurban regions were 46 and 67 ppb and 181 and 209 ppb, respectively (Wan et al. 2013). In the Yangtze delta region of China, Xu et al. (2008) observed a decrease in the average concentration, but an increase in the daily variations in diurnal O_3 concentration.

14.3.2 Ozone Concentration: An Indian Perspective

Rising O_3 concentration is also a threat for Indian subcontinent especially in Indo-Gangetic plains due to its subtropical location, which favors O_3 formation. Jain et al. (2005) showed variations in mean monthly O_3 concentrations between 62 and 95 ppb during summer, and 50–82 ppb were recorded during autumn at an urban site in New Delhi during 1997–2003 (Table 14.1). Singla et al. (2011) and Mahapatra et al. (2012) reported higher O_3 concentrations exceeding 50 ppb at Agra and Bhubaneswar. Ali et al. (2012) monitored O_3 concentrations from 1990 to 1999 at Pune and Delhi and reported maximum concentration during summer and minimum during monsoon season. Monitoring results carried out at Pune depicted that O_3 concentration varied from 17.5 to 43 ppb (Beig et al. 2007).

During growth period of wheat (November–March), seasonal mean O_3 concentration (12 h) was 40.1 ppb at a suburban site (Rai et al. 2007) and 36.7 ppb at a rural site during growth period of rice (July–October) (Rai and Agrawal 2008) (Table 14.1). Sarkar and Agrawal (2010a) reported mean O_3 concentrations of 45.3 and 47.3 ppb, respectively, at a rural site of Varanasi during 2007–2008 and 2008–2009. Tiwari et al. (2008) studied the seasonal variations between 2002 and 2006 based on 12 hourly O_3 monitoring and found that O_3 concentration was lowest during rainy season, followed by winter and highest during summer season (Table 14.1). Rai and Agrawal (2014) and Singh et al. (2014b) found high seasonal mean O_3 concentrations exceeding 50 ppb during winter. Ozone concentration data from Anantapur, a rural and semiarid place in south India, displayed minimum concentration in monsoon and maximum in summer (Reddy et al. 2012) (Table 14.1).

Table 14.1 Trends of tropospheric O_3 concentrations at different locations in India

City	Year/month/season	Ozone concentration (ppb)	References
Agra	2008–2009	55–65	Singla et al. (2011)
Allahabad	2002 monsoon	5.9–35.1	Agrawal et al. (2005)
Anantapur	2010		Reddy et al. (2012)
	Summer	70.2	
	Monsoon	20	
Bhubaneswar	2009–2011	50–78	Mahapatra et al. (2012)
Chennai	Summer 2005	2–53	Pulikesi et al.(2006)
New Delhi	1997–2003		Jain et al. (2005)
	Summer	62–95	
	Autumn	50–82	
New Delhi	2009–2010 winter	28	Singh et al. (2013)
	2010–2011 winter	33	
Mohali	2011–2014	46.5 (M12)	Sinha et al. (2015)
Pune	2003–2004	17.5–43	Beig et al. (2007)
Pune	1990–1999		Ali et al. (2012)
	Monsoon	6.7–13.8	
	Summer	9.3–32	
Delhi	Monsoon	5.6–26	
	Summer	4.7–34.7	
Varanasi	Winter 2004–2005	40.1 (M8)	Rai et al. (2007)
Varanasi	2002–2006		Tiwari et al. (2008)
	Summer	45.1–62.3	
	Rainy	24–43.85	
	Winter	28.5–44.2	
Varanasi	2005 monsoon	33.6	Rai and Agrawal (2008)
	2006 monsoon	36.7	
Varanasi	2007–2008 winter	45.3	Sarkar and Agrawal (2010a)
	2008–2009 winter	47.3	
Varanasi	2007 monsoon	49.3	Sarkar and Agrawal (2011)
Varanasi	November 2010–March 2011	49.4	Tripathi and Agrawal (2012)
Varanasi	2008–2009 winter	50.2	Rai and Agrawal (2014)
	2009–2010 winter	53.2	
Varanasi	2011–2012 winter	53.5	Singh et al. (2014a)
Varanasi	2010 summer	64.0	Chaudhary and Agrawal (2015)
	2011 summer	63.4	
Varanasi	2012 monsoon	42.4	Rai et al. (2015)

14.3.3 Future Projections in O_3 Concentrations

Based on modeling studies, Meehl et al. (2007) projected that if there are incessant emissions of O_3 precursors, then there would be an increase of 20–25 % in O_3 concentration by 2050 and 40–60 % by 2100. Morgan et al. (2006) also predicted that mean surface O_3 would increase by 23 % by 2050. A multimodal study of impacts of climate change on O_3 concentrations in Europe predicted an increase in the mean O_3 concentration in the range of 0.9–3.6 ppb for 2040–2049 compared to the concentrations that prevailed during 2000–2009 (Langner et al. 2012). These modeling-based studies in the past have shown that with unmitigated growth in O_3 precursor emissions, O_3 will present a serious global air pollution problem by the middle or end of the century. However, it is to be highlighted that due to the execution of O_3 precursor emission abatement policies, peak O_3 concentrations are gradually declining in Europe, North America, and Japan (Harmens 2014).

The US EPA has reported that emission reductions in O_3 precursors have been substantial over the past 29 years (US EPA 2009). With these emission reductions, many of the higher hourly average O_3 concentrations encountered during 1980s have been reduced, but the background O_3 levels have increased. Though emission abatement policies have been functional in developed countries, the scenario is entirely different in the developing economics of the world such as southern and central Asia where there are continued increase in the emissions of O_3 precursors (Harmens 2014). A report published by Royal Society (2008) emphasized that tropospheric O_3 is projected to increase in the regions where there are major sources of emission and where emission-curbing policies are currently weakest, such as Asia and Africa. Intergovernmental Panel on Climate Change (IPCC) Assessment Report Five (AR5) indicated that if current emission trend continues, O_3 may possibly rise by 20–25 % between 2015 and 2050 and further increase by 40–60 % by the end of this century (IPCC 2014).

14.4 Ozone-Generated Oxidative Stress and Related Responses

14.4.1 Ozone-Induced ROS Formation and Signal Transduction

The main route for O_3 entry into the leaves is via stoma, which is mainly governed by stomatal conductance (Ainsworth et al. 2012). After its entry into the substomatal chamber, O_3 does not persist in the apoplast for long and immediately breaks down or reacts with the compounds present in cell wall or apoplastic fluid to generate ROS such as superoxide radicals ($O_2^{\cdot -}$), hydrogen peroxide (H_2O_2), and hydroxyl radicals ($OH^{\cdot -}$) (Lasik et al. 1989). Staining by $CeCl_3$ revealed that extracellular H_2O_2 accumulation was one of the earliest detectable responses to O_3 in poplar leaves subjected to 150 ppb O_3 after 1 h exposure (Diara et al. 2005).

Among the ROS, $OH^{\cdot}$ radical is the most reactive of oxygen species causing serious damages (Iqbal et al. 1996). The ROS produced then act as early messenger molecules in signaling cascades, hence switching on the downstream signaling and also eliciting defense reactions in apoplast (Vainonen and Kangasjarvi 2014). These signaling molecules include ethylene (ET), salicylic acid (SA), jasmonic acid (JA), and NO as well as mitogen-activated protein kinases (MAP kinases) (Matyssek et al. 2008). Ozone is also known to induce Ca^{2+} influx within few seconds, which is required for the activation of MAP kinase and NADPH oxidase. The kinase (MAPK) cascade is one of the major pathways by which extracellular stimuli like O_3 stress is transuded into intracellular responses. The activated MAP kinase cascade is involved in upregulation of ET synthesis. In addition to ET, biosynthesis of SA is also induced which along with ET is required for the development of foliar injury or O_3-induced lesions (Vainonen and Kangasjarvi 2014). ET and NADPH oxidase spread the signal of oxidative burst from the site of lesion initiation to the surrounding cells and lead to cell death. When the cell death occurs, products of lipid peroxidation serve as the substrate for synthesis of JA, which acts antagonistically and decrease ET-dependent lesion production and hence spread of cell death (Vainonen and Kangasjarvi 2014). A dramatic increase in ET evolution was observed in the poplar clone "Eridana," which displayed sensitivity toward O_3 (Diara et al. 2005). The stimulatory roles of SA and ET and the prevention of lesion expansion by JA have been explicitly characterized by the deployment of mutants of *Arabidopsis thaliana* (Vainonen and Kangasjarvi 2014). ROS also mediate the abscisic acid (ABA)-induced stomatal closure response. In the guard cells of *Arabidopsis*, ROS stimulated ABA synthesis and induced stomatal closure via activation of plasma membrane calcium channels (Apel and Hirt 2004). Ethylene-dependent reductions in stomatal sensitivity to ABA have also been highlighted by Wilkinson and Davies (2010).

14.4.2 ROS Detoxification Mechanisms: Antioxidative Defense Responses

The stress condition generated by O_3 results in an oxidative burst due to enhanced production of ROS, which cause negative effects on cellular components leading to damage to the lipids (peroxidation of unsaturated fatty acids in membrane), proteins (denaturation), carbohydrates, and nucleic acids (Blokhina et al. 2003). To counteract the stress imposed by ROS, an array of antioxidant molecules are induced (Ashmore 2005; Caregnato et al. 2013) via nonenzymatic antioxidants such as ascorbic acid (AA), flavonoids, phenolics, vitamin E (tocopherol), peptides (glutathiones), carotenoids, polyamines, and organic buffering systems or through enzymatic antioxidants (Blokhina et al. 2003), viz., superoxide dismutase (SOD), ascorbate peroxidase (APX), glutathione reductase (GR), catalase (CAT), and various types of peroxidases (POD) (Caregnato et al. 2013).

Among the nonenzymatic antioxidants, ascorbic acid (AA) protects critical macromolecules from oxidative damage by directly reacting with $O_2^{\bullet-}$, H_2O_2, regenerating α-tocopherol from tocopheroxyl and removing H_2O_2 via AA-GSH cycle (Pinto et al. 2003). Increases in AA pool have been observed in various crops and tree species after O_3 exposure (Lu et al. 2009; Singh et al. 2010; Yan et al. 2010; Rai and Agrawal 2014). Rai et al. (2007) reported 11.2 % increase in AA content in wheat leaves under ambient O_3 pollution. Increments in mean value of AA by 40 % in 20 wheat cultivars grown in chambers receiving 82 ppb O_3 for 7 h day^{-1} have been reported by Biswas et al. (2008a). Upon O_3 exposure, differences in total AA concentrations in *Psidium guajava* based on the leaf rank were detected by Pina and Moraes (2010).

Higher AA content was observed in tolerant soybean cultivar PK 472 compared to sensitive cultivar Bragg at 70 and 100 ppb O_3 for 4 h from germination to maturity (Singh et al. 2010). Ozone-sensitive (NC-S) and ozone-resistant (NC-R) plants of *Trifolium repens* and *Centaurea jacea* exposed to moderate O_3 concentration in ambient air showed 50–70 % more AA in NC-R than NC-S (Severino et al. 2007). The level of total apoplastic AA correlates directly with O_3 tolerance in many plant species (Castagna and Ranieri 2009). But higher AA pool in sensitive varieties has also been reported in rice (Rai and Agrawal 2008) and wheat (Sarkar et al. 2010; Feng et al. 2010). D' Haese et al. (2005) reported that apoplastic AA alone cannot explain differential O_3 tolerance of *Trifolium* clones.

Padu et al. (2005) observed that ascorbate level in *Betula pendula* did not increase significantly under O_3 exposure even when stomata were fully open and O_3 flux to the mesophyll cells was at its maximum. Statistically insignificant results were obtained for the total ascorbate level and dehydroascorbate pool when *Pinus canariensis* was exposed to twice the level of ambient O_3 concentration (67 ppb). Hofer et al. (2008) found that ascorbate content was decreased in needle extract of *Picea abies* under twofold ambient O_3 concentrations. *Poa* plants in monoculture and in competition with *Vernonia* displayed a reduction in AA content by 21.3 % and 12.4 %, respectively (Scebba et al. 2006). Iglesias et al. (2006) exposed *Clementina mandarin* cv Marisol for 12 months to 30 and 65 ppb O_3 concentrations and detected a decrease in foliar ascorbate pool.

Exposure to ambient and elevated O_3 led to increases in the activities of SOD, APX, CAT, POD, and GR in wheat (Chen and Gallie 2005; Sarkar et al. 2010), rice (Rai and Agrawal 2008; Wang et al. 2013; Sarkar et al. 2015), maize (Singh et al. 2014b), and mung bean (Mishra and Agrawal 2015). Tree species have also depicted modulation in their enzymatic activities upon O_3 treatments. Activities of POD, CAT, APX, and MDHAR were higher in *Liriodendron tulipifera* under elevated O_3 treatment (Ryang et al. 2009).

Significant increases in the activities of various defense-related enzymes including SOD, CAT, APX, DHAR, MDHAR, and GR were also found in *Ginkgo biloba* under elevated O_3 (Lu et al. 2009; Feng et al. 2011a).

Sensitive variety of rice showed lower magnitude of increment in SOD activity compared to tolerant variety under elevated O_3 (Rai and Agrawal 2008); however, it did not vary between the filtered air and O_3-polluted environment in *Psidium guajava* (Pina and Moraes 2010). SOD activity increased significantly due to O_3 exposure in *Quercus mongolica* (Yan et al. 2010) and a sensitive birch clone (Toumainen et al. 1996). Biswas et al. (2008a) reported that there was a mean increase of 46 % in POD activity in 20 wheat varieties at 85 ppb O_3 concentration supplied 7 h day^{-1} for 21 days compared to filtered air. Rai et al. (2007) and Rai and Agrawal (2014) also reported higher POD activity in wheat plants under ambient O_3 compared to filtered air. In both *Achillea* and *Vernonia* grown in mixed culture, POD activity was increased by 54.4 % and 11.6 %, respectively, whereas reduced by 21.5 % and 27.7 %, respectively, when grown in monoculture (Scebba et al. 2006). Induction in GR activity was significantly higher in a sensitive birch clone (Toumainen et al. 1996). In *Fagus sylvatica* exposed to twofold ambient O_3 concentrations, glutathione content was significantly increased in seedlings as well as in mature trees compared to ambient levels (Herbinger et al. 2005). Increases in total and oxidized glutathione pool were also obtained in *Pinus canariensis* when exposed to 67 ppb O_3 concentration (Then et al. 2009). The constitutive APX activity measured in a resistant white clover clone with respect to a sensitive one was found to be more, suggesting its possible role in conferring higher tolerance toward O_3 stress (Nali et al. 2005).

Differences in the antioxidant defense response are often related with the differential O_3 sensitivity in various plants. Caregnato et al. (2013) have found that differences in O_3 sensitivity between the two varieties of *Phaseolus vulgaris* depended on the differences involved in maintenance of intracellular redox homeostasis. SoyFACE study done by Betzelberger et al. (2010) displayed cultivar variations in the antioxidant capacity of ten soybean cultivars and concluded that antioxidant capacity negatively correlated with photosynthesis and seed yield, hence, suggesting a trade-off between antioxidant metabolism and carbon gain. Zhang et al. (2012) also observed differences in the total antioxidative capacity of two deciduous (*Liriodendron chinense* and *Liquidambar formosana*) and six evergreen tree species (*Cinnamomum camphora*, *Cyclobalanopsis glauca*, *Schima superba*, *Ilex integra*, *Photinia×fraseri*, *Neolitsea sericea*) and found lowest value in *Liriodendron chinense* and highest in *Schima superba* contributing to their varying susceptibility toward O_3.

14.5 Physiological Responses

14.5.1 Photosynthetic Rate, Stomatal Conductance, and Photosynthetic Efficiency

A decline in the photosynthesis rate (Ps) of O_3-exposed plants is associated with reductions in photosynthetic pigments, structural damage to thylakoids, reduction in the efficiency of excitation energy captured, and negative effects on the electron transport system in photosystems (PSI and PSII) (Calatayud and Barreno 2001; Fiscus et al. 2005) as well as decline in the amount as well as the activity of ribulose-1,5-bisphosphate carboxylase/oxygenase (RuBisCO) (Agrawal et al. 2002). Reductions in Ps and stomatal conductance (g_s) of crops under variable O_3 concentrations have been widely reported in natural field conditions (Wahid 2006a, b; Rai et al. 2007; Rai and Agrawal 2008; Sarkar et al. (2010); Rai and Agrawal 2014; Mishra and Agrawal 2015) (Table 14.2). Meta-analysis study with wheat (Feng et al. 2008) and rice (Ainsworth 2008) showed varying degrees of negative response of Ps under O_3 exposure. Sarkar et al. (2010) found more reductions in Ps in wheat cultivar Sonalika, which was more sensitive to O_3 stress than less sensitive HUW234, due to higher depression in g_s suggesting reduced O_3 uptake. Rai and Agrawal (2008) working on two rice cultivars, however,

Table 14.2 Effects of O_3 on physiological parameters in various crop plants (↓) decrease, (↑) increase

Crops/cultivars	O_3 concentration (ppb)	Percent changes: Photosynthetic rate (Ps)	Stomatal conductance (g_s)	Fv/Fm ratio	References
Oryza sativa cv	Ambient O_3 (35 ppb 8 h)	29.3 (↓)	12.9 (↓)	19.4 (↓)	Rai and Agrawal (2008)
Saurabh 950		28.3 (↓)	18.8 (↓)	16.6 (↓)	
NDR 97					
Oryza sativa cv	1.5 times ambient (13.8–74.2) for 7 h	27.1 (↓)	33(↓)	NS	Pang et al. (2009)
SY 63		14.8 (↓)	NS (↓)	NS	
WYJ3					
Triticum aestivum cv	72 ppb (8 h)	20–22 (↓)	7–24 (↓)	9–17 (↓)	Wahid (2006a)
Inqilab-91, Punjab-96, Pasban-90					
Triticum aestivum	CF + O_3 (105 ppb for 8 h)	42.4 (↓)	–	–	Feng et al. (2007)
Triticum aestivum cv M 234	Ambient O_3 (42 ppb 8 h)	27 (↓)	20 (↓)	5.4 (↓)	Rai et al. (2007)
Triticum aestivum	82 ppb for 7 h	24 (↓)	8 (↓)	–	Biswas et al. (2008a)
20 cultivars					
Triticum aestivum	100 ppb for 7 h	36.9 (↓)	11.1 (↑)	–	Biswas et al. (2008b)
12 wild and cultivated cultivars					
Triticum aestivum	Elevated ozone (30–119, ≥120 ppb)	40 (↓)	31 (↓)	–	Feng et al. (2009)
Triticum aestivum cv	47.3 ppb for 12 h	31 (↓)	9.5 (↓)	11.1 (↓)	Sarkar et al. (2010)
M 510		15.5 (↓)	12 (↓)	5 (↓)	
Sonalika					
Triticum aestivum cv PBW 343	50.2 ppb (8 h)	19 (↓)	33.2 (↓)	5.9 (↓)	Rai and Agrawal (2014)
M 533		18.4 (↓)	43.6 (↓)	NS	
Hordeum vulgare cv Haider 93, Haider 91, Jou 87, Jou 85	71 ppb (6 h)	13–21 (↓)	6–12 (↓)	–	Wahid (2006b)
Glycine max	1.23× current concentration	NS	1 (↓)		Bernacchi et al. (2006)
Glycine max cv	70 and 100 ppb for 4 h	19.8 and 40.4 (↓)	21 and 26 (↓)	16.0 and 16.9 (↓)	Singh et al. (2009)
PK 472		25.6 and 31.6 (↓)	61 and 66 (↓)	2.5 and 5.8 (↓)	
Bragg					
Glycine max	2 times ambient O_3 (46 ppb)	11 (↓)	15 (↓)	–	Betzelberger et al. (2010)
10 cultivars					
Phaseolus vulgaris	60 ppb (12 h)	38 (↓)	52.6 (↓)	–	Flowers et al. (2007)
cv S156					
Phaseolus vulgaris cv Camellino	165 ppb for 3 h	36 (↓)	26 (↓)	–	Guidi et al. (2009)

(continued)

Table 14.2 (continued)

Crops/cultivars	O_3 concentration (ppb)	Percent changes: Photosynthetic rate (Ps)	Stomatal conductance (g_s)	Fv/Fm ratio	References
Vigna radiata cv HUM-2	+10 ppb above the ambient O_3 (63.4 ppb)	28.3 (↓)	38.8 (↓)	10.2(↓)	Mishra and Agrawal (2015)
HUM-6		19 (↓)	24.8 (↓)	2.3(↓)	
Brassica campestris cv Sanjukta	+10 ppb above the ambient O_3 (49.4 ppb)	49.3 (↓)	66.6 (↓)	45.6 (↓)	Tripathi and Agrawal (2012)
Vardan		46.6 (↓)	66 (↓)	33 (↓)	
Brassica napus	100 ppb (2 h)	21.3 (↓)	15.2 (↓)	NS	Feng et al. (2006)
Brassica oleracea	+20, 40 ppb for 8 h above ambient (30 ppb)	83.1 (↓)	79.7 (↓)	4.5 (↓)	De Bock et al. (2012)
Arachis hypogaea cv NC-VII	48 ppb for 12 h	21 (↓)	NS	–	Booker et al. (2007)
Gossypium hirsutum cv Giza 65	70 ppb for 10 h	18 (↓)	23 (↓)	–	Hassan and Tewfik (2006)

reported a contrasting response, when sensitive cultivar NDR 97 showed higher Ps along with reduction in g_s compared to tolerant cultivar Saurabh 950. Feng et al. (2011b) exposed two winter wheat cultivars Yanmai 16 and Yangfumai 2 at 127 % higher ambient O_3 and found more reductions in Ps rate and g_s in Yangfumai 2 (more sensitive), while initial chlorophyll fluorescence showed more decline in Yanmai 16 (less sensitive).

Responses of leaf Ps under different O_3 levels ranging from 37 to 116 ppb were investigated in soybean cultivars Dwight and IA3010 and it was found that Ps displayed linear reduction with increasing O_3 levels (Sun et al. 2014). Betzelberger et al. (2010) found that under elevated O_3 treatment, older leaves had a lower Ps rate on an average across six soybean cultivars during vegetative growth. Morgan et al. (2006) have also found that photosynthetic damage caused due to O_3 was more pronounced when plants and/or leaves were older. Significant reductions in Ps and g_s were recorded in an oil-yielding crop *Brassica napus* under O_3 exposure (Feng et al. 2006) (Table 14.2). Elevated O_3 led to an average reduction in net carbon assimilation by 11 % in ten test soybean cultivars (Betzelbetger et al. 2010) (Table 14.2). A meta-analysis applied on 39 different wheat varieties found that light-saturated photosynthesis rate (A_{sat}) reduced by 40 %, while g_s and chlorophyll content displayed reductions by 31 % and 46 %, respectively, under elevated O_3 (Feng et al. 2009). In a study performed by Hayes et al. (2009), reduced growth in *Trifolium repens* and *Lolium perenne* have been correlated with reduction in photosynthesis. Pleijel et al. (2006) and Biswas et al. (2008b) found that modern or cultivated species demonstrated higher O_3 flux as shown by increased g_s resulting in higher relative reduction in Ps than wild/old species of wheat.

Physiology of trees has also been found to be affected under O_3 exposure. Seedlings of *Populus nigra*, *Viburnum lantana*, and *Fraxinus excelsior* were subjected to charcoal-filtered and non-filtered air, and it was found that O_3 had significant effects on Ps and g_s of all the three species (Novak et al. 2005) (Table 14.3). Adverse effects of O_3 on physiology of *Populus tremuloides*, *Betula papyrifera*, and *Acer saccharum* were reported by Karnosky et al. (2005). Velikova et al. (2005) found that g_s of oak leaves reduced immediately after O_3 exposure. In a study on

Table 14.3 Effects of O_3 on physiological responses in various tree and grass species (↓) decrease, (↑) increase

Tree species	O_3 concentration	Percent change: Photosynthetic rate (Ps)/light-saturated photosynthetic CO_2 uptake (Asat)	Percent change: Stomatal conductance (gs)	Percent change: Fv/Fm ratio	References
Fagus sylvatica	Average 57 ppb, episodic O_3 <150 ppb	(↓)	(↓)	(↓)	Kitao et al. (2009)
Quercus ilex	300 ± 50 ppb (9 h day^{-1})	5 (↓)	No effect	No effect	Velikova et al. (2005)
Quercus pubescens		17.4 (↓)	23.2 (↓)	3.6 (↓)	
Populus nigra Viburnum lantana	Ambient O_3 47 ppb (12 h)	26 (↓)	28 (↓)	–	Novak et al. (2005)
		22 (↓)	24 (↓)		
Fraxinus excelsior		8 (↓)	20 (↓)		
Liriodendron tulipifera	70 ppb above ambient (30 ppb) for (5 h day^{-1})	57 (↓)	35 (↓)	–	Lombardozzi et al. (2012)
		74 (↓)	49 (↓)		
Quercus ilex	Elevated O_3 55.2 (12 h)	No change			Calatayud et al. (2011)
Quercus faginea		33 (↓)			
Quercus pyrenaica		64 (↓)	26 (↓)	16 (↓)	
Quercus robur		38 (↓)		16 (↓)	
Pinus canariensis	67 ppb (seasonal mean)	13.1 (↑)	1.3 (↑)	–	Then et al. (2009)
Quercus mongolica	80 ppb (9 h)	57.1 (↓)	62.5 (↓)	4.7 (↓)	Yan et al. (2010)
Fagus crenata	60 ppb (seasonal mean)	46 (↓)	18 (↓)	–	Watanabe et al. (2013)
Quercus mongolica var. crispula		15 (↓)	18 (↓)		
Betula pendula Roth	AOT 40 75.9 ppm.h			–	Riikonen et al. (2005)
	Mean of 3 year				
Clone 4		1(↑)	7 (↑)		
Clone 80		3(↓)	2 (↑)		
Liriodendron chinense	150 ppb (8 h)	42 (↓)	25 (↓)	–	Zhang et al. (2012)
Liquidambar formosana		36 (↓)	15 (↑)		
Cinnamomum camphora		27 (↓)	6 (↑)		
Cyclobalanopsis glauca		29 (↓)	19 (↓)		
Schima superba		32 (↓)	14 (↓)		
Ilex integra		3 (↓)	15 (↓)		
Photinia × fraseri		6 (↓)	1 (↑)		
Neolitsea sericea		16 (↓)	12 (↓)		
Grass species					
Ilex aquifolium	70 ppb for 7 h	–	(↑)	–	Ranford and Reiling (2007)
Trifolium pratense	32.4 ppb for 14 h	21–23 (↑)	–	–	Saviranta et al. (2010)
Trifolium alexandrinum 6 cultivars	+10 ppb above the ambient O_3	12.7–31.5 (↓)	5–29.4(↓)	16.6–19.2 (↓)	Chaudhary and Agrawal (2013)
Ranunculus acris	16.2–89.5 ppb 24 h mean	–	6.1 (↑)	–	Wagg et al. (2013)
Dactylis glomerata			26.7 (↑)		

effects of O_3 on maple, decrease in CO_2 assimilation was observed (Calatayud et al. 2007). In a FACE setup with beech and oak, elevated O_3 concentrations caused reductions in Ps and g_s, and it was found that photosynthetic activity of beech was more sensitive to O_3 than oak (Watanabe et al. 2013) (Table 14.3).

Wittig et al. (2007) reviewed 348 measurements across 61 studies on Ps and 266 measurements across 55 studies on g_s and documented that mean elevation of O_3 to 47 ppb since the industrial revolution reduced A_{sat} by 11 %, while average elevated O_3 concentrations of 42 ppb reduced g_s by 13 % (Wittig et al. 2007). There were also variations among the species reflecting that A_{sat} and g_s were significantly reduced in *Fraxinus*, *Populus*, *Prunus*, and *Viburnum* species, while no changes were recorded in *Picea*, *Pinus*, and *Quercus* species. In a similar meta-analysis, it was found that angiosperms were affected more than gymnosperms and younger trees were affected less than the older tress under O_3 stress. One evergreen (*Quercus ilex*) and three deciduous (*Quercus faginea*, *Quercus pyrenaica*, and *Quercus robur*) tree species, when exposed to elevated O_3 levels in OTCs, showed reductions in Ps except in *Quercus ilex*, which displayed no change (Calatayud et al. 2011) (Table 14.3).

Seedlings of *Populus nigra*, *Viburnum lantana*, and *Fraxinus excelsior* when grown under non-filtered chambers displayed reductions in net photosynthesis (Pn) and g_s compared to filtered chambers (Novak et al. 2005). Tulip poplar seedlings when exposed to two elevated O_3 levels displayed reductions in Ps rate (Lombardozzi et al. 2012) (Table 14.3). Seedlings of *Metasequoia glyptostroboides*, a living fossil, when exposed to elevated O_3, showed reductions in photosynthetic pigments and gaseous exchange (Zhang et al. 2014a). Within-canopy variations in leaf gas exchange were also observed in *Fagus sylvatica* exposed to twofold ambient O_3 concentrations. Reduction in Ps of sun-exposed leaves was more pronounced than the leaves under shaded condition, while a reverse trend was followed for g_s (Herbinger et al. 2005). In a similar experiment, gas exchange parameters were more affected in adult trees than the juvenile ones (Herbinger et al. 2005). It was found that Ps and g_s increased though being nonsignificant while quantum yield of PSII activity decreased when *Pinus canariensis* was grown in a FACE environment at elevated O_3 (67 ppb) (Then et al. 2009). Elevated O_3 concentration of 32.4 ppb (14 h) increased the Pn of red clover by approximately 21–23 % before appearance of visible injuries (Saviranta et al 2010) (Table 14.3).

Reductions were found in g_s when Timothy grass (*Phleum pratense*) was grown at 60 ppb O_3 for 12 h (Danielsson et al. 2013). Mills et al. (2009) stated that upon chronic O_3 exposure, reduction in sensitivity of stomatal cells to abscisic acid could be a reason that some grass species displayed increased g_s. The g_s were higher in *Dactylis glomerata* and *Leontodon hispidus* when grown under O_3 concentrations varying between 21.4 and 102.5 ppb (Mills et al. 2009). Stomatal conductance increased under O_3 stress in *Ilex aquifolium* (Ranford and Reiling 2007), *Ranunculus acris*, and *Dactylis glomerata* (Wagg et al. 2013) (Table 14.3).

Some species experienced closure of stomata in order to decrease the g_s and thus O_3 diffusion reduces into stomatal chamber/leaf mesophyll (Madkour and Laurence 2002). However, it may happen that stomata get impaired by chronic exposure of O_3 and are unable to close (Hoshika et al. 2013). Some of the studies have focused on stomatal acclimation and stomatal sluggishness upon O_3 exposure. Darbah et al. (2010) showed no evidence of stomatal acclimation in two aspen clones (42 E and 271) even after continuous O_3 exposure for a decade. Stomatal sluggishness was observed by Paoletti and Grulke (2010) in *Quercus kelloggii* and *Quercus douglasii* when exposed to 70 ppb O_3 for 8 h day^{-1} in OTCs. Stomatal sluggishness increased the time of stomata to open and close limiting the CO_2 uptake, hence affecting the photosynthesis.

Ratio of photosynthetic efficiency (Fv/Fm) is a measure of maximum efficiency of PS II, and its reduction suggests decrease in linear flow of electron through PS II, leading to lesser electrons available for Calvin cycle and water–water cycle. Lowering of Fv/Fm ratio under O_3 exposure was observed in wheat (Sarkar and Agrawal 2010a;

Rai and Agrawal 2014), rice (Sarkar et al. 2015), soybean (Singh et al. 2009; Betzelberger et al. 2010; Zhang et al. 2014b), and mung bean (Mishra and Agrawal 2015) (Table 14.3). Fv/Fm ratio reduced by 12 % in white clover sensitive clone (NC-S) at 200 ppb O_3 for 5 h day^{-1} (Francini et al. 2007), by 9.3 % in snap bean cv S 156 at 60 ppb O_3 (Flowers et al. 2007) and by 5.4 % in wheat cv M 234 at mean ambient concentration of O_3 42.4 ppb (Rai et al. 2007). A decrease in Fv/Fm ratio with increasing O_3 exposure has been demonstrated in some tree species (Bussotti et al. 2011), and out of 78 studies, about 52 % showed significant differences in Fv/Fm ratio between O_3-treated and control plants. In *Achillea* and *Vernonia*, Fv/Fm ratio was negatively affected under O_3 treatment (non-filtered air with added 50 ppb O_3) when grown in monoculture, while there was an increment in this ratio by 27.5 % when grown in mixed culture with *Poa pratensis*. *Fagus sylvatica* displayed reduction of 18.2 % in Fv/Fm ratio under twice-ambient O_3 concentration (Gilen et al. 2007).

14.5.2 Photosynthetic Pigments

Ozone-generated ROS are known to alter membrane-bound organelles such as chloroplast, which leads to destruction of photosynthetic pigments (Rai and Agrawal 2008, 2012). Significant reductions of 27 % and 44 %, respectively, in total chlorophyll content of rice plants under ambient and elevated dose of O_3 were reported (Rai and Agrawal 2008; Sarkar et al. 2015). In a study with 20 cultivars of wheat, Biswas et al. (2008a) found 24–35 % reductions in total chlorophyll content in recent cultivars and 3–12 % in older ones at 82 ppb O_3 given for 7 h day^{-1}. Reductions in chlorophyll by 6.4 % and carotenoids by 5.7 % at each 10 ppb increase in mean O_3 were observed in two soybean cultivars grown under O_3 levels varying between 37 and 116 ppb (Sun et al. 2014). Feng et al. (2008) in their meta-analytical study also reported about 40 % decrease in chlorophyll content in wheat plants under O_3 exposure.

Elevated O_3 decreased total chlorophyll content in subtropical broad-leaved tree species *Cinnamomum camphora* and *Cyclobalanopsis glauca* (Zhang et al. 2012) and in *Citrus clementina* (Iglesias et al. 2006). Gilen et al. (2007) exposed *Fagus sylvatica* to twice-ambient O_3 level and detected 15.9 % reduction in total chlorophyll content. Riikonen et al. (2005) reported 5 and 19 % reductions in total chlorophyll content in two European silver birch (*Betula pendula*) clones 4 and 80, respectively, upon O_3 treatment. In six *Trifolium alexandrinum* cultivars, total chlorophyll showed reductions varying from 13.1 % to 57.3 % and carotenoids by 9.4–39.2 % under O_3 dose ambient +10 ppb (Chaudhary and Agrawal 2013). Total chlorophyll content increased in DHM117 and HQPM1 cultivars of maize grown at lower O_3 dose (ambient +15 ppb), while a reduction was observed in plants exposed with higher concentration (ambient +30 ppb). Similarly total chlorophyll content increased by 13, 37 and 5 % and decreased by 47 % in maize plants at elevated levels of O_3, i.e., +20,+40,+60, and +80, respectively, relative to non-filtered chambers (Leitao et al. 2007). *Pinus canariensis* exposed to elevated O_3 concentration (67 ppb) displayed an increment of 14.3 % in photosynthetic pigments though insignificant (Then et al. 2009). In *Vernonia*, also, O_3 exposure increased the total chlorophyll content (Scebba et al. 2006).

Carotenoids are vital photoprotective agents, which prevent photooxidative chlorophyll destruction (Singh et al. 2010, 2014b). Changes in carotenoids content under O_3 treatment may lead to a modification in their capacity to protect photosystem against photooxidation. Several studies have noticed O_3-induced reduction (Rai and Agrawal 2008; Sarkar et al. 2010; Sarkar and Agrawal 2011; Mishra and Agrawal 2015) or induction (Leitao et al. 2007; Singh et al. 2014b) in carotenoid content. The effects of O_3 on 3-year-old *Clementina mandarin* trees were studied at two O_3 concentrations, and reductions in total chlorophyll as well as carotenoids pools in leaves were recorded (Iglesias et al. 2006) (Table 14.4).

Table 14.4 Summary of yield responses of selected agricultural crops and growth responses of tree species at varying concentrations of O_3

Study site	Crops and cultivars	O_3 concentrations (ppb)	Parameters assessed	Response (%) /↑↓	References
Crops					
Lahore, Pakistan	Barley: Haider 93, Haider 91, Jou 87, Jou 85	71	Wt. of grains plant^{-1}	↓12.9–43.9 %	Wahid (2006b)
Lahore, Pakistan	Wheat: Inquilab 91 Punjab 96, Pasban 96	72	Wt. of grains plant^{-1}	↓18–43 %	Wahid (2006a)
Varanasi, India	Wheat: M 234, PBW 343, M 533	40–48	Wt. of grains plant^{-1}	↓20.7, 16, 14 %	Rai et al. (2007)
Varanasi, India	Rice: NDR 97, Saurabh 950	35.5	Wt. of grains plant^{-1}	↓14.5 and 15 %	Rai et al. (2010)
Varanasi, India	Wheat: Sonalika, M 510	45.3	Wt. of grains plant^{-1}	↓11–20 %	Sarkar and Agrawal (2010a)
Varanasi, India	Rice: Shivani Malviyadhan 36	51.3	Wt of grains m^{-2}	↓13–15 %	Sarkar and Agrawal (2011)
Varanasi, India	Mustard Ashirwad and Kranti	44.6	Wt. of grains plant^{-1}	↓7–19.3 %	Singh et al. (2012)
Varanasi, India	Wheat cvLok-1	58.2	Wt. of grains plant^{-1}	↓16 %	Singh et al. (2014c)
Varanasi, India	HQPM1, DHM117	55.6	Wt. of kernels plant^{-1}	↓4–5.5 %	Singh et al. (2014a)
Varanasi, India	Mung cultivars	64	Wt. of grains plant^{-1}	↓15.4–9.8 %	Chaudhary and Agrawal (2015)
US	Soybean	37.9–46	Yield	↓11-3–36.8 %	Betzelberger et al. (2010)
China	Rice: SY63, WYJ 63	74.2	Wt. of grains plant^{-1}	↓20.7–6.3 %	Pang et al. (2009)
Forest					
US Great Smoky Mountains	Eastern hardwood forest	>60 ppb	Net primary productivity	↓3–16	Karnosky et al. (2007)
Northern Wisconsin, USA	Blackcherry and aspen	>60 ppb	Growth	↓20 %	Karnosky et al. (2005)
New Jersey, USA	Southern pines	45	Growth	↓10 %	Karnosky et al. (2007)
Northern Wisconsin, USA	Aspen FACE	50–60	Growth	↓10–15 %	Karnosky et al. (2007)
Mediterranean Forest, Italy	*Fagus sylvatica*	AOT 40=45 ppm.h	Crown transparency	↑39 %	Paoletti (2006)
Mediterranean Forest, Italy	Deciduous oak	AOT 40 = 45 ppm.h	Basal area increment	↓	Paoletti (2006)
France	Hybrid poplar	200 ppb	Cambial growth	↓	Richet et al. (2011)

(↓) decrease; (↑) increase

14.5.3 RuBisCO Content

The decline in RuBisCO content upon O_3 exposure has a direct effect in form of significant reduction in photosynthetic capacity (Wittig et al. 2009). Ozone-induced decline in RuBisCO quantity may possibly be due to inhibition of synthesis and/or its increased degradation (Eckardt and Pell 1994). Sarkar and Agrawal (2010b) reported that elevated O_3 caused degradation of large subunit (LSU) and small subunit (SSU) of RuBisCO in rice. Similar results were obtained in wheat (Sarkar et al. 2010), maize (Singh et al. 2014a), and mung bean (Mishra and Agrawal 2015). Amount of RuBisCO increased by 10, 20, and 17 % and decreased by 7 % in O_3 atmospheres of +20, +40, +60, and +80, respectively, compared to non-filtered chambers (Leitao et al. 2007). RuBisCO content was also reduced by 35.9 % in beech, while increased by 23.0 % in oak at 60 ppb O_3 concentration. Riikonen et al. (2005) found respective reductions of 11 and 7 % in RuBisCO content in two European silver birch (*Betula pendula*) clones 4 and 80 under elevated O_3 exposure.

14.5.4 Secondary Metabolism

In plants, phenylpropanoid metabolism is induced as a general response to stress, and enhancement of key enzyme activities and accumulation of secondary metabolites occur early after stress in order to improve the resistance/tolerance to stresses. Ozone exposure elevates the levels of secondary metabolites through the phenylpropanoid pathway, thereby supplying carbon skeletons for secondary metabolites (Fig. 14.1) (Santos and Furlan 2013). The enhancement of phenylpropanoid biosynthesis by O_3 is documented (Iriti and Faoro 2009). Phenyl ammonium lyase (PAL) is a key enzyme of phenylpropanoid pathway and is considered as a biochemical marker indicating the activation of plant defense, which includes the synthesis of both structural and protective compounds. Gene encoding PAL was induced rapidly but transiently in both clones of Birch during O_3 exposure (Tuomainen et al. 1996). Long-term O_3 exposure can lead to increase in levels of phenolic acids, flavonoids, and related compounds (Booker and Miller 1998). In Arabidopsis (*Arabidopsis thaliana*), PAL mRNA is rapidly and transiently induced within 3 h of O_3 treatment (300 ppb daily for 6 h), reaching a threefold higher level than control plants (Vainonen and Kangasjarvi 2014). Interestingly, the increase in activity of this enzyme has been associated with higher lignin content in O_3-exposed leaves (Richet et al. 2011) and the newly synthesized lignin structurally differed from the control lignin.

14.6 Secondary Responses

Ozone has a potential to affect natural and seminatural vegetation due to differential O_3 sensitivity of plant species, leading to changes in the floristic composition of plant communities (Fuhrer et al. 1994).

14.6.1 Herbs, Grasses, and Native Species

In case of crops, yield of the plant has been set as an important criterion to determine O_3 sensitivity, while it is more difficult to decide criteria to adopt with most of the natural vegetation. Annual or monocarpic species must produce seeds to survive, so seed output is an obvious criteria to use, but perennials pose a difficult problem due to their long life.

The most common parameters studied to measure O_3 sensitivity are visible symptoms and growth. Visible symptoms of O_3 were first recorded in crops, ornamentals, and wild species in California, USA, during 1940s (Thomas 1951). However, there was a weak correlation between visible symptoms and biomass in three grass species (*Dactylis glomerata*, *D. ascherso-niana*, and *Phleum alpinum*) (Pleijel and Dannielsson 1997). These grasses showed visible symptoms, but no effect on growth was recorded. Some studies have, however, reported significant

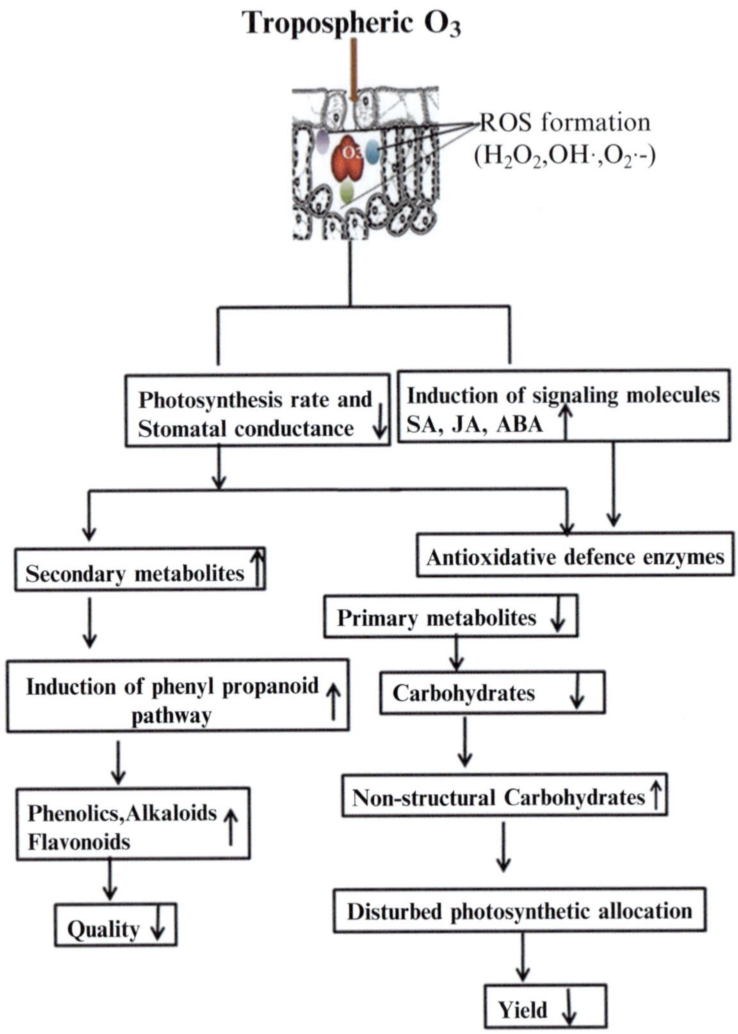

Fig. 14.1 Generalized diagram showing effect of ozone on plant metabolism, physiology, yield, and quality

effect on growth without visible symptoms of injury (Davison and Barnes 1998; Reiling and Davison 1992; Bermejo et al. 2003).

Visible symptoms varied between different plant species. Bermejo et al. (2003) conducted a screening study in OTCs to assess the O_3 sensitivity of 22 representative therophytes of grassland ecosystems based on the appearance and extent of foliar injury. A distinction was made between specific O_3 injury and non-specific O_3 discoloration at three O_3 treatments (charcoal-filtered air, non-filtered air, and non-filtered air supplemented with 40 ppb O_3). The results showed that Papilionaceae species were more sensitive than Poaceae. O_3-induced injury in Papilionaceae species were consisted of brown-reddish necrotic spots that were associated with foliar chlorosis. While in Poaceae, apical necrosis and reddish-brown spots to interveinal symptoms were recorded. *Trifolium* was the most sensitive among Papilionaceae family and *Avena barbata* in Poaceae.

Bergmann et al. (1995) reported wide range of visible injury symptoms in native herbaceous species of *Rumex obtusifolius*, *Cirsium arvense*, *Atriplex patula*, *Chamomilla recutita*, *Chenopodium album*, *Cirsium arvense*, *Galinsoga parviflora*, *Malva sylvestris*,

Matricaria discoidea, *Senecio vulgaris*, and *Solanum nigrum* grown at three O_3 treatments, viz., charcoal-filtered air (CF), CF+30 ppb as background level of O_3, and CF+70 ppb O_3 for 8 h day^{-1}. The most dominant symptoms were premature leaf senescence. Reddish pigmentation was observed in *Rumex obtusifolius*, *Senecio vulgaris*, and *Sonchus asper*, while necrotic spots were observed in *Malva sylvestris* and *Cirsium arvense*.

Other traits in grassland species affected under O_3 exposure were stomatal control, specific leaf area, and biochemical defense capacity (Bender et al. 2006). These traits were correlated with visible symptoms and growth of plants. The cumulative dose of O_3 taken up by leaves as determined by both g_s and O_3 concentration at leaf level was found to be a key factor influencing O_3 damage to the plants (Mills et al. 2009). Detoxification processes are based on a generic response to oxidative stress, induced by increased levels of free radicals (Fuhrer and Booker 2003). The leaf area-based antioxidant content has been found to increase with age. In addition, physical defense in the form of cell wall thickening has been observed in a variety of species (Bass et al. 2006; Bussotti et al. 2005). Both biochemical and physical responses require energy for regeneration and transport of antioxidants; thus, O_3 tolerance could depend on the amount of carbohydrates available and on the energy supplied from photosynthesis (Fuhrer and Booker 2003). High specific O_3 sensitivity relates with thinner palisade mesophyll layers and a high ratio of spongy to palisade mesophyll cells (Bennett et al. 1992). Such leaf traits decrease mesophyll resistance to O_3 diffusion and in turn increase the cumulative dose per cell. O_3 sensitivity may increase with increasing specific leaf area (Körner 2003).

Ozone sensitivity also varies with different phenological stages, but studies are limited for herbs and grasses. *Plantago major* was found to be more sensitive in the seedling stage (Davison and Barnes 1998; Reiling and Davison 1992). *Centaurea jacea* showed enhanced sensitivity during the reproductive phase (Bassin et al. 2004), but it was not possible to distinguish whether this was due to physiological changes, or due to more exposure period to O_3. In perennials, which typically dominate temperate grasslands, three phases can be separated: establishment phase, vegetative growth phase, and recurring generative growth phase. Establishment and generative stages were identified to be O_3 sensitive (Bassin et al. 2007). The possible reason is that initial establishment phase is characterized by expansion with high relative growth rate (RGR), large carbon investments in roots and shoots, and high g_s which make the plant species highly sensitive to ozone. While in generative phase, specific leaf area (SLA) is very high and has less detoxification capacity and altered carbon allocation more toward developing reproductive structures and less toward root and leaf. During the vegetative phase, growth rates are reduced as compared to the establishment phase and carbon is mainly invested in growth of storage organs but also in structures and biochemical defense (Ye et al. 2000). Therefore, during this phase, plants may be more sensitive to O_3 than during the vegetative phase.

Another most important factor is nutrient availability in modifying O_3 response of semi-natural vegetation (Davison and Barnes 1998). Increasing nutrient supply could have opposing effects: increasing sensitivity due to stimulated growth and higher SLA or alternatively increasing tolerance due to enhanced detoxification capacity through stimulated photosynthesis. Several studies have been carried out in OTCs using a cross-factorial design with more O_3 treatments and nutrient levels. Whitfield et al. (1998) found that high nutrient supply protected *Plantago major* against O_3 damage, but similar response was not confirmed in *Trifolium subterraneum*, where fertilization (only N) was reduced.

14.6.2 Plant Communities

Effects of O_3 may occur at various levels of organization, i.e., from the cellular level through the level of individual organs and plants to the level of plant communities and ecosystems (Fuhrer 2009). Ozone affects at community level by affecting species richness and diversity (Pfleeger et al. 2010). The existence of phenotypic variability in species sensitivity to O_3 suggests a potential capacity of O_3 for driving species evolutionary changes (Anav et al. 2011). To measure O_3 sensitivity of a community, it is assumed that sensitivity of the component species determines its sensitivity, however, in communities dominated by species with traits, which are associated with high sensitivity to O_3, may lead to a shift in species cover or frequency, or loss of productivity. Hayes et al. (2007a, b) conducted a meta-analysis on 83 seminatural vegetation species based on field experimental facilities, OTCs, field release systems, and solardomes, with exposure duration of at least 3 weeks and mean maximum hourly O_3 concentration of less than 100 ppb. Meta-analysis showed that approximately one-third of the species in this database showed reductions in aboveground biomass above 10 % at 15 ppm.h AOT-40 compared to 3 ppm.h considered as sensitive to O_3 (Hayes et al. 2007a, b). About 15 species showed 5 % stimulation of aboveground biomass at 15 ppm.h O_3 compared to 3 ppm.h and were classified as moderately sensitive, and the remaining 41 species were considered insensitive to ozone. Meta-analysis data also showed that Fabaceae was more sensitive to O_3 than plants of the families Asteraceae, Caryophyllaceae, and Poaceae; however, a species of Asteraceae, namely, *Leontodon hispidus*, was reported highly sensitive to O_3. The most possible reason for Fabaceae to show greater sensitivity may be due to differences in the detoxification of O_3 and ROS production rather than differences in O_3 uptake (Hayes et al. 2007a, b).

An investigation to establish relationship between sensitivity to O_3 and Raunkiaer life-form showed that the therophytes were more sensitive than chamaephytes and hemichamaephytes (Hayes et al. 2007a, b). In therophytes, higher sensitivity was observed due to disturbance in partitioning of photosynthates to their seeds at the expense of maintaining other aerial biomass. On the basis of meta-analysis, traits were identified which correlated with O_3 sensitivity like species with higher leaf N was more sensitive than with those having less N content. Reich et al. (1998) suggested that the combination of high mass-based leaf N concentration, dark respiration, and maximum photosynthetic rates in leaves entails specific trade-offs, since such leaves are fragile (high specific area and low toughness) and less well physically defended against biotic (herbivory) and abiotic stresses (including O_3).

Leaves of fast-growing species generally show the above characteristics, which implies that fast-growing species are more sensitive to O_3 than slow-growing species. Ellenberg ecological habitat scores showed that light-loving species were more sensitive than shade-loving species to O_3. Mills et al. (2007) mapped O_3-sensitive communities using available European land-cover datasets: the SEI land-cover dataset and the European Environment Agency Corine land-cover dataset. Mapping was done on the basis of the European Nature Information System (EUNIS), which was used for mapping impacts of pollutants in Europe. Using the EUNIS approach, O_3-sensitive communities identified were alpine and sun-alpine grasslands, Woodland fringed and clearings, and E1 Dry grasslands.

14.6.3 Change in Competitive Hierarchies

In non-managed systems, competitive mechanisms are among the most determining factors of shifts in the floristic composition in response to changing environmental conditions (Bishop and Cook 1981). Neighboring individuals compete for resources both below- and aboveground, and any changes in resource availability, resulting from effects of O_3 on one of the competing species, may lead to adaptations in other competing species. Rooting pattern is strongly influenced by the presence of neighbors, and in dense stands, competition for light causes changes in shoot growth and architecture (Turkington et al.

1994). Thus, it can be hypothesized that the sensitivity to O_3 depends on whether the plant is grown in isolation, in a monoculture, or in intraspecific competition. In a competitive environment, growth may be enhanced or reduced depending on the relative responses to O_3 of the target species and its neighbors. Ozone impacts on these interactions are important, but little is known about interspecific interactions under O_3 stress (Fuhrer and Booker 2003). Blackberry (*Rubus cuneifolius*) under high O_3 became dominant in an early successional community previously dominated by sumac (*Rhus copallina*) despite the great sensitivity of blackberry to O_3 (Barbo et al. 1998).

Bender et al. (2006) reported that interspecific plant competition was altered under O_3 exposure. Using a phytometer approach, O_3 effect on growth of *Poa pratensis* was observed. *Poa pratensis* was grown in monoculture and in mixed cultures with four competitor plant species (*Anthoxanthum odoratum*, *Achillea millefolium*, *Rumex acetosa*, and *Veronica chamaedrys*). Mesocosms were exposed to charcoal-filtered air +25 ppb O_3 as background levels and nonfiltered air +50 ppb O_3 (elevated O_3). A result showed that biomass of *Poa pratensis* was highest in monoculture, but lowered in mixtures due to competition with other species. Among plant competitor species, *Veronica chamaedrys* was the strongest competitor. Foliar injury symptoms were observed in both the monoculture and mixed cultures of *Poa pratensis*. *Poa pratensis* was the most sensitive when grown all alone, while competing species reduced the extent of injured leaves, indicating that O_3 effects in a competitive situation differ from effects observed in monoculture. A possible reason for the reduced O_3 sensitivity of *Poa pratensis* in competition with other species might be alteration in plant canopy structure leading to reduction in O_3 flux. Nussbaum et al. (2000) showed that competition changed the species reaction to O_3 in an experiment with binary mixtures. *Trisetum flavescens* planted with either *Trifolium repens* or *Centaura jacea* in different mixing ratios exhibited a significant $O_3 \times$ mixing ratio interaction with *T. repens*, but not with other species (Nussbaum et al. 2000). In a study of four wetland species grown in competition with *Agrostis capillaris*, O_3 had no impact on species productivity, but competition substantially influenced the biomass of the target species (Tonneijck et al. 2004).

14.6.4 Changes in Forage Quality and Herbivory Pattern

Effects of O_3 on plant growth and productivity can vary depending upon timing of exposure and whether plants are growing alone or in competition with other plants for growth resources such as light, water, nutrients, etc. (Davison and Barnes 1998). Most of the studies on seminatural grasslands are related to shift in species composition; therefore, resistant grasses are benefitted at the expense of sensitive grasses. Few studies addressing the effects of ambient and rising levels of O_3 on pasture communities revealed significant reduction in forage nutritive quality, a factor critical for animal production and is also commercially important. Ozone exposure can directly influence plant nutritive quality for herbivores as a result of altered concentrations of minerals, protein, and carbohydrates (Fuhrer 2009). Fuhrer and Booker (2003) found that elevated O_3 exposure reduced clover biomass and thus increased the grass/clover ratio. Blum et al. (1982) reported that O_3 exposure reduced total nonstructural carbohydrate and increased mineral concentrations in ladino clover (*Trifolium repens*), while Rebbeck et al. (1988) found increase in starch with no change in soluble sugar content under O_3 exposure.

Changes in forage quality under elevated O_3 are due to direct effect on secondary metabolism, or a change in plant development (Fernández et al. 2008). Increase in levels of phenolic acids, flavonoids, and related compounds under long-term O_3 exposure (Booker and Miller 1998) may negatively affect ruminant microorganisms and enzyme systems. One of the most common effects of O_3 is to promote leaf senescence. Thus, in pastures or other types of grasslands exposed to O_3, the fraction of senescing tissues may increase (Fuhrer et al. 2003). Because of increase in lignification, and decrease in leaf/stem ratio, forage digestibility decline under O_3 stress (Fernández et al. 2008). In grass-clover forage, leaf in vitro dry matter disappearance (IVDMD) and N content in leaves decreased and neutral detergent fiber (NDF) increased in *Trifolium repens* under ambient O_3 (50 ppb) compared with charcoal-filtered air (Burns et al. 1997). Bender et al. (2006) also reported decrease in nutritive quality of *Poa pratensis* in terms of relative food value (RFV) and increase in NDF and acid detergent fiber (ADF) due to higher lignin content under elevated O_3, suggesting significant nutritional implications for its utilization by ruminants.

A study was conducted to assess the impact of O_3 and its interactions with soil-nutrient regime on the productivity and nutritive quality of ryegrass/clover sward exposed to ambient O_3 (nonfiltered air, NFA), NFA+ 25 ppb, NFA +40 ppb, and NFA+ 55 ppb (Fernández et al. 2008). Results showed deterioration in forage quality of ryegrass and clover in form of reductions RFV and consumable food value (CFV). Increases in NDF, IVDMD, and ADL were observed in clover (Fernández et al. 2008). Changes in forage quality may result from shift in species composition. Differential O_3 sensitivities between grasses and legumes have been found to cause a shift in the grass/legume ratio in favor of grasses, hence, modifying the protein concentration and other quality traits relevant for animal nutrition (Rebbeck et al. 1988).

A number of studies have documented the negative impact of O_3 on nutritive quality of forage crops and crop residues for ruminant herbivores as determined by in vitro incubation experiments and/or chemical composition using wet-chemistry methods (Gonźalez-Fernandeź et al. 2009; Frei et al. 2011). Gilliland et al. (2012) conducted a digestibility experiment with a mixture of common Southern Piedmont (USA) grassland species (*Lolium arundinacea*, *Paspalum dilatatum*, *Cynodon dactylon*, *and Trifolium repens*) by exposing at ambient O_3 (non-filtered; NF) and elevated O_3 (twice-ambient (2X) concentration) and then fed to individually caged New Zealand white rabbits (*Oryctolagus cuniculus*). Forages and feed refusals were analyzed for concentrations of total cell wall constituents, lignin, crude protein, and soluble and hydrolyzable phenolic fractions. Neutral detergent fiber and acid detergent fiber digestibility by rabbits were significantly lower at 2X O_3 than NF forage. Decreased digestibility could not be attributed to lignin concentration but was associated with increases in concentrations of acid-hydrolyzable and saponifiable phenolics (Gilliland et al. 2012). Exposure of forage to elevated O_3 resulted in decrease in digestible dry matter intake by rabbits. The negative impact on forage quality under elevated O_3 resulted in decreased nutrient utilization by mammalian herbivores in Southern Piedmont grasslands under projected future climate scenarios of IPCC (Ainsworth et al. 2012).

14.6.5 Induction and Suppression of Pathogenicity

Ozone can have secondary effects on crops by affecting the incidence of pests and diseases and by altering crop–weed competition. The effects of O_3 on plants lead to altered disease susceptibility, but the effects are variable (Sandermann 2000). As observed by von Tiedemann et al. (1991), powdery mildew (*Erysiphe graminis*), leaf spot disease (*Septoria nodorum*), and spot blotch following inoculation with *Bipolaris sorokiniana* were significantly enhanced under exposure of wheat flag leaves to O_3. Conversely, Plazek et al. (2001) found a positive effect of O_3 on resistance of barley and fescue to *Bipolaris sorokiniana* and in rapeseed to *Phoma lingam*. Leaf rust disease (*Puccinia recondita*) on wheat leaves was strongly inhibited by O_3 (von Tiedemann and Firsching 2000).

In the field, the incidence of powdery mildew was reduced because of negative effects of O_3 on canopy structure resulting in a drier canopy microclimate, while infections caused by facultative pathogens were generally increased (Sandermann 2000). This suggests that under altered climatic conditions favoring infection pressure, plants weakened by O_3 stress become particularly susceptible. But the interaction between O_3 and pathogens may be determined primarily by the timing of O_3 exposure relative to the presence of the inoculation. Sandermann (2000) suggested that O_3 stress may induce a burst of active oxygen, which triggers the plant defense system in the leaves (systemic acquired resistance, hypersensitive response). Thus the outcome of plant–pathogen interactions may strongly vary with timing, stage of plant development, predisposing factors, and environmental conditions.

Investigations on the effects of elevated O_3 on the virus–plant system were conducted by Ye et al. (2012), and a susceptible cultivar Yongding of tobacco and Vam, a resistant cultivar to Potato virus Y petiole necrosis strain (PVYN) infection, were grown in OTCs under ambient and elevated O_3 concentrations. Results showed that under ambient O_3, the resistant cultivar possessed greater biomass and a lower C/N ratio after infection than the susceptible cultivar; however, under elevated O_3, the resistant cultivar lost its biomass advantage but maintained a lower C/N ratio. Chlorophyll content remained constant in the resistant cultivar but decreased significantly in the susceptible cultivar. This study indicated that a virus-resistant tobacco cultivar showed increased sensitivity to elevated O_3 compared to a virus-sensitive cultivar.

14.7 Yield Response

Tropospheric O_3 was found to adversely affect the growth and yield of a variety of agricultural plants. It also reduced the marketable yield of a range of crop species even in the absence of visible injury, primarily through its effects in reducing photosynthetic rates and accelerating leaf senescence (Rai et al. 2015). Yield reductions of 29–47 % were reported for six cultivars of barley, 37–46 % for three cultivars of wheat (Wahid 2006a, b) in Pakistan, and 10–25 % in four cultivars of rice in India (Rai et al. 2010; Sarkar and Agrawal 2011). Feng and Kobayashi (2009) in a meta-analysis showed maximum reductions in yield of soybean (40–60 %), followed by wheat (20–40 %), rice (10–20 %), and minimum in barley. Pang et al. (2009) reported that the grain yield was reduced by 20.7 and 6.3 % in rice cultivars SY63 and WYJ 63 in a FACE experiment at 74.2 ppb O_3. Ozone exposure of 70 and 100 ppb for 4 h day^{-1} for 70 days led to reductions in yield by 13.9 % and 10 % and 33.5 % and 25 % in soybean cultivars PK 472 and Bragg, respectively (Singh et al. 2010). Sarkar and Agrawal (2010a) found reductions of 7, 16.7, and 22 % in wheat cultivars Sonalika and 8.4, 18.5, and 25 % in cultivar HUW 510 grown at 45.3, 50.4, and 55.6 ppb O_3 compared to FCs. In SoyFACE experiment, ten soybean cultivars were exposed to ambient (46.3 and 37.9 ppb) and 20 % elevated O_3 concentrations in 2007–2008 and yield reductions varied from 11.3–36.8 % in 2007 to 7.5–16 % in 2008. The yield response relationships also indicated that Loda and Pana were tolerant and IA 3010 was sensitive (Betzelbergler et al. 2010).

Zhu et al. (2011) exposed four winter wheat cultivars (Yannog 19, Yangmai 16, Yangmain 15, and Yangfumai 2) under elevated O_3 using a FACE system from 2007 to 2009 with mean O_3 levels of 56.9 ppb for 7 h in 2006–2007, 57.6 ppb in 2007–2008, and 57.3 ppb for 2008–2009. The grain yield reductions recorded were 18.7, 34.7, and 10.1 % in Y 19 in three consecutive years of O_3 exposure. Singh et al. (2012) reported reductions in mustard yield by 19.3 and 7 % in Aashirwad and Vardan cultivars grown in OTCs receiving mean O_3 concentration of 44.6 ppb. Pandey et al. (2015) reported yield loss of 25–48 % in 18 rice cultivars receiving 83 ppb O_3. Singh et al. (2014c) reported yield reductions of 16 % in wheat cultivar Lok-1 grown at mean concentration of 58.2 ppb O_3. Yield losses recorded in six cultivars of mung bean ranged between 9.8 % and 15.4 % under mean ambient O_3 concentration of 64 ppb (Chaudharyand and Agrawal 2015).

Decline in yield of maize cultivars HQPM1 (4 %) and DHM 117 (5.5 %) exposed to ambient O_3 concentration of 52 ppb (Singh et al. 2014b).

Ozone also causes reductions in growth and biomass of forest tree species (Paoletti 2009). O_3 induced 10–15 % reductions in height and diameter growth of aspen (Paoletti 2009). Chappelka and Samuelson (1998) reported growth losses of 20 % for >50 % of blackcherry and aspen exposed to elevated O_3. Ollinger et al. 1997 used a canopy to stand model to predict forest response to O_3. Results of modeling study showed decrease of 3–16 % in net primary productivity. Mean annual production of forest decreased by 3–22 % under O_3 exposure. Ozone also affects wood composition and commercial value. Studies have shown that O_3 reduced xylem volume and increased lignin concentration in aspen and poplars (Richet et al. 2011; Kostiainen et al. 2008). Another changes associated with wood structure were vessel lumen diameter which decreased leading to reduction in water transport efficiency in the xylem (Karnosky et al. 2005).

Grassland communities have also shown reduction in aboveground biomass (Fernandez et al. 2008; Pleijel and Dannielson 1997). Harvest index (percentage aboveground biomass found in the reproductive structures at final harvest) did not decrease in 22 herbs exposed to 49 and 73 ppb O_3 (Pleijel and Dannielson 1997). In several studies, O_3-induced growth reductions in herbaceous species have been measured in terms of relative growth rate (RGR). Bender et al. (2006) found no changes in total aboveground biomass production of *Poa pratensis* either as primary growth or in regrowth at the end of season under O_3 exposure. This suggests that grasses are relatively insensitive to O_3 compared to forbs and legumes (Fuhrer et al. 1994; Davison and Barnes 1998; Gimeno et al. 2004). In a similar exposure study with wet-grassland species, Tonneijck et al. (2004) also reported only minimal O_3 effects on biomass production of the investigated grass species at comparable O_3 doses.

14.8 IPCC Projections for O_3 Concentrations and Yield Losses

Global demand for food is expected to increase by at least 50 % from 2010 to 2050 mainly as a result of population growth. Tai et al. (2014) quantified individual and combined effects of mean temperature and O_3 pollution on the global production of wheat, rice, maize, and soybean from 2000 to 2050. Quantification was done with Community Earth System Model (CESM) to simulate present-day (2000) and derive future (2050) projections of hourly temperature and O_3 concentrations consistent with the representative concentration pathways (RCPs) represented in the Intergovernmental Panel on Climate Change Fifth Assessment Report (AR5) (Wise et al. 2009; Meinshausen et al. 2011). Future O_3 projections not only follow trends in anthropogenic emissions of precursor gases but also include the effects of climate and land use change as these factors significantly affect future projections.

Two IPCC scenarios, RCP4.5, representing an intermediate pathway with a global reduction in surface O_3 due to pollution control measures worldwide (except in South Asia), and RCP8.5, representing a more "pessimistic," energy-intensive pathway with a worldwide increase in O_3 except in the USA and around Japan, were considered. The two scenarios represent a range of policy options regarding O_3 regulation. Both scenarios project a global increase in surface temperature with negative effects on crop production. Ozone formation is strongly correlated with temperature, so the observed crop–temperature relationship may arise in part from O_3 damage at high temperature. Results showed that on a global scale, more severe O_3 pollution is expected for RCP8.5 leading to substantial crop damage (except in the USA and South Korea), thus reducing global total crop production by 3.6 %, but aggressive pollution control worldwide expected for RCP4.5 led to substantial gain in crop production in many regions (except South

Asia) with an overall 3.1 % increase in global production (Tai et al. 2014). It was suggested that O_3 pollution control as represented in RCP4.5 has the potential to partially offset the negative impact of climate change, leading to a smaller combined decrease of 9.0 % compared with RCP8.5, where O_3 pollution and climate change combined to reduce global crop production by 15 %.

Further evaluation made on how combined changes of O_3 and temperature projections may shift in current (2000) distribution of per capita food consumption in developing countries, leading to a change in the rate of undernourishment as a proxy for the potential societal impact, and the current undernourishment rate of 18 % was used as the baseline for estimation, and no accounts for agricultural advances, land use change, and international politics were included. For RCP8.5, more serious O_3 pollution worldwide and climate change combined to increase undernourishment rate in developing countries by 49 % (from a rate of 18–27 %). For RCP4.5, undernourishment rate may increase by 27 % because O_3 regulation may partially offset the warming effect, suggesting the importance of air quality management in devising strategies for food security. Wheat in all major producing regions is mostly sensitive to O_3 and its effect is much larger compared to temperature effect. Ozone regulation as represented in RCP4.5 has the potential to completely reverse the warming impact and lead to substantial gain in wheat production in the USA and China. In South Asia where O_3 pollution is projected to worsen in both scenarios, wheat production will be reduced up to 40 %. Results of rice and maize production in China were also found sensitive to O_3 pollution.

The study suggests that greater collaboration between farmers, agricultural policymakers, and air quality managers is needed to achieve coordinated goals concerning public health and food security.

14.9 Conclusions

Tropospheric ozone has been identified as a phytotoxic secondary air pollutant and a green house gas. Indiscriminate anthropogenic activities had increased its concentration tremendously, leading to potential threats to crops, pastures, and forests. Ozone causes visible injury, impairs photosynthesis, alters stomatal behavior, decreases growth and productivity, and alters the metabolic pathway of plants. Sensitivity of crops, grasses, and tree species depends on exposure, leaf uptake, plant defense capacity, and developmental stages. Fast-growing species are found to be more sensitive than slow-growing species. Ozone also alters interspecific plant competition as grasses grown in monoculture are more sensitive than in mixed culture. Ozone increases incidence of pest and disease attacks on plants. O_3 induces crop yield and quality losses. Changes in forage quality through shift in pasture communities leading to change in herbivory pattern are also found under ambient levels of O_3. Ozone reduces commercial value of wood. Quantification of effects of O_3 on nutritive quality of herbaceous vegetation and on animal nutrition is lacking. Tropospheric O_3 has damaged sensitive forest species of the USA and Europe and change in wood quality and reduced commercial value of wood.

There is a dearth of field data on estimating the impact of rising O_3 concentration on natural and seminatural ecosystems in developing countries where highest O_3 concentrations are projected in all future scenarios. In view of the study, there is a need of greater collaboration between farmers, agricultural policymakers, and air quality managers to achieve coordinated goals concerning public health, food security, food availability, and forest sustainability under projected scenarios of elevated ozone and temperature.

References

Agathokleous E, Saitanis CJ, Koike T (2015) Tropospheric O_3, the nightmare of wild plants: a review study. J Agr Meteorol 71:142–152

Agrawal GK, Rakwal R, Yonekura M, Kubo A, Saji H (2002) Proteome analysis of differentially displayed proteins as a tool for investigating ozone stress in rice (*Oryza sativa* L.) seedlings. Proteomics 2:947–959

Agrawal SB, Singh A, Rathore D (2005) Role of ethylene diurea (EDU) in assessing impact of ozone on *Vigna radiata* L. plants in a suburban area of Allahabad (India). Chemosphere 61:218–228

Ainsworth EA (2008) Rice production in a changing climate: a metaźanalysis of responses to elevated carbon dioxide and elevated ozone concentration. Global Chang Biol 14:1642–1650

Ainsworth EA, Yendrek CR, Sitch S, Collins WJ, Emberson LD (2012) The effects of tropospheric ozone on net primary productivity and implications for climate change. Annu Rev Plant Biol 63:637–661

Ali K, Inamdar SR, Beig G, Ghude S, Peshin S (2012) Surface ozone scenario at Pune and Delhi during the decade of 1990s. J Earth Syst Sci 121:373–383

Anav A, Menut L, Khvorostyanov D, Viovy N (2011) Impact of tropospheric ozone on the Euro-Mediterranean vegetation. Glob Chang Biol 17:2342–2359

Apel K, Hirt H (2004) Reactive oxygen species: metabolism, oxidative stress, and signal transduction. Ann Rev Plant Biol 55:373–399

Ashmore MR (2005) Assessing the future global impacts of ozone on vegetation. Plant Cell Environ 28:949–964

Barbo DN, Chappelka AH, Somers GL, Miller- Goodman MS, Stolt K (1998) Diversity of an early successional plant community as influenced by ozone. New Phytol 138:653–662

Bass DJ, Barnes JD, Lyons T, Mills G (2006) The impact of tropospheric ozone on semi- natural vegetation. PhD thesis, Newcastle University, UK

Bassin S, Kolliker R, Cretton C, Bertossa M, Widmer F, Bungener P, Fuhrer J (2004) Intra- specific variability of ozone sensitivity in *Centaurea jacea* L., a potential bioindicator for elevated ozone concentrations. Environ Pollut 131:1–12

Bassin S, Volk M, Fuhrer J (2007) Factors affecting the ozone sensitivity of temperate European grasslands: an overview. Environ Pollut 146:678–691

Beig G, Gunthe S, Jadhav DB (2007) Simultaneous measurements of ozone and its precursors on a diurnal scale at a semi urban site in India. J Atmos Chem 57:239–253

Bender J, Muntiferang RB, Lin JC, Weigel HJ (2006) Growth and nutritive quality of *Poa pratensis* as influenced by ozone and fumigation. Environ Pollut 142:109–115

Bennett JP, Rassat P, Berrang P, Karnosky DF (1992) Relationships between leaf anatomy and ozone sensitivity of *Fraxinus pennsylvanica* Marsh and *Prunus serotina* Ehrh. Environ Exp Bot 32:33–41

Bergmann E, Bender J, Weigel HJ (1995) Growth responses and foliar sensitivities of native herbaceous species to ozone exposure. Water Air Soil Pollut 85:1437–1442

Bermejo V, Gimeno BS, Sanz J, de la Torre D, Gil JM (2003) Assessment of the ozone sensitivity of 22 native plant species from Mediterranean annual pastures based on visible injury. Atmos Environ 37:4667–4677

Bernacchi CJ, Leakey AD, Heady LE, Morgan PB, Dohleman FG, McGrath JM, Ort DR (2006) Hourly and seasonal variation in photosynthesis and stomatal conductance of soybean grown at future CO_2 and ozone concentrations for 3 years under fully openźair field conditions. Plant Cell Environ 29:2077–2090

Betzelberger AM, Gillespie KM, Mcgrath JM, Koester RP, Nelson R, Ainsworth EA (2010) Effects of chronic elevated ozone concentration on antioxidant capacity, photosynthesis and seed yield of 10 soybean cultivars. Plant Cell Environ 33:1561–1589

Bishop JA, Cook LM (1981) Genetic consequences of man made change. Academic, London

Biswas DK, Xu H, Li YG, Sun JZ, Wang XZ, Han XG, Jiang GM (2008a) Genotypic differences in leaf biochemical, physiological and growth responses to ozone in 20 winter wheat cultivars released over the past 60 years. Global Chang Biol 14:46–59

Biswas DK, Xu H, Li YG, Liu MZ, Chen YH, Sun JZ, Jiang GM (2008b) Assessing the genetic relatedness of higher ozone sensitivity of modern wheat to its wild and cultivated progenitors/relatives. J Exp Bot 59:951–963

Blokhina O, Virolainen E, Fagerstedt KV (2003) Antioxidants, oxidative damage and oxygen deprivation stress: a review. Ann Bot Lond 91:179–194

Blum U, Smith GR, Fites RC (1982) Effects of multiple O_3 exposures on carbohydrates and mineral contents of ladino clover. Environ Exp Bot 22:143–154

Booker FL, Miller JE (1998) Phenylpropanoid metabolism and phenolic composition of soybean [*Glycine max* (L.) Merr] leaves following exposure to ozone. J Exp Bot 49:1191–1202

Booker FL, Burkey KO, Pursley WA, Heagle AS (2007) Elevated carbon dioxide and ozone effects on peanut: I. Gas-exchange, biomass, and leaf chemistry. Crop Sci 47:1475–1487

Booker F, Muntifering R, McGrath M, Burkey K, Decoteau D, Fiscus E, Manning W, Sagar K, Chappelka A, Grantz D (2009) The ozone component of global change: potential effects on agricultural and horticultural plant yield, product quality and interactions with invasive species. J Integr Plant Biol 51:337–351

Burns JC, Heagle AS, Fisher DS (1997) Nutritive value of ozone sensitive and resistant Ladino white clover clones after chronic ozone and carbon dioxide exposure. In: Advances in carbon dioxide effects research. ASA Spec Publ Madison (WI): American Society of Agronomy 61: 153–167

Bussotti F, Agati G, Desotgiu R, Matteini P, Tani C (2005) Ozone foliar symptoms in woody plant species assessed with ultrastructural and fluorescence analysis. New Phytol 142:283–293

Bussotti F, Desotgiu R, Cascio C, Pollastrini M, Gravano E, Gerosa G, Strasser RJ (2011) Ozone stress in woody plants assessed with chlorophyll a fluorescence. A critical reassessment of existing data. Environ Exp Bot 73:19–30

Calatayud A, Barreno E (2001) Chlorophyll a fluorescence, antioxidant enzymes and lipid peroxidation in tomato in response to ozone and benomyl. Environ Pollut 115:283–289

Calatayud V, Cerveró J, Sanz MJ (2007) Foliar, physiological and growth responses of four maple species exposed to ozone. Water Air Soil Pollut 185:239–254

Calatayud V, Cerveró J, Calvo E, García-Breijo FJ, Reig-Armiñana J, Sanz MJ (2011) Responses of evergreen and deciduous *Quercus* species to enhanced ozone levels. Environ Pollut 159:55–63

Caregnato FF, Bortolin RC, Junior AMD, Moreira JCF (2013) Exposure to elevated ozone levels differentially affects the antioxidant capacity and the redox homeostasis of two subtropical *Phaseolus vulgaris* L. varieties. Chemosphere 93:320–330

Castagna A, Ranieri A (2009) Detoxification and repair process of ozone injury: from O_3 uptake to gene expression adjustment. Environ Pollut 157:1461–1469

Chappelka AH, Samuelson LJ (1998) Ambient ozone effects on forest trees of the eastern United States: a review. New Phytol 139:91–108

Chaudhary N, Agrawal SB (2013) Intraspecific responses of six Indian clover cultivars under ambient and elevated levels of ozone. Environ Sci Pollut R 20:5318–5329

Chaudhary N, Agrawal SB (2015) The role of elevated ozone on growth, yield and seed quality amongst six cultivars of mung bean. Ecotoxcol Environ Safe 111:286–294

Chen Z, Gallie DR (2005) Increasing tolerance to ozone by elevating foliar ascorbic acid confers greater protection against ozone than increasing avoidance. Plant Physiol 138:1673–1689

Chevalier A, Gheusi F, Delmas R, Ordonez C, Sarrat C, Zbinden R, Thouret V, Athie G, Cousin JM (2007) Influence of altitude on ozone levels and variability in the lower troposphere: a ground-based study for western Europe over the period 2001–2004. Atmos Chem Phys 7:4311–4326

Cooper OR, Parrish DD, Stohl A, Trainer M, Nédélec P, Thouret V, Cammas JP, Oltmans SJ, Johnson BJ, Tarasick D, Leblanc T, McDermid IS, Jaffe D, Gao R, Stith J, Ryerson T, Aikin K, Campos T, Weinheimer A, Avery MA (2010) Increasing springtime ozone mixing ratios in the free troposphere over western North America. Nature 463:344–348

D'Haese D, Vandermeiren K, Asard H, Horemans N (2005) Other factors than apoplastic ascorbate contribute to the differential ozone tolerance of two clones of *Trifolium repens* L. Plant Cell Environ 28:623–632

Danielsson H, Karlsson PE, Pleijel H (2013) An ozone response relationship for four *Phleum pratense* genotypes based on modelling of the phytotoxic ozone dose (POD). Environ Exp Bot 90:70–77

Darbah JN, Kubiske ME, Nelson N, Kets K, Riikonen J, Sober A, Karnosky DF (2010) Will photosynthetic capacity of aspen trees acclimate after long-term exposure to elevated CO_2 and O_3? Environ Pollut 158:983–991

Davison AW, Barnes JD (1998) Effects of ozone on wild plants. New Phytol 139:135–151

De Bock M, Ceulemans R, Horemans N, Guisez Y, Vandermeiren (2012) Photosynthesis and crop growth of spring oilseed rape and broccoli under elevated tropospheric ozone. Environ Exp Bot 82:28–36

Derwent RG, Simmonds PG, Manning AJ, Spain TG (2007) Trends over a 20-year period from 1987 to 2007 in surface ozone at the atmospheric research station, Mace Head, Ireland. Atmos Environ 41:9091–9098

Diara C, Castagna A, Baldan B, Mensuali Sodi A, Sahr T, Langebartels C, Sebastiani L, Ranieri A (2005) Differences in the kinetics and scale of signaling molecule production modulate the ozone sensitivity of hybrid poplar clones: the roles of H_2O_2, ethylene and salicylic acid. New Phytol 168:351–364

EANET (2006) Data report on acid deposition on East Asia Region 2005. Network Centre of EANET, Japan. http://www.eanet.cc/. Accessed 16 August 2015

Eckardt NA, Pell EJ (1994) O_3źinduced degradation of Rubisco protein and loss of Rubisco mRNA in relation to leaf age in *Solanum tuberosum* L. New Phytol 127:741–748

Emberson LD, Buker P, Ashmore MR, Mills G, Jackson LS, Agrawal M, Atikuzzaman MD, Cinderby S, Engardt M, Jamir C, Kobayshi K, Oanh OTR, Quadir QF, Wahid A (2009) A comparison of North-American and Asian exposure-response data for ozone effects on crop yields. Atmos Environ 43:1945–1953

Feng Z, Kobayashi K (2009) Assessing the impacts of current and future concentrations of surface ozone on crop yield with meta-analysis. Atmos Environ 43:1510–1519

Feng Z, Wang X, Zheng Q, Feng Z, Xie J, Chen Z (2006) Response of gas exchange of rape to ozone concentration and exposure regime. Acta Ecol Sin 26:823–829

Feng ZZ, Yao FF, Chen Z, Wang XK, Zheng QW, Feng ZW (2007) Response of gas exchange and yield components of field grown *Triticum aestivum* L. to elevated ozone in China. Photosynthetica 45:441–446

Feng Z, Kobayashi K, Ainsworth EA (2008) Impact of elevated ozone concentration on growth, physiology, and yield of wheat (*Triticum aestivum* L.): a metaźanalysis. Global Chang Biol 14:2696–2708

Feng Z, Kobayashi K, Wang X, Feng Z (2009) A meta-analysis of responses of wheat yield formation to elevated ozone concentration. Chin Sci Bull 54:249–255

Feng Z, Pang J, Nouchi I, Kobayashi K, Yamakawa T, Zhu J (2010) Apoplastic ascorbate contributes to the differential ozone sensitivity in two varieties of winter wheat under fully open-air field conditions. Environ Pollut 158:3539–3545

Feng Z, Niu J, Zhang W, Wang X, Yao F, Tian Y (2011a) Effects of ozone exposure on sub-tropical evergreen *Cinnamomum camphora* seedlings grown in different nitrogen loads. Trees 25:617–625

Feng Z, Pang J, Kobayashi K, Zhu J, Ort DR (2011b) Differential responses in two varieties of winter wheat to elevated ozone concentration under fully openźair field conditions. Global Chang Biol 17:580–591

Fernandeź IG, Bass D, Muntifering R, Mills G, Barnes J (2008) Impacts of ozone pollution on productivity and forage quality of grass/clover swards. Atmos Environ 42:8755–8769

Fiscus EL, Booker FL, Burkey KO (2005) Crop responses to ozone: uptake, modes of action, carbon assimilation and partitioning. Plant Cell Environ 28:997–1011

Flowers MD, Fiscus EL, Burkey KO, Booker FL, Dubois JJB (2007) Photosynthesis, chlorophyll fluorescence, and yield of snap bean (*Phaseolus vulgaris* L.) genotypes differing in sensitivity to ozone. Environ Exp Bot 61:190–198

Francini A, Nali C, Picchi V, Lorenzini G (2007) Metabolic changes in white clover clones exposed to ozone. Environ Exp Bot 60:11–19

Frei M, Kohno Y, Wissuwa M, Makkar HPS, Becker KL (2011) Negative effects of tropospheric ozone on the feed value of rice straw are mitigated by an ozone tolerance QTL. Glob Chang Biol 17:2319–2329

Fuhrer J (2009) Ozone risk for crops and pastures in present and future climates. Naturwissenschaften 96:173–194

Fuhrer J, Booker F (2003) Ecological issues related to ozone: agricultural issues. Environ Int 29:141–154

Fuhrer J, Shariat- Madari H, Perler R, Tschannew W, Grub A (1994) Effects of ozone on managed pasture II. Yield species composition canopy structure and forage quality. Environ Pollut 97:91–106

Fuhrer J, Ashmore MR, Mills G, Hayes F, Davison A (2003) Critical levels for semi-natural vegetation. In: Karlsson PE, Sellden G, Pleijel H (eds) Establishing ozone critical levels II. IVL, Stockholm, pp 183–198

Gielen B, Löw M, Deckmyn G, Metzger U, Franck F, Heerdt C, Ceulemans R (2007) Chronic ozone exposure affects leaf senescence of adult beech trees: a chlorophyll fluorescence approach. J Exp Bot 58:785–795

Gilliland NJ, Chappelka AH, Muntifering RB, Booker FL, Ditchkoff SS (2012) Digestive utilization of ozone-exposed forage by rabbits (*Oryctolagus cuniculus*). Environ Pollut 163:281–286

Gimeno BS, Bermejo V, Sanz J, de la Torre D, Elvira S (2004) Growth response to ozone of annual species from Mediterranean pastures. Environ Pollut 132:297–306

Gonźalez-Fernandeź I, Bass D, Muntifering R, Mills G, Barnes J (2009) Impacts of ozone pollution on productivity and forage quality of grass/ clover swards. Atmos Environ 42:8755–8769

Guidi L, Degl'Innocenti E, Martinelli F, Piras M (2009) Ozone effects on carbon metabolism in sensitive and insensitive *Phaseolus* cultivars. Environ Exp Bot 66:117–125

Harmens, H (2014) Air pollution and vegetation: ICP Vegetation annual report 2013/2014 Type of book: monografija Formal editor/s: Harmens, Harry; Mills, Gina; Hayes, Felicity; Sharps, Katrina, Frontasyeva, Marina

Hassan IA, Tewfik I (2006) CO_2 photo assimilation, chlorophyll fluorescence, lipid peroxidation and yield in cotton (*Gossypium hirsutum* L. cv Giza 65) in response to O_3. World Rev Sci Techno Sust Dev 3:70–78

Hayes F, Jones MLM, Mills G, Ashmore M (2007a) Meta-analysis of the relative sensitivity of semi-natural vegetation species to ozone. Environ Pollut 146:754–762

Hayes F, Mills G, Harmens H, Norris D (2007b) Evidence of widespread ozone damage to vegetation in Europe (1990–2006). ICP Vegetation Programme Coordination Centre, CEH Bangor, UK

Hayes F, Mills G, Ashmore M (2009) Effects of ozone on inter-and intra-species competition and photosynthesis in mesocosms of *Lolium perenne* and *Trifolium repens*. Environ Pollut 157:208–214

Herbinger K, Then C, Löw M, Haberer K, Alexous M, Koch N, Wieser G (2005) Tree age dependence and within-canopy variation of leaf gas exchange and antioxidative defence in *Fagus sylvatica* under experimental free-air ozone exposure. Environ Pollut 137:476–482

Hofer N, Alexou M, Heerdt C, Löw M, Werner H, Matyssek R, Haberer K (2008) Seasonal differences and within-canopy variations of antioxidants in mature spruce (*Picea abies*) trees under elevated ozone in a free-air exposure system. Environ Pollut 154:241–253

Hoshika Y, Omasa K, Paoletti E (2013) Both ozone exposure and soil water stress are able to induce stomatal sluggishness. Environ Exp Bot 88:19–23

Iglesias DJ, Calatayud A, Primo-Millo EBE, Talon M (2006) Responses of citrus plants to ozone: leaf biochemistry, antioxidant mechanisms and lipid Peroxidation. Plant Physiol Biochem 44:125–131

IPCC (2013) Climate change 2013. The physical science basis. www.ipcc.ch

IPCC (2014) Pachauri RK, Allen MR, Barros VR, Broome J, Cramer W, Christ R, … & Vuuren D Climate Change 2014: Synthesis Report. Contribution of Working Groups I, II and III to the Fifth Assessment Report of the Intergovernmental Panel on Climate Change

Iqbal M, Abdin M, Mahmooduzzafar Z, Yunus M, Agrawal M (1996) Resistance mechanisms in plants against air pollution. In: Iqbal M, Yunus M (eds) Plant response to air pollution. Wiley, New York, pp 195–240

Iriti M, Faoro F (2009) Chemical diversity and defence metabolism: how plants cope with pathogens and ozone pollution. Int J Mol Sci 10:3371–3399

Jaffe D, Ray J (2007) Increase in surface ozone at rural sites in the western US. Atmos Environ 41:5452–5463

Jain SL, Arya BC, Kumar A, Ghude SD, Kulkarni PS (2005) Observational study of surface ozone at New Delhi, India. Int J Remote Sens 26:3515–3524

Jenkin ME (2008) Trends in ozone concentration distributions in the UK since 1990: local, regional and global influences. Atmos Environ 42:5434–5445

Karnosky DF, Pregitzer KS, Zak DR, Kubiske ME, Hendrey GR, Weinstein D, Nosal M, Percy KE (2005) Scaling ozone responses of forest trees to the ecosystem level in a changing climate. Plant Cell Environ 28:965–981

Karnosky DF, Skelly JM, Percy KE, Chappelka AH (2007) Perspectives regarding 50 years of research on effects of tropospheric ozone air pollution on US forests. Environ Pollut 147:489–506

Kitao M, Löw M, Heerdt C, Grams TE, Häberle KH, Matyssek R (2009) Effects of chronic elevated ozone exposure on gas exchange responses of adult beech trees (*Fagus sylvatica*) as related to the within-canopy light gradient. Environ Pollut 157:537–544

Körner C (2003) Alpine plant life, 2nd edn. Springer, Berlin

Kostiainen K, Kaakinen S, Warsta E, Kubiske ME, Nelson ND, Sober J, Karnosky DF, Saranpaa P, Vapaavuori E (2008) Wood properties of trembling aspen and paper birch after 5 years of exposure to elevated concentrations of CO_2 and O_3. Tree Physiol 28:805–813

Krupa SV, Manning WJ (1988) Atmospheric ozone: formation and effects on vegetation. Environ Pollut 50:101–137

Laisk A, Kull O, Moldau H (1989) Ozone concentration in leaf intercellular air spaces is close to zero. Plant Physiol 90:1163–1167

Langner J, Engardt M, Baklanov A, Christensen JH, Gauss M, Geels C, Hedegaard GB, Nuterman R, Simpson D, Soares J, Sofiev M, Wind P, Zakey A (2012) A multi-model study of impacts of climate change on surface ozone in Europe. Atmos Chem Phys 12:10423–10440

Leitao L, Delacôte E, Dizengremel P, Le Thiec D, Biolley JP (2007) Assessment of the impact of increasing concentrations of ozone on photosynthetic components of maize (*Zea mays* L.), a C 4 plant. Environ Pollut 146:5–8

Lombardozzi D, Sparks JP, Bonan G, Levis S (2012) Ozone exposure causes a decoupling of conductance and photosynthesis: implications for the Ball-Berry stomatal conductance model. Oecologia 169:651–659

Lu T, He X, Chen W, Yan K, Zhao T (2009) Effects of elevated O_3 and/or elevated CO_2 on lipid peroxidation and antioxidant systems in Ginkgo biloba leaves. B Environ Contam Tox 83:92–96

Madkour SA, Laurence JA (2002) Egyptian plant species as new ozone indicators. Environ Pollut 120:339–353

Mahapatra PS, Jena J, Moharana S, Srichandan H, Das T, Chaudhury GR, Das SN (2012) Surface ozone variation at Bhubaneswar and intra-corelationship study with various parameters. J Earth Syst Sci 121:1163–1175

Matyssek R, Sandermann H, Wieser G, Booker F, Cieslik S, Musselman R, Ernst D (2008) The challenge of making ozone risk assessment for forest trees more mechanistic. Environ Pollut 156:567–582

Meehl GA, Stocker TF, Collins WD, Friedlingstein P, Gaye AT, Gregory JM, Kitoh A, Knutti R, Murphy JM, Noda A, Raper SCB, Watterson IG, Weaver AJ, Zhao ZC (2007) Global climate projections. In: Solomon S, Qin D, Manning M, Chen Z, Marquis M, Averyt KB, Miller HL, Tignor M (eds) Climate change 2007: the physical basis contribution of working group I to fourth assessment report of IPCC on climate change. Cambridge University Press, Cambridge

Meinshausen M, Smith SJ, Calvin K, Daniel JS, Kainuma MLT, Lamarque J-F, Malsumoto K, Sa M, Raper SCB, Riahi K, Thomson A, Velders GJM, van Vuuren DPP (2011) The RCP greenhouse gas concentrations and their extensions from 1765 to 2300. Clim Chang 109:213–241

Mills G, Hayes F, Jones MLM, Cinderby S (2007) Identifying ozone sensitive communities of (semi-) natural vegetation suitable for mapping exceedance of critical levels. Environ Pollut 146:736–743

Mills G, Hayes F, Wilkinson S, Davies WJ (2009) Chronic exposure to increasing background ozone impairs stomatal functioning in grassland species. Global Chang Biol 15:1522–1533

Mishra AK, Agrawal SB (2015) Biochemical and physiological characteristics of tropical mung bean (*Vigna radiata* L.) cultivars against chronic ozone stress: an insight to cultivar-specific response. Protoplasma 252:797–811

Mittal ML, Hess PG, Jain SL, Arya BC, Sharma C (2007) Surface ozone in the Indian region. Atmos Environ 41:6572–6584

Monks PS (2005) Gas phase chemistry in the troposphere. Chem Soc Rev 34:376–395

Morgan PB, Mies TA, Bollero GA, Nelson RL, Long SP (2006) Season-long elevation of ozone concentration to projected 2050 levels under fully open-air conditions substantially decreases the growth and production of soybean. New Phytol 170:333–343

Nali C, Pucciariello C, Mills G, Lorenzini G (2005) On the different sensitivity of white clover clones to ozone: physiological and biochemical parameters in a multivariate approach. Water Air Soil Pollut 164:137–153

Novak K, Schaub M, Fuhrer J, Skelly JM, Hug C, Landolt W, Kräuchi N (2005) Seasonal trends in reduced leaf gas exchange and ozone-induced foliar injury in three ozone sensitive woody plant species. Environ Pollut 136:33–45

Nussbaum S, Bungener P, Geissman M, Fuhrer J (2000) Plant-plant interactions and soil moisture might be important in determining ozone impacts on grasslands. New Phytol 147:327–335

Ollinger SV, Aber JD, Reich PB (1997) Simulating ozone effects on forest productivity: interactions among leaf-, and standlevel processes. Ecol Appl 7:1237–1251

Padu E, Kollist H, Tulva I, Oksanen E, Moldau H (2005) Components of apoplastic ascorbate use in *Betula pendula* leaves exposed to CO_2 and O_3 enrichment. New Phytol 165:131–142

Pandey AK, Majumder B, Keski-Saari S, Kontunen SS, Mishra A, Sahu N, Pandey V, Oksanen E (2015) Searching for common responsive parameters for ozone tolerance in 18 rice cultivars in India: results from ethylene diurea studies. Sci Total Environ 532:230–238

Pang J, Kobayashi K, Zhu J (2009) Yield and photosynthetic characteristics of flag leaves in Chinese rice (*Oryza sativa* L.) varieties subjected to free-air release of ozone. Agric Ecosyst Environ 132:203–211

Paoletti E (2006) Impact of ozone on Mediterranean forests: a review. Environ Pollut 144:463–474

Paoletti E (2009) Ozone and urban forests in Italy. Environ Pollut 157:1506–1512

Paoletti E, Grulke NE (2010) Ozone exposure and stomatal sluggishness in different plant physiognomic classes. Environ Pollut 158:2664–2671

Paoletti E, Contran N, Bernasconi P, Günthardt-Goerg MS, Vollenweider P (2010) Erratum to Structural and physiological responses to ozone in Manna ash (*Fraxinus ornus* L.) leaves of seedlings and mature trees under controlled and ambient conditions. Sci Tot Environ 408:2014–2024

Pellegrini E, Francini A, Lorenzini G, Nali C (2011) PSII photochemistry and carboxylation efficiency in Liriodendron tulipifera under ozone exposure. Environ Exp Bot 70:217–226

Percy KE, Nosal M, Heilman W, Dann T, Sober J, Legge AH, Karnosky DF (2007) New exposure-based metric approach for evaluating O_3 risk to North American aspen forests. Environ Pollut 147:554–566

Pfleeger TG, Plocher M, Bichel P (2010) Response of pioneer plant communities to elevated ozone exposure. Agric Ecosyst Environ 138:116–126

Pina JM, Moraes RM (2010) Gas exchange, antioxidants and foliar injuries in saplings of a tropical woody species exposed to ozone. Environ Exp Bot 73:685–691

Pinto E, Sigaudźkutner T, Leitao MA, Okamoto OK, Morse D, Colepicolo P (2003) Heavy metal–induced oxidative stress in algae. J Phycol 39:1008–1018

Plazek A, Hura K, Rapacz H (2001) The influence of ozone fumigation on metabolic efficiency and plant resistance to fungal pathogens. J Appl Bot 75:8–13

Pleijel H, Dannielsson H (1997) Growth of 27 herbs and grasses in relation to ozone exposure and plant strategy. New Phytol 135:361–367

Pleijel H, Eriksen AB, Danielsson H, Bondesson N, Selldén G (2006) Differential ozone sensitivity in an old and a modern Swedish wheat cultivar—grain yield and quality, leaf chlorophyll and stomatal conductance. Environ Exp Bot 56:63–71

Pulikesi M, Baskaralingam P, Rayudu VN, Elango D, Ramamurthi V, Sivanesan S (2006) Surface ozone measurements at urban coastal site Chennai, in India. J Hazard Mater 137:1554–1559

Rai R, Agrawal M (2008) Evaluation of physiological and biochemical responses of two rice (*Oryza sativa* L.) cultivars to ambient air pollution using open top chambers at a rural site in India. Sci Tot Environ 407:679–691

Rai R, Agrawal M (2012) Impact of tropospheric ozone on crop plants. Proc Nat Acad Sci India Sect B Biol Sci 82:241–257

Rai R, Agrawal M (2014) Assessment of competitive ability of two Indian wheat cultivars under ambient O_3 at different developmental stages. Environ Sci Pollut R 21:1039–1053

Rai R, Agrawal M, Agrawal SB (2007) Assessment of yield losses in tropical wheat using open top chambers. Atmos Environ 41:9543–9554

Rai R, Agrawal M, Agrawal SB (2010) Threat to food security under current levels of ground level ozone: a case study for Indian cultivars of rice. Atmos Environ 44:4272–4282

Rai R, Agrawal M, Choudhary KK, Agrawal SB, Emberson L, Büker P (2015) Application of ethylene diurea (EDU) in assessing the response of a tropical soybean cultivar to ambient O_3: nitrogen metabolism, antioxidants, reproductive development and yield. Ecotox Environ Safe 112:29–38

Ranford J, Reiling K (2007) Ozone induced leaf loss and decreased leaf production of European Holly (*Ilex aquifolium* L.) over multiple seasons. Environ Pollut 145:355–364

Rebbeck J, Blum U, Heagle AS (1988) Effects of ozone on the regrowth and energy reserves of a ladino-clover- tall fescue pasture. J Appl Ecol 25:659–681

Reddy BSK, Kumar KR, Balakrishnaiah G, Gopal KR, Reddy RR, Sivakumar V, Lal S (2012) Analysis of diurnal and seasonal behavior of surface ozone and its precursors (NOx) at a semi-arid rural site in Southern India. Aerosol Air Qual Res 12:1081–1094

Reich PB, Walters MB, Ellsworth DS, Vose JM, Volin JC, Greshaam C, Bowman WD (1998) Relationships of leaf dark respiration to leaf nitrogen, specific leaf area and leaf life-span: a test across biomes and functional groups. Oecologia 114:471–482

Reiling K, Davison AW (1992) Effects of a short ozone exposure given at different stages in the development of *Plantago major* L. New Phytol 12:643–647

Richet N, Afif D, Huber F, Pollet B, Banvoy J, El Zein R, Lapierre C, Dizengremel P, Perre' P, Cabane M (2011) Cellulose and lignin biosynthesis is altered by ozone in wood of hybrid poplar (*Populus tremula* alba). J Exp Bot 62(10):3575–3586

Riikonen J, Holopainen T, Oksanen E, Vapaavuori E (2005) Leaf photosynthetic characteristics of silver birch during three years of exposure to elevated concentrations of CO_2 and O_3 in the field. Tree Physiol 25:621–632

Royal Society (2008) Ground-level ozone in the 21st century: future trends, impacts and policy implications, Science Policy report 15/08. The Royal Society, London

Ryang SZ, Woo SY, Kwon SY, Kim SH, Lee SH, Kim KN, Lee DK (2009) Changes of net photosynthesis, antioxidant enzyme activities, and antioxidant contents of *Liriodendron tulipifera* under elevated ozone. Photosynthetica 47:19–25

Sadanaga Y, Shibata S, Hamana M, Takenaka N, Bandow H (2008) Weekday/weekend difference of ozone and its precursors in urban areas of Japan, focusing on

nitrogen oxides and hydrocarbons. Atmos Environ 42:4708–4723
Saitanis CJ, Panagopoulous G, Dasopoulou V, Agathokleous E, Papatheohari Y (2015) Integrated assessment of ambient ozone phytotoxicity in Greece's Tripolis Plateau. J Agr Meteorol 71:55–64
Sandermann JH (2000) Ozone/biotic disease interactions: molecular biomarkers as a new experimental tool. Environ Pollut 108:327–332
Santos ACDR, Furlan CM (2013) Levels of phenolic compounds in *Tibouchina pulchra* after fumigation with ozone. Atmos Pollut Res 4:250–256
Sarkar A, Agrawal SB (2010a) Elevated ozone and two modern wheat cultivars: an assessment of dose-dependent sensitivity with respect to growth, reproductive, and yield parameters. Environ Exp Bot 69:328–337
Sarkar A, Agrawal SB (2010b) Identification of ozone stress in Indian rice through foliar injury and differential protein profile. Environ Monit Assess 161:283–302
Sarkar A, Agrawal SB (2011) Evaluating the response of two high yielding Indian rice cultivars against ambient and elevated levels of ozone using open top chambers. J Environ Man 95:S19–S24
Sarkar A, Rakwal R, Agrawal SB, Shibato J, Ogawa Y, Yoshida Y, Agrawal GK, Agrawal M (2010) Investigating the impact of elevated levels of O_3 on tropical wheat using integrated phenotypical, physiological, biochemical and proteomics approaches. J Proteome Res 9:4565–4584
Sarkar A, Singh AA, Agrawal SB, Ahmad A, Rai SP (2015) Cultivar specific variations in antioxidative defense system, genome and proteome of two tropical rice cultivars against ambient and elevated ozone. Ecotox Environ Safe 115:101–111
Saviranta NM, Julkunen-Tiitto R, Oksanen E, Karjalainen RO (2010) Leaf phenolic compounds in red clover (*Trifolium pratense* L.) induced by exposure to moderately elevated ozone. Environ Pollut 158:440–446
Scebba F, Canaccini F, Castagna A, Bender J, Weigel HJ, Ranieri A (2006) Physiological and biochemical stress responses in grassland species are influenced by both early-season ozone exposure and interspecific competition. Environ Pollut 142:540–548
Severino JF, Stich K, Soja G (2007) Ozone stress and antioxidant substances in *Trifolium repens* and *Centaurea jacea* leaves. Environ Pollut 146:707–714
Sicard P, De Marco A, Troussier F, Renou C, Vas N, Paoletti E (2013) Decrease in surface ozone concentrations at Mediterranean remote sites and increase in the cities. Atmos Environ 79:705–715
Simmonds PG, Derwent RG, Manning AL, Spain G (2004) Significant growth in surface ozone at Mace Head, Ireland, 1987–2003. Atmos Environ 38:4769–4778
Singh E, Tiwari S, Agrawal M (2009) Effects of elevated ozone on photosynthesis and stomatal conductance of two soybean varieties: a case study to assess impacts of one component of global climate change. Plant Biol 11:101–108
Singh E, Tiwari S, Agrawal M (2010) Variability in antioxidant and metabolite levels, growth and yield of two soybean varieties: an assessment of anticipated yield losses under projected elevation of ozone. Agri Ecosyst Environ 135:168–177
Singh P, Singh S, Agrawal SB, Agrawal M (2012) Assessment of the interactive effects of ambient O_3 and NPK levels on two tropical mustard varieties (*Brassica campestris* L.) using open-top chambers. Environ Monit Assess 184(10):5863–5874
Singh S, Bhatia A, Tomer R, Kumar V, Singh B, Singh SD (2013) Synergistic action of tropospheric ozone and carbon dioxide on yield and nutritional quality of Indian mustard (*Brassica juncea* (L.) Czern.). Environ Monit Assess 185:6517–6529
Singh AA, Agrawal SB, Shahi JP, Agrawal M (2014a) Assessment of growth and yield losses in two *Zea mays* L. cultivars (quality protein maize and nonquality protein maize) under projected levels of ozone. Environ Sci Pollut Res 21:2628–2641
Singh AA, Agrawal SB, Shahi JP, Agrawal M (2014b) Investigating the response of tropical maize (*Zea mays* L.) cultivars against elevated levels of O_3 at two developmental stages. Ecotoxicology 23:1447–1463
Singh P, Agrawal M, Agrawal SB, Singh S, Singh A (2014c) Genotypic differences in utilization of nutrients in wheat under ambient ozone concentrations: growth, biomass and yield. Agric Ecosys Environ 199:26–33
Singh AA, Singh S, Agrawal M, Agrawal SB (2015) Assessment of ethylene diurea-unduced protection in plants against ozone phytotoxicity. DM Whitacre (Ed.) Rev Environ Cont Toxicol 233:129–184
Singla V, Satsangi A, Pachauri T, Lakhani A, Kumari KM (2011) Ozone formation and destruction at a suburban site in North Central region of India. Atmos Res 101:373–385
Sinha B, Singh Sangwan K, Maurya Y, Kumar V, Sarkar C, Chandra BP, Sinha V (2015) Assessment of crop yield losses in Punjab and Haryana using two years of continuous in-situ ozone measurements. Atmos Chem Phys Discuss 15:2355–2404
Sun J, Feng Z, Ort DR (2014) Impacts of rising tropospheric ozone on photosynthesis and metabolite levels on field grown soybean. Plant Sci 226:147–161
Tai APK, Martin MV, Heald CL (2014) Threat to future global food security from climate change and ozone air pollution. Nat Clim Chang 4:817–821
Tang H, Takigawa M, Liu G, Zhu J, Kobayashi K (2013) A projection of ozoneźinduced wheat production loss in China and India for the years 2000 and 2020 with exposureźbased and fluxźbased approaches. Global Chang Biol 19:2739–2752
Then C, Herbinger K, Luis VC, Heerdt C, Matyssek R, Wieser G (2009) Photosynthesis, chloroplast pigments, and antioxidants in *Pinus canariensis* under free-air ozone fumigation. Environ Pollut 157:392–395

Thomas MD (1951) Gas damage to plants. Ann Rev Plant Physiol 2:293–322

Tiedemann AV, Firsching KH (2000) Interactive effects of elevated ozone and carbon dioxide on growth and yield of leaf rust-infected versus non-infected wheat. Environ Pollut 108:357–363

Tiwari S, Rai R, Agrawal M (2008) Annual and seasonal variations in tropospheric ozone concentrations around Varanasi. Int J Remote Sens 29:4499–4514

Tonneijck AEG, Franzaring J, Brouwer G, Metselaar K, Dueck TA (2004) Does interspecific competition alter effects of early season ozone exposure on plants from wet grasslands? Results of a three- year experiment in open-top chambers. Environ Pollut 131:205–213

Toumainen J, Pellinen R, Roy S, Kiiskinen M, Eloranta T, Karjalainen R, Kangasjärvi J (1996) Ozone affect birch (*Betula pendula* Roth) phenylpropanoid, polyamine and reactive oxygen detoxifying pathways at biochemical and gene expression levels. J Plant Physiol 148:179–188

Tripathi R, Agrawal SB (2012) Effects of ambient and elevated level of ozone on *Brassica campestris* L. with special reference to yield and oil quality parameters. Ecotox Environ Safe 85:1–12

Turkington R, Klein E, Maze J (1994) Conditioning effects by neighbours on the growth and form of Trifolium repens. Can J Bot 72:783–787

U.S. Environmental Protection Agency (US EPA) (2006) Air quality criteria for resistance and a possible metric. Atmos Environ 38:2323–2337

U.S. Environmental Protection Agency US EPA (2009) 1980–2008 Average annual emissions, all criteria pollutants in MS Excel. National Emissions Trend Data, Office of Air Quality planning and Standards

Vainonen JP, Kangasjärvi J (2014) Plant signalling in acute ozone exposure. Plant Cell Environ. doi:10.1111/pce.12273

Van Dingenen R, Dentener FJ, Raes F, Krol MC, Emberson L, Cofala J (2009) The global impact of ozone on agricultural crop yields under current and future air quality legislation. Atmos Environ 43:604–618

Van Tienhoven AM, Zunckel M, Emberson L, Koosailee A, Otter L (2006) Preliminary assessment of risk of ozone impacts to maize (*Zea mays*) in southern Africa. Environ Pollut 140:220–230

Velikova V, Tsonev T, Pinelli P, Alessio GA, Loreto F (2005) Localized ozone fumigation system for studying ozone effects on photosynthesis, respiration, electron transport rate and isoprene emission in field-grown Mediterranean oak species. Tree Physiol 25:1523–1532

von Tiedemann A, Weigel H, Jäger HJ (1991) Effects of open-top chamber fumigations with ozone on three fungal leaf diseases of wheat and the mycoflora of the phyllosphere. Environ Pollut 72(3):205–224

Wagg S, Mills G, Hayes F, Wilkinson S, Davies WJ (2013) Stomata are less responsive to environmental stimuli in high background ozone in *Dactylis glomerata* and *Ranunculus acris*. Environ Pollut 175:82–91

Wahid A (2006a) Influence of atmospheric pollutants on agriculture in developing countries: a case study with three new varieties in Pakistan. Sci Total Environ 371:304–313

Wahid A (2006b) Productivity losses in barley attributable to ambient atmospheric pollutants in Pakistan. Atmos Environ 40:5342–5354

Wan WX, Xia YJ, Zhang HX, Wang J, Wang XK (2013) The ambient ozone pollution and foliar injury of the sensitive woody plants in Beijing exurban region. Acta Ecol Sin 33:109

Wang T, Wei XL, Ding AJ, Poon CN, Lam KS, Li YS, Chan LY, Anson M (2009) Increasing surface ozone concentrations in the background atmosphere of Southern China, 1994–2007. Atmos Chem Phys 9:6217–6227

Wang J, Zeng Q, Zhu J, Liu G, Tang H (2013) Dissimilarity of ascorbate–glutathione (AsA–GSH) cycle mechanism in two rice (Oryza sativa L.) cultivars under experimental free-air ozone exposure. Agr Ecosyst Environ 165:39–49

Watanabe M, Hoshika Y, Inada N, Wang X, Mao Q, Koike T (2013) Photosynthetic traits of Siebold's beech and oak saplings grown under free air ozone exposure in northern Japan. Environ Pollut 174:50–56

Whitfield CP, Davison AW, Ashendra TW (1998) The effects of nutrient limitation on the response of *Plantago major* to ozone. New Phytol 140:219–230

Wilkinson S, Davies WJ (2010) Drought, ozone, ABA and ethylene: new insights from cell to plant to community. Plant Cell Environ 33:510–525

Wise M, Calvin K, Thomson A, Clarke L, Lamberty BB, Sands R, Smith SJ, Janetos A, Edmonds J (2009) Implications of limiting CO_2 concentrations for land use and energy. Science 324:1183–1186

Wittig VE, Ainsworth EA, Long SP (2007) To what extent do current and projected increases in surface ozone affect photosynthesis and stomatal conductance of trees? A metaźanalytic review of the last 3 decades of experiments. Plant Cell Environ 30:1150–1162

Wittig VE, Ainsworth EA, Naidu SL, Karnosky DF, Long SP (2009) Quantifying the impact of current and future tropospheric ozone on tree biomass, growth, physiology and biochemistry: a quantitative metaźanalysis. Global Chang Biol 15:396–424

World Health Organization (2006) WHO ambient air quality guidelines. http://w3.whosea.org/techinfo/air.htm

Xu X, Lin W, Wang T, Yan P, Tang J, Meng Z, Wang Y (2008) Long term trend of surface ozone at a regional background station in eastern China 1991–2006: enhanced variability. Atmos Chem Phys 8:215–243

Yamaji K, Ohara T, Uno I, Tanimoto H, Kurokawa JI, Akimoto H (2006) Analysis of the seasonal variation of ozone in the boundary layer in East Asia using the community multi-scale air quality model: what controls surface ozone levels over Japan? Atmos Environ 40:1856–1868

Yan K, Chen W, He X, Zhang G, Xu S, Wang L (2010) Responses of photosynthesis, lipid peroxidation and antioxidant system in leaves of *Quercus mongolica* to elevated O_3. Environ Exp Bot 69:198–204

Ye ZZ, Rodriguez R, Tran A, Hoang H, De Los Santos D, Brown S, Vellanoweth RL (2000) The developmental transition to flowering represses ascorbate peroxidase activity and induces enzymatic lipid peroxidation in leaf tissue in *Arabidopsis thaliana*. Plant Sci 158:115–127

Ye L, Fu X, Ge F (2012) Enhanced sensitivity to higher ozone in a pathogen-resistant tobacco cultivar. J Exp Bot 63:1341–1347

Zeng G, Pyle JA, Young PJ (2008) Impact of climate change on tropospheric ozone and its global budgets. Atmos Chem Phys 8:369–387

Zhang W, Feng Z, Wang X, Niu J (2012) Responses of native broadleaved woody species to elevated ozone in subtropical China. Environ Pollut 163:149–157

Zhang W, Feng Z, Wang X, Niu J (2014a) Impacts of elevated ozone on growth and photosynthesis of *Metasequoia glyptostroboides* Hu et Cheng. Plant Sci 226:182–188

Zhang W, Wang G, Liu X, Feng Z (2014b) Effects of elevated O_3 exposure on seed yield, N concentration and photosynthesis of nine soybean cultivars (*Glycine max* (L.) Merr.) in Northeast China. Plant Sci 226:172–181

Zhao C, Wang Y, Zeng T (2009) East China plains: a "basin" of ozone pollution. Environ Sci Technol 43:1911–1915

Zhu X, Feng Z, Sun T, Liu X, Tang H, Zhu J, Guo W, Kobayashi K (2011) Effects of elevated ozone concentration on yield of four Chinese cultivars of winter wheat under fully open-air field conditions. Global Chang Biol 17:2697–2706

MIX
Papier aus verantwortungsvollen Quellen
Paper from responsible sources
FSC® C105338

If you have any concerns about our products,
you can contact us on
ProductSafety@springernature.com

In case Publisher is established outside the EU,
the EU authorized representative is:
Springer Nature Customer Service Center GmbH
Europaplatz 3, 69115 Heidelberg, Germany

Printed by Libri Plureos GmbH
in Hamburg, Germany